Lecture Notes in Computer Science

Lecture Notes in Artificial Intelligence   14375

Founding Editor

Jörg Siekmann

Series Editors

Randy Goebel, *University of Alberta, Edmonton, Canada*
Wolfgang Wahlster, *DFKI, Berlin, Germany*
Zhi-Hua Zhou, *Nanjing University, Nanjing, China*

The series Lecture Notes in Artificial Intelligence (LNAI) was established in 1988 as a topical subseries of LNCS devoted to artificial intelligence.

The series publishes state-of-the-art research results at a high level. As with the LNCS mother series, the mission of the series is to serve the international R & D community by providing an invaluable service, mainly focused on the publication of conference and workshop proceedings and postproceedings.

Van-Nam Huynh · Bac Le · Katsuhiro Honda ·
Masahiro Inuiguchi · Youji Kohda
Editors

Integrated Uncertainty in Knowledge Modelling and Decision Making

10th International Symposium, IUKM 2023
Kanazawa, Japan, November 2–4, 2023
Proceedings, Part I

Springer

Editors
Van-Nam Huynh ⓘ
Japan Advanced Inst of Science
and Technology
Nomi, Ishikawa, Japan

Katsuhiro Honda ⓘ
Osaka Metropolitan University
Sakai, Osaka, Japan

Youji Kohda ⓘ
Japan Advanced Institute of Science
and Technology
Nomi, Ishikawa, Japan

Bac Le ⓘ
Vietnam National University
Ho Chi Minh City, Vietnam

Masahiro Inuiguchi ⓘ
Osaka University
Toyonaka, Osaka, Japan

ISSN 0302-9743   ISSN 1611-3349 (electronic)
Lecture Notes in Artificial Intelligence
ISBN 978-3-031-46774-5   ISBN 978-3-031-46775-2 (eBook)
https://doi.org/10.1007/978-3-031-46775-2

LNCS Sublibrary: SL7 – Artificial Intelligence

This Springer imprint is published by the registered company Springer Nature Switzerland AG
The registered company address is: Gewerbestrasse 11, 6330 Cham, Switzerland

Paper in this product is recyclable.

Preface

This volume contains the papers that were accepted for presentation at the 10th International Symposium on Integrated Uncertainty in Knowledge Modelling and Decision Making (IUKM 2023), held in Ishikawa, Japan, November 2–4, 2023.

The IUKM conference aims to provide a forum for exchanges of research results and ideas and practical experiences among researchers and practitioners involved with all aspects of uncertainty modelling and management. Previous editions of the conference were held in Ishikawa, Japan (2010), Hangzhou, China (2011), Beijing, China (2013), Nha Trang, Vietnam (2015), Da Nang, Vietnam (2016), Hanoi, Vietnam (2018), Nara, Japan (2019), Phuket, Thailand (2020), Ishikawa (online), Japan (2022) and their proceedings were published by Springer in AISC 68, LNAI 7027, LNAI 8032, LNAI 9376, LNAI 9978, LNAI 10758, LNAI 11471, LNAI 12482, and LNAI 13199, respectively.

The IUKM 2023 was jointly organized by Japan Advanced Institute of Science and Technology (JAIST), University of Science – Vietnam National University Ho Chi Minh City, Vietnam, and Osaka University, Japan. During IUKM 2023, a special event was also organized for honoring Hung T. Nguyen and Sadaaki Miyamoto for their contributions to the field of uncertainty theories.

This year, the conference received 107 submissions from 16 different countries. Each submission was peer single-blind review by at least two members of the Program Committee. After a thorough review process, 58 papers (54.2%) were accepted for presentation and inclusion in the LNAI proceedings. In addition to the accepted papers, the conference program also included 10 short presentations and featured the following four keynote talks:

- Hung T. Nguyen (New Mexico State University, USA; Chiang Mai University, Thailand), On Uncertainty in Partially Identified Models
- Sadaaki Miyamoto (University of Tsukuba, Japan), Two Classes of Fuzzy Clustering: Their Theoretical Contributions
- Hyun Oh Song (Seoul National University, Korea), Contrastive Discovery of Hierarchical Achievements in Reinforcement Learning
- Thierry Denoeux (Université de Technologie de Compiègne, France), Random Fuzzy Sets and Belief Functions: Application to Machine Learning

The conference proceedings are split into two volumes (LNAI 14375 and LNAI 14376). This volume contains four keynote abstracts, and 28 papers related to Uncertainty Management and Decision Making, Optimization and Statistical Methods, and Economic Applications.

As a follow-up of IUKM 2023, a special issue of the journal *Annals of Operations Research* is anticipated to include a small number of extended papers selected from the conference as well as other relevant contributions received in response to subsequent open calls. These journal submissions will go through a fresh round of reviews in accordance with the journal's guidelines.

IUKM 2023 was partially supported by JAIST Research Fund and the U.S. Office of Naval Research Global (Award No. N62909-23-1-2105). We are very thankful to the local organizing team from Japan Advanced Institute of Science and Technology for their hard working, efficient services, and wonderful local arrangements.

We would like to express our appreciation to the members of the Program Committee for their support and cooperation in this publication. We are also thankful to the staff of Springer for providing a meticulous service for the timely production of this volume. Last, but certainly not the least, our special thanks go to all the authors who submitted papers and all the attendees for their contributions and fruitful discussions that made this conference a great success.

November 2023

Van-Nam Huynh
Bac Le
Katsuhiro Honda
Masahiro Inuiguchi
Youji Kohda

Organization

General Chairs

Masahiro Inuiguchi   Osaka University, Japan
Youji Kohda   Japan Advanced Institute of Science and
Technology (JAIST), Japan

Advisory Board

Michio Sugeno   European Center for Soft Computing, Spain
Hung T. Nguyen   New Mexico State University, USA; Chiang Mai
University, Thailand
Sadaaki Miyamoto   University of Tsukuba, Japan
Akira Namatame   AOARD/AFRL and National Defense Academy
of Japan, Japan

Program Chairs

Van-Nam Huynh   JAIST, Japan
Bac Le   University of Science, VNU-Ho Chi Minh City,
Vietnam
Katsuhiro Honda   Osaka Metropolitan University, Japan

Special Sessions and Workshop Chair

The Dung Luong   Academy of Cryptography Techniques, Vietnam

Publicity Chair

Toan Nguyen-Mau   University of Information Technology, VNU-Ho
Chi Minh City, Vietnam

Program Committee

Yaxin Bi	University of Ulster, UK
Matteo Brunelli	University of Trento, Italy
Tru Cao	University of Texas Health Science Center at Houston, USA
Tien-Tuan Dao	Centrale Lille Institut, France
Yong Deng	University of Electronic Science and Technology of China, China
Thierry Denoeux	University of Technology of Compiègne, France
Sebastien Destercke	University of Technology of Compiègne, France
Zied Elouedi	LARODEC, ISG de Tunis, Tunisia
Tomoe Entani	University of Hyogo, Japan
Katsushige Fujimoto	Fukushima University, Japan
Lluis Godo	IIIA - CSIC, Spain
Yukio Hamasuna	Kindai University, Japan
Katsuhiro Honda	Osaka Metropolitan University, Japan
Tzung-Pei Hong	National University of Kaohsiung, Taiwan
Jih Cheng Huang	Soochow University, Taiwan
Van-Nam Huynh	JAIST, Japan
Hiroyuki Inoue	University of Fukui, Japan
Atsuchi Inoue	Eastern Washington University, USA
Masahiro Inuiguchi	Osaka University, Japan
Radim Jirousek	Prague University of Economics and Business, Czech Republic
Jung Sik Jeong	Mokpo National Maritime University, South Korea
Yuchi Kanzawa	Shibaura Institute of Technology, Japan
Yasuo Kudo	Muroran Institute of Technology, Japan
Vladik Kreinovich	University of Texas at El Paso, USA
Yoshifumi Kusunoki	Osaka Metropolitan University, Japan
Bac Le	University of Science, VNU-Ho Chi Minh City, Vietnam
Churn-Jung Liau	Academia Sinica, Taiwan
Marimin Marimin	Bogor Agricultural University, Indonesia
Luis Martinez	University of Jaen, Spain
Radko Mesiar	Slovak Univ. of Tech. in Bratislava, Slovakia
Tetsuya Murai	Chitose Institute of Science and Technology, Japan
Michinori Nakata	Josai International University, Japan
Canh Hao Nguyen	Kyoto University, Japan

Duy Hung Nguyen   Sirindhorn International Institute of Technology,
                               Thailand
Vu-Linh Nguyen   University of Technology of Compiègne, France
Akira Notsu   Osaka Metropolitan University, Japan
Vilem Novak   Ostrava University, Czech Republic
Sa-Aat Niwitpong   King Mongkut's University of Technology North
                               Bangkok, Thailand
Warut Pannakkong   Sirindhorn International Institute of Technology,
                               Thailand
Irina Perfilieva   Ostrava University, Czech Republic
Frédéric Pichon   Artois University, France
Zengchang Qin   Beihang University, China
Jaroslav Ramik   Silesian University in Opava, Czech Republic
Hiroshi Sakai   Kyushu Institute of Technology, Japan
Hirosato Seki   Osaka University, Japan
Kao-Yi Shen   Chinese Culture University, Taiwan
Dominik Slezak   University of Warsaw, Poland
Roman Slowinski   Poznan University of Technology, Poland
Kazuhiro Takeuchi   Osaka Electro-Communication University, Japan
Yongchuan Tang   Zhejiang University, China
Roengchai Tansuchat   Chiang Mai University, Thailand
Phantipa Thipwiwatpotjana   Chulalongkorn University, Thailand
Vicenc Torra   University of Skövde, Sweden
Seiki Ubukata   Osaka Metropolitan University, Japan
Guoyin Wang   Chongqing Univ. of Posts and Telecom., China
Woraphon Yamaka   Chiang Mai University, Thailand
Xiaodong Yue   Shanghai University, China
Chunlai Zhou   Renmin University of China

Local Arrangement Team

Van-Nam Huynh   JAIST, Japan
Yang Yu   JAIST, Japan
Trong-Hiep Hoang   JAIST, Japan
Leelertkij Thanapat   JAIST, Japan
Aswanuwath Lalitpat   JAIST, Japan
Supsermpol Pornpawee   JAIST, Japan
Dang-Man Nguyen   JAIST, Japan
Chemkomnerd Nittaya   JAIST, Japan
Xuan-Thang Tran   JAIST, Japan
Jiaying Ni   JAIST, Japan
Xuan-Truong Hoang   JAIST, Japan

Keynote Lectures

On Uncertainty in Partially Identified Models

Hung T. Nguyen

New Mexico State University, USA; Chiang Mai University, Thailand

Abstract. This talk is a panorama of uncertainty topics that I have been involved in from Generalized Information Theory to Ambiguity, focusing on current interests in using Random Sets as statistical tools for investigating prediction and decision in the face of ambiguity, especially in partially identified econometrics models.

Overall the talk spells out how random sets will become the main tools to investigate uncertainty in the sense of ambiguity in general decision theory, and in particular in partially identified statistical models in econometrics and finance. The talk evokes the current interests in connections with Optimal Transport Theory in general mathematics.

Two Classes of Fuzzy Clustering: Their Theoretical Contributions

Sadaaki Miyamoto

University of Tsukuba, Japan

Abstract. Two classes of fuzzy clustering are overviewed with suggestions for future studies. They are known as fuzzy equivalence relations and fuzzy c-means. By people in non-fuzzy fields, they are sometimes regarded as the classical single linkage and a variation of EM algorithm: the former is correct while the latter is wrong.

A fundamental question is what real and useful contributions of them. The contribution of fuzzy c-means seems obvious to us in view of their good performance and robustness, while people in other fields may not appreciate such properties and still oppose the use of this method. We will consider here the contribution of fuzzy c-means regarding their ability of generalizations and flexibility as well as their theoretical soundness, showing different variations of fuzzy c-means with current problems to be solved. The other class of fuzzy equivalence corresponds to the well-known method of the single linkage in agglomerative hierarchical clustering. There appears to be no real contribution to the traditional theory. We will show, however, a class of generalizations here related to current non-fuzzy algorithms with future research possibilities. These two classes appear to have no relation, but can effectively be used together in real clustering problems. An example of their relation will be shown.

Contrastive Discovery of Hierarchical Achievements in Reinforcement Learning

Hyun Oh Song

Seoul National University, Korea

Abstract. Discovering hierarchical achievement structures on procedurally generated environments is a challenging task. This requires that agents possess a wide range of skills, including generalization and long-term reasoning. Many previous approaches have relied on model-based or hierarchical methods, assuming that incorporating an explicit module for long-term planning would enhance the learning of hierarchical achievements. However, these approaches suffer from the drawbacks of requiring a high number of interactions with the environment or large model sizes, making them less practical. In this study, we first show that proximal policy optimization (PPO), a simple and general model-free algorithm, outperforms previous methods when incorporating recent implementation techniques. Furthermore, we find that the PPO agent is capable of predicting the next achievement to a certain extent, albeit with low confidence. Building on this observation, we propose a new contrastive learning technique, named achievement distillation, which strengthens the agent's ability to forecast subsequent achievements. Our method exhibits a remarkable performance for discovering hierarchical achievements and achieves state-of-the-art performance on the challenging Crafter environment, with fewer model parameters in a sample-efficient manner.

Random Fuzzy Sets and Belief Functions: Application to Machine Learning

Thierry Denoeux

Université de Technologie de Compiègne

Abstract. The theory of belief functions is a powerful formalism for uncertain reasoning, with many successful applications to knowledge representation, information fusion, and machine learning. Until now, however, most applications have been limited to problems (such as classification) in which the variables of interest take values in finite domains. Although belief functions can, in theory, be defined in infinite spaces, we lacked practical representations allowing us to manipulate and combine such belief functions. In this talk, I show that the theory of epistemic random fuzzy sets, an extension of Possibility and Dempster-Shafer theories, provides an appropriate framework for evidential reasoning in general spaces. In particular, I introduce Gaussian random fuzzy numbers and vectors, which generalize both Gaussian random variables and Gaussian possibility distributions. I then describe an application of this new formalism to nonlinear regression.

Contents – Part I

Economic Applications

Contents – Part II

Pattern Classification and Data Analysis

Security and Privacy in Machine Learning

Uncertainty Management and Decision Making

Many-Valued Judgment Aggregation – Some New Possibility Results

Sebastian Uhl and Christian G. Fermüller[✉]

Institue of Logic and Computation, TU Wien, 1040 Vienna, Austria
chrisf@logic.at

Abstract. Judgment aggregation (JA) poses the problem of finding a consistent collective judgment for a set of logically related propositions, based on judgments of individuals. We present two corresponding possibility results for Kleene-Zadeh logic and Gödel logic, respectively. Moreover, we introduce median aggregation, which can be applied to arbitrary many-valued logics.

Keywords: Judgment aggregation · Kleene-Zadeh logic · Gödel logic

1 Introduction

Judgment aggregation (JA) is a topic in the intersection of economy, logic, and mathematics, closely related to social choice and voting theory (see, e.g., [6,11]). JA is concerned with the aggregation of individual judgments on a set of logically connected propositions into a collective judgment (valuation). Reasonable aggregation rules should yield consistent collective judgments and satisfy certain constraints. In particular, the aggregation rule should not be dictatorial; i.e., it should not ignore the opinions of all but one of the agents. Classical impossibility results state that there are no reasonable non-dictatorial aggregation rules that yield a consistent aggregation of arbitrary consistent individual judgments for certain types of agendas (see, e.g., [2,3,6]). This triggers the question of whether we can improve that situation if we allow for some type of uncertainty—here modeled by various fuzzy logics—in the individual and collective judgments.[1] At first sight, the answer is negative since classical impossibility results can be generalized to large families of many-valued logics as demonstrated, e.g., in [4,7]. This motivates the search for criteria that guarantee that natural aggregation rules yield consistent collective judgments, at least for certain types of agendas and judgment profiles. Classical possibility results of this kind are summarized in [6]. Recently Fermüller [5] presented some positive results concerning average aggregation for Kleene-Zadeh logic and Łukasiewicz logic. Our main goal here is to extend the range of aggregation functions as well as of underlying fuzzy logics for further possibility results.

---

[1] One sometimes speaks of 'attitude aggregation' rather than judgment aggregation if the underlying logic is fuzzy. Here, we prefer the term 'many-valued judgment aggregation' to emphasize the relation to classical results.

© The Author(s), under exclusive license to Springer Nature Switzerland AG 2023
V.-N. Huynh et al. (Eds.): IUKM 2023, LNAI 14375, pp. 3–14, 2023.
https://doi.org/10.1007/978-3-031-46775-2_1

After reviewing basic notions for classical and many-valued JA in Sect. 2, we provide possibility results for order restricted profiles for Kleene-Zadeh logic KZ and for Gödel logic G, in Sect. 3 and Sect. 4, respectively. In Sect. 5, we introduce median aggregation and generalize a classical possibility result for unidimensionally aligned profiles to arbitrary many-valued logics, before concluding in Sect. 6.

2   Classical and Many-Valued Judgment Aggregation

To understand the challenges for judgment aggregation it is useful to revisit the so-called *doctrinal paradox*, due to the legal scholars Kornhauser and Sager [8]. (Our presentation follows [14].) A plaintiff has brought a civil suit against a defendant, alleging a breach of contract between them. Three judges (J_1, J_2, J_3) have to determine whether the defendant must pay damages to the plaintiff (d or $\neg d$). The case brings up two issues, namely, whether the contract was valid (v or $\neg v$), and whether the defendant was in breach of it (b or $\neg b$). Contract law stipulates that the defendant must pay damages if and only if the contract was valid and he was in breach of it ($d \leftrightarrow v \wedge b$). Suppose that the judges have the following views on the relevant issues. (Acceptance of a proposition is denoted by an assignment of 1 to it and rejection by an assignment of 0.)

	v	b	d
J_1:	1	0	0
J_2:	0	1	0
J_3:	1	1	1

The paradox arises from two contradictory legal doctrines that might apply:

(1) Since the majority of the judges reject d, no damage is to be paid.
(2) Since each of the two conditions, v and b, are judged to hold by a majority, the defendant must pay damages.

Pettit [15] introduced a general version of the above scenario, dubbed *discursive dilemma*. His presentation shifts the focus away from the conflict between the two methods of finding an adequate overall judgment and instead makes the logical inconsistency of the set of propositions that are accepted by a majority of the judges explicit. In the above example, this inconsistent set is $\{v, b, d \leftrightarrow v \wedge b, \neg d\}$.

To ease the transition from the classical to the many-valued scenario, we will identify individual as well as collective judgments with assignments of values to the propositions of the agenda. In the classical case, the possible values are 0 and 1; in the many-valued case, this is generalized to values from the unit interval $[0, 1]$.

Definition 1. *A propositional language $\mathcal{L}$ is the set of formulas built up from atoms (propositional variables) in At using connectives in $Op_\mathcal{L}$ as follows: (1) At $\subseteq \mathcal{L}$, (2) if $\phi_1, \ldots, \phi_n \in \mathcal{L}$, then $\circ(\phi_1, \ldots, \phi_n) \in \mathcal{L}$ for every n-ary $\circ \in Op_\mathcal{L}$.*
A many-valued logic Λ over the set of truth values $[0, 1]$ is specified by associating with each n-ary $\circ \in Op_\mathcal{L}$ a corresponding truth function $\widetilde{\circ} : [0, 1]^n \to [0, 1]$.

A Λ-valuation V of a set of formulas $X \subseteq \mathcal{L}$ is an assignment of truth values to the formulas in X that respects the truth functions associated with Λ. An arbitrary assignment of values in $[0,1]$ to formulas in X is called Λ-consistent if it is a Λ-valuation; otherwise, it is called Λ-inconsistent.

Definition 2. *A judgment aggregation problem (JA problem) over a propositional language $\mathcal{L}$ is a pair $\langle N, \mathcal{A} \rangle$, where $N = \{1, \ldots, n\}$ is the set of individuals (judges) and $\mathcal{A}$ is a finite subset of $\mathcal{L}$, called agenda.*

Throughout the paper, we speak only of *closed* agendas, by which we mean that $\phi \in \mathcal{A}$ implies that also the subformulas of ϕ are in $\mathcal{A}$. This does not amount to a substantial restriction since one may simply ignore values assigned to subformulas if they are not deemed relevant.

Definition 3. *Given a many-valued logic Λ, a Λ-judgment $J : \mathcal{A} \to [0,1]$ is a Λ-valuation of $\mathcal{A}$. The set of all Λ-judgments is denoted by $\mathbf{J}_\Lambda$. A corresponding judgment profile $P = \langle J_i \rangle_{i \in N} \in \mathbf{J}_\Lambda^n$ over Λ is an n-tuple of Λ-judgments, where J_i denotes the judgment of agent $i \in N$. $\mathbf{P}_\Lambda$ denotes the set of all corresponding judgment profiles over Λ.*

Note that classical JA arises as a special case, where judgments assign only the values 0 or 1, according to the classical truth functions.

Definition 4. *A (many-valued) aggregation function (rule) for a given JA problem $\langle N, \mathcal{A} \rangle$ is of type $F : \mathbf{P}_\Lambda \to [0,1]^{\mathcal{A}}$ for some logic Λ. F maps any given Λ-judgment profile $P \in \mathbf{P}_\Lambda$ into the collective (aggregated) judgment $F(P)$. If $F(P)$ is a Λ-valuation, then the collective judgment is called Λ-consistent or rational (with respect to the given logic).*

Remark 1. Note that an aggregated judgment may well be Λ-inconsistent; i.e., in general, $F(P)$ is not a Λ-valuation and hence not a Λ-judgment.

We refer to [6] and [9] for an overview of various types of classical aggregation rules. An important example is the *majortiy rule F_{maj}*, defined by

$$F_{maj}(P)(\phi) = \begin{cases} 1 & \sum_{i \in N} J_i(\phi) > |N|/2 \\ 0 & \text{otherwise} \end{cases}$$

In a many-valued setting, *average aggregation*, defined by

$$F_{avg}(P)(\phi) = \sum_{i \in N} J_i(\phi)/|N|$$

is an example of a natural aggregation rule.

We specify a few important possible properties of aggregation rules.

Definition 5. *Given a JA problem $\langle N, \mathcal{A} \rangle$ and the corresponding set* **P** *of judgment profiles, an aggregation rule $F : \mathbf{P} \rightarrow 2^{\mathcal{A}}$ is*

- *unanimous iff $\forall \phi \in \mathcal{A} \; \forall P \in \mathbf{P}$: $[\forall i \in N \; J_i(\phi) = 1]$ implies $F(P)(\phi) = 1$;*
- *anonymous iff for any permutation $\sigma : N \rightarrow N$ s.t. $\forall \langle J_i \rangle_{i \in N} \in \mathbf{P} \; \forall \phi \in \mathcal{A}$: $F(\langle J_i \rangle_{i \in N})(\phi) = F(\langle J_{\sigma(i)} \rangle_{i \in N})(\phi)$;*
- *monotonic iff $\forall \phi \in \mathcal{A} \; \forall \langle J_i \rangle_{i \in N}, \langle J_i' \rangle_{i \in N} \in \mathbf{P}$: $[\forall i \in N : J_i(\phi) \leq J_i'(\phi)]$ implies $[F(\langle J_i \rangle_{i \in N})(\phi) \leq F(\langle J_i' \rangle_{i \in N})(\phi)]$;*
- *strongly monotonic iff $\forall \phi \in \mathcal{A} \; \forall \langle J_i \rangle_{i \in N}, \langle J_i' \rangle_{i \in N} \in \mathbf{P}$: $[\forall i \in N : J_i(\phi) < J_i'(\phi)]$ implies $[F(\langle J_i \rangle_{i \in N})(\phi) < F(\langle J_i' \rangle_{i \in N})(\phi)]$;*
- *systematic iff $\exists f : [0,1]^n \rightarrow [0,1]$ s.t. $\forall \langle J_i \rangle_{i \in N} \in \mathbf{P} \; \forall \phi \in \mathcal{A} : F(\langle J_i \rangle_{i \in N})(\phi) = f(J_1(\phi), \dots, J_n(\phi))$;*
- *dictatorial iff $\exists i \in N$ s.t. $\forall P \in \mathbf{P}$: $F(P) = J_i$.*

Note that F_{maj} as well as F_{avg} are unanimous, anonymous, monotonic and systematic, but not dictatorial. Arguably, these are *desiderata* for any reasonable form of judgment aggregation. Unfortunately, various classical *impossibility results* (see, e.g., [6,12,13]) show that there is no non-dictatorial aggregation function that satisfies these desiderata and yields (classically) consistent collective judgments for all profiles over certain types of non-trivial agenda, like the one for the discursive dilemma. Some of these results have been generalized to many-valued JA. In particular, Herzberg [7] proved that all systematic aggregation functions for non-trivial JA problems based on Łukasiewicz logic are dictatorial. Recently Esteban *et al.* [4] generalized that result to a wide range of logics, covering, in particular, all many-valued logics.

3   Possibility for Logic KZ

Fermüller [5] provides a possibility theorem for Kleene-Zadeh logic KZ using average aggregation. We generalize this result to a wider range of aggregation functions.

Definition 6 (Kleene-Zadeh logic KZ). *Like Boolean formulas, KZ-formulas are built up from atoms At using the connectives $\wedge$, $\vee$, and $\neg$. The semantics of KZ is given by extending any assignment (valuation) $v : At \rightarrow [0,1]$ from atoms to arbitrary formulas as follows: $v(\neg \phi) = 1 - v(\phi)$, $v(\phi \wedge \psi) = \min(v(\phi), v(\psi))$, $v(\phi \vee \psi) = \max(v(\phi), v(\psi))$.*

Definition 7 (Order compatible profiles [5]). *Let $\mathcal{J} = \langle N, \mathcal{A} \rangle$ be a JA problem over a closed agenda $\mathcal{A}$. A profile $P = \langle J_i \rangle_{i \in N} \in \mathbf{P}$ for $\mathcal{J}$ is order compatible iff there exists an enumeration $\langle p_1, \dots, p_n \rangle$ of all atoms occurring in the agenda $\mathcal{A}$ such that $\forall i \in N$: $J_i(p_1) \leq \dots \leq J_i(p_n)$.*

Note that the profile of the discursive dilemma presented in Sect. 2 is not order compatible. However, the following variation of it is order compatible:

	v	b	$v \wedge b$
$J_1 :$	1	1	1
$J_2 :$	0	1	0
$J_3 :$	0	0	0

Theorem 1. *Let $\langle N, \mathcal{A} \rangle$ be a JA problem, where $\mathcal{A}$ is closed and contains only negation-free formulas. Then every systematic and monotonic aggregation function F yields a* KZ-*consistent collective judgment $F(P)$ for every order compatible* KZ-*judgment profile $P \in \mathbf{P}$.*

Proof. If all formulas in $\mathcal{A}$ are atomic, then the claim holds trivially since every assignment of values in $[0,1]$ to propositional variables constitutes a valuation over any logic. We proceed inductively. Since the formulas in $\mathcal{A}$ are negation-free, there are only two cases to consider. $\phi \wedge \psi$: Let P be an order compatible KZ-judgment profile over the agenda $\mathcal{A}$. By simply ignoring the values for $\phi \wedge \psi$, P induces an order compatible judgment profile $P' = \langle J_i' \rangle_{i \in N}$ over $\mathcal{A}' = \mathcal{A} - \{\phi \wedge \psi\}$. The induction hypothesis states that $F(P')$ amounts to a KZ-consistent valuation of the propositions in $\mathcal{A}'$. In particular, by systematicity, we have $F(P')(\phi) = f(J_1'(\phi), \ldots, J_n'(\phi))$ and $F(P')(\psi) = f(J_1'(\psi), \ldots, J_n'(\psi))$.

By order compatibility we know that either (1) $J_i(\phi) \leq J_i(\psi)$ for every $i \in N$ or (2) $J_i(\psi) \leq J_i(\phi)$ for every $i \in N$. Hence, we have the following two cases: (1) $J_i(\phi \wedge \psi) = J_i(\phi)$ for all $i \in N$, or (2) $J_i(\phi \wedge \psi) = J_i(\psi)$ for all $i \in N$.

P and P' are identical for all formulas in $\mathcal{A}'$, hence in particular $J_i(\phi) = J_i'(\phi)$ and $J_i(\psi) = J_i'(\psi)$. Thus it follows that if (1), then $F(P)(\phi \wedge \psi) = f(J_1(\phi \wedge \psi), \ldots, J_n(\phi \wedge \psi)) = f(J_1(\phi), \ldots, J_n(\phi)) = F(P)(\phi)$. Analogously if (2), then $F(P)(\phi \wedge \psi) = F(P)(\psi)$.

Combining the cases using monotonicity we get that

$$F(P)(\phi \wedge \psi) = f(J_1(\phi \wedge \psi), \ldots, J_n(\phi \wedge \psi))$$
$$= \min(f(J_1(\phi), \ldots, J_n(\phi)), f(J_1(\psi), \ldots, J_n(\psi)))$$
$$= \min(F(P)(\phi), F(P)(\psi))$$

$\phi \vee \psi$: The argument is analogous to that for $\phi \wedge \psi$.   □

Theorem 1 only applies to negation-free formulas. In order to fully generalize the result of [5], let us call an agenda *internally positive* if the negation sign can only occur at the top-most level in its formulas. Moreover, we introduce the following negation-oriented property of aggregation functions.

Definition 8 (Linear systematicity). *An aggregation function F for a JA-problem $\langle N, \mathcal{A} \rangle$ with $N = \{1, \ldots, n\}$ is linear systematic iff there is some function f with $f(1 - x_1, \ldots, 1 - x_n) = 1 - f(x_1, \ldots, x_n)$ for any $\langle x_1, \ldots, x_n \rangle$ such that $\forall P = \langle J_1, \ldots, J_n \rangle \in \mathbf{P} \; \forall \phi \in \mathcal{A} : F(P)(\phi) = f(J_1(\phi), \ldots, J_n(\phi))$.*

Remark 2. Regarding Definition 8 of linear systematicity, we observe that every linear systematic aggregation function is also systematic, as the decision function is only restricted to a subset of systematic functions.

Corollary 1. *Let $\langle N, \mathcal{A} \rangle$ be a JA problem, where $\mathcal{A}$ is closed and internally positive. Then every linear systematic aggregation function F yields a* KZ-*consistent collective judgment $F(P)$ for every order compatible* KZ-*judgment profile $P \in \mathbf{P}$.*

Proof. The inductive part for closed and positive agendas $\mathcal{A}$ has already been shown in Theorem 1.

Hence, we only have to show that negation works for internally positive formulas in the agenda. In particular, for negated formulas in $\phi \in \mathcal{A}$ by Definition 8 of linear systematicity, we have that there is a decision function f such that $F(P)(\neg\phi) = f(1 - J_1(\phi), \ldots, 1 - J_n(\phi)) = 1 - f(J_1(\phi), \ldots, J_n(\phi)) = 1 - F(P)(\phi).$

This means that the KZ-valuation provided by F remains KZ-consistent if extended from negation-free formulas to negations of such formulas if F is linear systematic.

Note that at first appearance, it is not obvious that Corollary 1 allows more aggregation functions than the average rule F_{av} which satisfies the condition of linear systematicity. However, observe that the constant aggregation function $F_{0.5}(P)(\phi) = 0.5$ for each $P \in \mathbf{P}$ and every formula $\phi \in \mathcal{A}$ satisfies both anonymity and linear systematicity. Another example satisfying both anonymity and linear systematicity is the *majority rule* F_{maj}.

Example 1 (Linear systematicity of the majority rule). Let $\langle N, \mathcal{A} \rangle$ be a judgment aggregation problem with a closed agenda $\mathcal{A}$ and odd $|N|$. Let $P \in \mathbf{P}$ be any KZ-judgment profile and $\phi \in \mathcal{A}$ any formula. Assume that $F_{maj}(P)(\phi) = 1$. By Definition of the majority rule, it holds that $F_{av}(P)(\phi) = \sum_{i \in N} J_i(\phi)/|N| > 0.5$. Moreover, the average rule is linearly systematic. Hence we get $F_{av}(P)(\neg\phi) = 1 - F_{av}(P)(\phi) < 0.5$. Since $F_{av}(P)(\neg\phi) < 0.5$ it follows that $F_{maj}(P)(\neg\phi) = 0$. Hence, it holds that $F_{maj}(P)(\neg\phi) = 0 = 1 - F_{maj}(P)(\phi)$. Thus (at least for odd $|N|$), the majority rule is linearly systematic.

4   Possibility for Logic G

One of the most important t-norm based fuzzy logics is Gödel(-Dummett) logic G (see, e.g., [1]), which features the same semantics for conjunction $\wedge$ and disjunction $\vee$ as Kleene-Zadeh logic KZ. Additionally, we also have an implication with the following semantics:

$$v(\phi \to \psi) = \begin{cases} 1 & \text{if } v(\phi) \le v(\psi) \\ v(\psi) & \text{otherwise.} \end{cases}$$

Moreover, negation is defined by $\neg\phi = \phi \to \bot$.

To adapt the proof of Theorem 1 to Gödel logic, we have to strengthen order compatibility of profiles and require that the individuals agree for each pair of atoms, whether they should be judged identically or whether one should receive a strictly smaller value than the other.

Definition 9 (Strictly order compatible profiles). *Let $\langle N, \mathcal{A} \rangle$ be a JA problem where $\mathcal{A}$ is closed. The profile $P = \langle J_i \rangle_{i \in N} \in \mathbf{P}$ is strictly order compatible iff there exists an enumeration $\langle p_1, \ldots, p_n \rangle$ of all atoms occurring in the agenda $\mathcal{A}$ such that $\forall i \in N : J_i(p_1) \trianglelefteq \cdots \trianglelefteq J_i(p_n)$, where $\trianglelefteq \in \{=, <\}$.*

Theorem 2. *Let $\langle N, \mathcal{A} \rangle$ be a JA problem for a closed agenda $\mathcal{A}$. Then every systematic, strongly monotonic, and unanimous aggregation function F yields a G-consistent collective judgment $F(P)$ for every strictly order compatible G-judgment profile $P \in \mathbf{P}$.*

Proof. Like for Theorem 1, the claim trivially holds if all propositions in $\mathcal{A}$ are atomic. For non-atomic propositions we distinguish the following cases.

$\phi \wedge \psi$, $\phi \vee \psi$: These cases are analogous to the proof of Theorem 1.

$\phi \rightarrow \psi$: Let P be a strictly order compatible G-judgment profile over the agenda $\mathcal{A}$. By simply ignoring the values for $\phi \rightarrow \psi$, P induces a strictly order compatible judgment profile $P' = \langle J'_i \rangle_{i \in N}$ over $\mathcal{A}' = \mathcal{A} - \{\phi \rightarrow \psi\}$. The induction hypothesis states that $F(P')$ amounts to a G-consistent valuation of the propositions in $\mathcal{A}'$. In particular, by systematicity, we have

$$F(P')(\phi) = f(J'_1(\phi), \ldots, J'_n(\phi)),$$
$$F(P')(\psi) = f(J'_1(\psi), \ldots, J'_n(\psi)) \text{ and}$$
$$F(P')(\phi \rightarrow \psi) = f(J'_1(\phi \rightarrow \psi), \ldots, J'_n(\phi \rightarrow \psi)).$$

By strict order compatibility we know that either (1) $J_i(\phi) < J_i(\psi)$ for every $i \in N$ or $J_i(\phi) = J_i(\psi)$ for every $i \in N$, or (2) $J_i(\psi) < J_i(\phi)$ for every $i \in N$.

Thus, we have the following cases:

(1) $J_i(\phi \rightarrow \psi) = 1$ for all $i \in N$, or
(2) $J_i(\phi \rightarrow \psi) = J_i(\psi)$ for all $i \in N$.

P and P' are identical for all formulas in $\mathcal{A}'$, hence in particular $J_i(\phi) = J'_i(\phi)$ and $J_i(\psi) = J'_i(\psi)$. In case (1) and by unanimity of F we have

$$F(P)(\phi \rightarrow \psi) = f(J_1(\phi \rightarrow \psi), \ldots, J_n(\phi \rightarrow \psi)) = f(1, \ldots, 1) = 1.$$

Moreover, by monotonicity of F it follows that

$$F(P)(\phi) = f(J_1(\phi), \ldots, J_n(\phi)) \leq f(J_1(\psi), \ldots, J_n(\psi)) = F(P)(\psi).$$

Hence, for case (1) the semantics of implication in G is also satisfied on the collective level.

Moreover, if (2) it follows that

$$F(P)(\phi \rightarrow \psi) = f(J_1(\phi \rightarrow \psi), \ldots, J_n(\phi \rightarrow \psi))$$
$$= f(J_1(\psi), \ldots, J_n(\psi)) = F(P)(\psi).$$

Moreover, by strong monotonicity of F, we get that

$$F(P)(\psi) = f(J_1(\psi), \ldots, J_n(\psi)) < f(J_1(\phi), \ldots, J_n(\phi)) = F(P)(\phi).$$

Hence, in both cases (1) and (2) the semantics of implication in the logic G is satisfied.

$\neg \phi$: Since $\neg \phi = \phi \rightarrow \bot$ this case is covered by the one for implication.

5   Possibility for Arbitrary Logics via Median Aggregation

In Sects. 3 and 4, we have established possibility results for Kleene-Zadeh logic KZ and Gödel logic G, respectively. These results are specific to the mentioned logics and do not hold, e.g., for Łukasiewicz logic Ł.[2] In this section, we will formulate a profile condition that guarantees the consistency of *(propositionwise) median aggregation* for arbitrary many-valued logics. Our result generalizes a well-known possibility result for classical JA by List that shows that the majority judgment is consistent for unidimensionally aligned profiles [10].

We assume that there are at least two individuals in every JA problem.

Definition 10 (Median Judgments). *Let $P = \langle J_i \rangle_{i \in N} \in \mathbf{P}$ be a profile for a many-valued logic L over $[0,1]$. $J^{md} : \mathcal{A} \to [0,1]$ is a median judgment for $\phi \in \mathcal{A}$ in P if there exists a partition of N into $N = N_\downarrow \cup N_\uparrow$, $N_\downarrow \cap N_\uparrow = \emptyset$, such that $|N_\downarrow| = |N_\uparrow|$ or $|N_\downarrow| = |N_\uparrow| + 1$ or $|N_\downarrow| = |N_\uparrow| - 1$ and the following holds: $\forall i \in N_\downarrow : J_i(\phi) \leq J^{md}(\phi)$ and $\forall i \in N_\uparrow : J_i(\phi) \geq J^{md}(\phi)$. If $J^{md}(\phi)$ coincides with J_i for some $i \in N$, we call i a median judge for ϕ in P.*

J^{md} is a (propositionwise) median judgment for P and $\mathcal{A}$ if J^{md} is a median judgment for every $\phi \in \mathcal{A}$. If J^{md} coincides with J_i for every $\phi \in \mathcal{A}$ then i is a median judge in P for $\mathcal{A}$.

Definition 11 (Median Aggregation Functions). *Let $\langle N, \mathcal{A} \rangle$ be an L-JA problem for a many-valued logic L over $[0,1]$. Then a function $F^m : \mathbf{P} \to [0,1]^{\mathcal{A}}$ is a median aggregation function if, for every $P \in \mathbf{P}$, $F^m(P)$ is a median judgment in P for $\mathcal{A}$.*

Note that median judges exist for each proposition $\phi \in \mathcal{A}$ in every profile $\langle J_i \rangle_{i \in N}$. However, in general, there is no median judge for the whole agenda $\mathcal{A}$. If one orders the individuals such that the judgment values $J_i(\phi)$ appear in either ascending or descending order, then both $J_m(\phi)$ and $J_{m+1}(\phi)$, are median judgments (and hence m and $m + 1$ are median judges) for ϕ if $|N| = 2m$. If $|N| = 2m + 1$ one even obtains three median judgments: $J_m(\phi)$, $J_{m+1}(\phi)$ and $J_{m+2}(\phi)$. This implies that a median judgment is only useful if there are more than three individuals.[3] On the other hand, if there are a lot of voters and the individual values attached to propositions are widespread, then median judgments may well amount to representative aggregations. One may argue that determining the average values, rather than picking a median value, is even more informative. However, note that for every proposition ϕ there exists only a single average value, whereas, in general, there are intervals of possible median judgments for ϕ. In the above scenario, where m and $m + 1$ and, for the case $|N| = 2m + 1$, also $m + 2$ are median judges for ϕ, we may set $J^{md}(\phi) =$

---

[2] For other recent possibility results specific to KZ and Ł see [5].

[3] In particular, for the special case of the discursive dilemma (see Sect. 2), every individual judgment constitutes a median judgment for the given profile. While this hardly amounts to a resolution of the dilemma, it reflects the fact that any of the three individual judgments seems equally representative in this case.

x for any $x \in [J_m(\phi), J_{m+1}(\phi)]$ or $x \in [J_m(\phi), J_{m+2}(\phi)]$, respectively. Thus uncountably many different median judgments may arise. As a consequence, median aggregation is more flexible than average aggregation in general.

Example 2. Consider the following profile P for a JA-problem $\langle N, \mathcal{A} \rangle$, where $N = \{1, \ldots, 6\}$ and $\mathcal{A} = \{\phi_1, \phi_2, \phi_3, \phi_4\}$.

	ϕ_1	ϕ_2	ϕ_3	ϕ_4
1 :	0	1	1	0
2 :	0.2	0.7	0.7	0.2
3 :	0.3	0.5	0.5	0.3
4 :	0.4	0.4	0.4	0.4
5 :	0.8	0.3	0.8	0.3
6 :	1	0	1	0

Note that for every $i \in N$ we have $J_i(\phi_3) = \max(J_i(\phi_1), J_i(\phi_2))$ and $J_i(\phi_4) = \min(J_i(\phi_1), J_i(\phi_2))$. Hence, we have L-consistent individual judgments if $\phi_3 = \phi_1 \vee \phi_2$ and $\phi_4 = \phi_1 \wedge \phi_2$ if L is a logic where disjunction is modeled by minimum and conjunction by maximum, like, for example, in Kleene-Zadeh logic KZ or Gödel logic G.

Any median judgment J^{md} for P and $\mathcal{A}$ has to satisfy the following constraints: $J^{md}(\phi_1) \in [0.3, 0.4]$, $J^{md}(\phi_2) \in [0.4, 0.5]$, $J^{md}(\phi_3) \in [0.7, 0.8]$, and $J^{md}(\phi_4) \in [0.2, 0.3]$. If we assume that $\phi_3 = \phi_1 \vee \phi_2$ and $\phi_4 = \phi_1 \wedge \phi_2$, then there is no KZ- or G-consistent median judgment for $\mathcal{A}$, since the possible median values for $\phi_3 = \phi_1 \vee \phi_2$ are strictly higher than those for ϕ_1 and ϕ_2. However, if we delete ϕ_3 from the agenda, we can satisfy the constraints. Indeed, individual 3 is a median judge for $\mathcal{A} - \{\phi_3\}$.

If we change $J_2(\phi_2)$ and $J_2(\phi_3)$ to 0.5 then we obtain the constraint $J^{md}(\phi_3) \in [0.5, 0.8]$ for $\phi_3 = \phi_1 \vee \phi_2$. In that case, individual 3 is a median judge for the whole agenda $\mathcal{A}$.

Median judgments are related to majority judgments. For classical JA, where $J_i(\phi) = 0$ or $J_i(\phi) = 1$ for every $i \in N$ and every $\phi \in \mathcal{A}$, median judgments are non-discriminating, i.e. $J^{md}(\phi)$ can be 0 as well as 1, for a proposition ϕ, if and only if (almost) as many judges evaluate ϕ to 1 and to 0, respectively. More precisely, let $n^+(\phi) = |\{i \in N \mid J_i(\phi) = 1\}|$ and $n^-(\phi) = |\{i \in N \mid J_i(\phi) = 0\}|$. Then $J^{md}(\phi)$ can take both values if $n^+(\phi) = n^-(\phi)$ (for even $|N|$) or $|n^+(\phi) - n^-(\phi)| = 1$ (for uneven $|N|$). In all other cases, the median judgment coincides with the majority judgment.

Also in the general, many-valued scenario, median aggregation has attractive features. Clearly, median aggregation functions are non-dictatorial. By definition, median judgments keep a balance between sets of individuals that value a given proposition at least as high or at least as low, respectively, as some intermediary value. Median aggregation returns such an intermediary *median value* as the collective value for each proposition of the agenda. In contrast to the *average* of the individuals' values, median values are not unique, as we have seen above. The possibility to choose between different median values implies that one can, in principle, define median aggregation functions that are neither anonymous nor systematic. However, if one keeps the choice of the median value independent from the identity of individuals and uniform for all propositions, then one obtains an anonymous and systematic median aggregation function. For example, we may stipulate that the aggregation function always assigns the lowest (or highest) possible median judgment value to a given proposition.

As we have seen in Example 2, median aggregation, just like average aggregation, does not yield consistent collective judgments in general. However, the following restriction on profiles, which generalizes a notion defined in [10] for classical JA, provides a sufficient condition for the existence of a median judge for the whole agenda.

Definition 12 (Unidimensional Alignment). *Let $P = \langle J_i \rangle_{i \in N} \in \mathbf{P}$ be a profile for an L-JA problem $\langle N, \mathcal{A} \rangle$. P is unidimensionally aligned with respect to a (strict) linear ordering $\prec$ of the individuals N if, for every $\phi \in \mathcal{A}$ one of the following holds.*

(1) $\forall i, j \in N \; i \prec j \Rightarrow J_i(\phi) \leq J_j(\phi)$, or
(2) $\forall i, j \in N \; i \prec j \Rightarrow J_i(\phi) \geq J_j(\phi)$.

In case (1) ϕ is upwards aligned in P with respect to $\prec$. In case (2) ϕ is downwards aligned in P with respect to $\prec$.

A profile P is unidimensionally alignable if there exists a linear ordering $\prec$ of N such that P is unidimensionally aligned with respect to $\prec$.

Note that in contrast to (strictly) order compatible profiles (see Sects. 3 and 4), unidimensionally aligned profiles are ordered along individuals, not along propositions.

Example 3. We revisit the profile P of Example 2. Proposition ϕ_1 is upwards aligned and proposition ϕ_2 is downwards aligned in P with respect to the natural order relation ($<$) on $N = \{1, \ldots, 6\}$. But ϕ_3 and ϕ_4 are not unidimensionally aligned with respect to $<$. We can of course, downwards or upwards align those propositions individually with respect to some other ordering of N. However, there is no ordering $\prec$ of N such that all four propositions in $\mathcal{A}$ are unidimensionally aligned with respect to $\prec$. Hence P is not unidimensionally alignable. In fact, even if we delete either ϕ_3 or ϕ_4 from $\mathcal{A}$ and correspondingly from P, the two resulting profiles are not unidimensionally alignable. On the other hand, the profile restricted to the agenda $\mathcal{A}' = \{\phi_1, \phi_2\}$ is already unidimensionally aligned.

Theorem 3. *There exists a median judge for every unidimensionally alignable profile for any JA problem $\langle N, \mathcal{A} \rangle$ with respect to any many-valued logic L.*

Proof. Let $P = \langle J_i \rangle_{i \in N} \in \mathbf{P}$ be a unidimensionally alignable profile for the JA problem $\langle N, \mathcal{A} \rangle$. By Definition 12, there is a linear ordering $\prec$ of N such that P is unidimensionally aligned with respect to $\prec$. In particular, every $\phi \in \mathcal{A}$ is either upwards aligned or downwards aligned in P with respect to $\prec$.

We distinguish two cases, according to whether $|N|$ is odd or even.

$|N|$ *is Odd:* Since N is linearly ordered by $\prec$, there is a midpoint $m \in N$ with respect to $\prec$. This means that we can partition N into $N_1 = \{i \mid i \prec m\}$ and $N_2 = \{i \mid m \prec i\} \cup \{m\}$ such that $|N_1| = |N_2| - 1$. Note that, since $\prec$ is a strict ordering we have $N = N_1 \cup N_2$ and $N_1 \cap N_2 = \emptyset$. If $\phi \in \mathcal{A}$ is upwards aligned in P with respect to $\prec$ then $\forall i \in N_1 \; J_i(\phi) \leq J_m(\phi)$ and $\forall i \in N_2 \; J_i(\phi) \geq J_m(\phi)$. Hence J_m is a median judgment for ϕ according to Definition 10, where N_1 instantiates $N_\downarrow$ and N_2 instantiates $N_\uparrow$. If, on the other hand, ϕ is downwards aligned in P with respect to $\prec$ then $\forall i \in N_1 \; J_i(\phi) \geq J_m(\phi)$ and $\forall i \in N_2 \; J_i(\phi) \leq J_m(\phi)$. Again J_m is a median judgment for ϕ, where now N_2 instantiates $N_\downarrow$ and N_1 instantiates $N_\uparrow$ in the defining condition. To sum up, J_m is a median judgment for every $\phi \in \mathcal{A}$, and hence m is a median judge for the profile P.

$|N|$ *is even*: The argument is analogous to the previous case. We now obtain two adjacent midpoints m and m' with respect to $\prec$. We partition N into $N_1 = \{i \mid i \prec m\} \cup \{m\}$ and $N_2 = \{i \mid m' \prec i\} \cup \{m'\}$ and obtain that both, m and m' are median judges for P.

Corollary 2. *For every JA problem $\langle N, \mathcal{A} \rangle$ there is a median aggregation function F^{md} that yields an L-consistent collective judgment for every unidimensionally alignable profile for $\langle N, \mathcal{A} \rangle$.*

Proof. By definition, every individual judgment in a profile is L-consistent. Hence the statement follows immediately from Definition 11 and Theorem 3.

To sum up: if the agenda is unidimensionally aligned, then a median judgment coincides with the judgment of one of the individuals (the median judge) for the whole agenda and hence is L-consistent by definition.

We have already indicated above how average aggregation, which is in the focus of [5], compares with median aggregation, introduced here. We emphasize that median aggregation leaves more room for consistent collective judgments than average aggregation. For example, for any classical (0/1-valued) profile where $|N| = 3$, every individual is a median judge and hence provides a consistent collective median judgment. In contrast, the doctrinal paradox (see Sect. 2) presents a classical profile that cannot be aggregated consistently using the average rule, even if we allow for intermediary values in the collective judgment. The greater 'liberality' of median aggregation is related to the fact that it only rests on comparisons of individual judgments, whereas computing the average requires that arithmetic operations are applied to individual judgments.

6 Conclusion

We have shown three possibility results for many-valued judgment aggregation. The first two results, referring to Kleene-Zadeh logic KZ and Gödel logic G, respectively, apply to judgment profiles where the individuals agree on the relative order of values that the atomic propositions contained in the agenda should receive. The third result introduces median aggregation for arbitrary many-valued logics and generalizes a classical possibility result of List [10] for unidimensionally alignable profiles.

Our results, jointly with those of [5], should be considered only as starting points for a wider exploration of possibilities for fuzzy logic based judgment aggregation. In particular, the newly introduced median aggregation deserves further attention, since the consistency criterion of Theorem 3 (unidimensional alignability) is not at all tight, thus leaving room for other types of profiles that can be consistently aggregated in this manner.

References

1. Baaz, M., Preining, N.: Gödel-Dummett logics. In: Cintula, P., Hájek, P., Noguera, C. (eds.) Handbook of Mathematical Fuzzy Logic - Volume 2, pp. 585–625. College Publications (2015)
2. Dietrich, F., List, C.: Arrow's theorem in judgment aggregation. Soc. Choice Welfare **29**, 19–33 (2007)
3. Dokow, E., Holzman, R.: Aggregation of binary evaluations with abstentions. J. Econ. Theory **145**(2), 544–561 (2010)
4. Esteban, M., Palmigiano, A., Zhao, Z.: An abstract algebraic logic view on judgment aggregation. In: van der Hoek, W., Holliday, W.H., Wang, W. (eds.) LORI 2015. LNCS, vol. 9394, pp. 77–89. Springer, Heidelberg (2015). https://doi.org/10.1007/978-3-662-48561-3_7
5. Fermüller, C.G.: Some consistency criteria for many-valued judgment aggregation. In: 53rd IEEE International Symposium on Multiple-Valued Logic, ISMVL 2023, Matsue, Japan, May 22–24, 2023, pp. 215–220. IEEE (2023)
6. Grossi, D., Pigozzi, G.: Judgment Aggregation: A Primer. Springer, Cham (2014). https://doi.org/10.1007/978-3-031-01568-7
7. Herzberg, F.S.: Universal algebra for general aggregation theory: many-valued propositional-attitude aggregators as MV-homomorphisms. J. Log. Comput. **25**(3), 965–977 (2013)
8. Kornhauser, L.A., Sager, L.G.: The one and the many: adjudication in collegial courts. Calif. Law Rev. **81**(1), 1–59 (1993)
9. Lang, J., Slavkovik, M.: Judgment aggregation rules and voting rules. In: Perny, P., Pirlot, M., Tsoukiàs, A. (eds.) ADT 2013. LNCS (LNAI), vol. 8176, pp. 230–243. Springer, Heidelberg (2013). https://doi.org/10.1007/978-3-642-41575-3_18
10. List, C.: A possibility theorem on aggregation over multiple interconnected propositions. Math. Soc. Sci. **45**(1), 1–13 (2003)
11. List, C.: Social choice theory. In: Zalta, E.N. (ed.) The Stanford Encyclopedia of Philosophy. Stanford University, Metaphysics Research Lab (2022)
12. List, C., Pettit, P.: Aggregating sets of judgments: an impossibility result. Econ. Philos. **18**, 89–110 (2002)
13. List, C., Pettit, P.: Aggregating sets of judgments: two impossibility results compared. Synthese **140**, 207–235 (2004)
14. Mongin, P.: Judgment aggregation. In: Hansson, S.O., Hendricks, V.F. (eds.) Introduction to Formal Philosophy. SUTP, pp. 705–720. Springer, Cham (2018). https://doi.org/10.1007/978-3-319-77434-3_38
15. Pettit, P.: Deliberative democracy and the discursive dilemma. Philos. Issues **11**(1), 268–299 (2001)

Free Quantification in Four-Valued and Fuzzy Bilattice-Valued Logics

Libor Běhounek$^{(\boxtimes)}$ [ID], Martina Daňková [ID], and Antonín Dvořák [ID]

University of Ostrava, IRAFM, 30. dubna 22, 701 03 Ostrava, Czech Republic
{libor.behounek,martina.dankova,antonin.dvorak}@osu.cz

Abstract. We introduce a variant of free logic (i.e., a logic admitting terms with nonexistent referents) that accommodates truth-value gluts as well as gaps. Employing a suitable expansion of the Belnap–Dunn four-valued logic, we specify a dual-domain semantics for free logic, in which propositions containing non-denoting terms can be true, false, neither true nor false, or both true and false. In each model, the dual domain semantics separates existing and non-existing objects into two subdomains, making it possible to quantify either over all objects or existing objects only. We also outline a fuzzy variant of the dual-domain semantics, accommodating non-denoting terms in fuzzy contexts that can be partially indeterminate or inconsistent.

Keywords: Free logic · Belnap–Dunn logic · Bi-lattice logic · Fuzzy logic · Dual-domain semantics

1 Introduction

The four-valued bilattice logic BD (Belnap–Dunn logic, also known as FDE) has been designed to deal with underdetermined (i.e., neither true nor false) as well as overdetermined (i.e., both true and false) modes of truth and falsity besides the usual truth values *true* and *false* [4,9,14]. Its generalizations to $[0,1]^2$-valued logics have been used for reasoning under indeterminacy and uncertainty [6,7,12]. Here we employ a suitable expansion of the first-order logic BD (in particular, the logic BDΔ of [15]) and its $[0,1]^2$-valued generalization over Łukasiewicz fuzzy logic Ł$_\Delta$ as the background machinery to develop logics of the so-called free quantification in the presence of (either crisp or graded) indeterminacy and overdeterminacy of truth and falsity.

Free logic, or 'logic free of existential assumptions' [5], is a field of logic that studies quantification over terms that may be non-denoting. Such terms naturally occur in various situations, e.g., in logical analysis of natural language (definite descriptions), fictional discourse, and predicate modal logics (incl. epistemic and temporal). When terms over which quantification is carried include non-denoting ones, some standard quantification laws, such as universal instantiation (also known as specification) and its existential dual, are no longer valid. Naturally, the invalidity of these laws should be reflected in the semantics of free logics.

© The Author(s), under exclusive license to Springer Nature Switzerland AG 2023
V.-N. Huynh et al. (Eds.): IUKM 2023, LNAI 14375, pp. 15–26, 2023.
https://doi.org/10.1007/978-3-031-46775-2_2

Free logics are developed in several variants (see, e.g., [13]). One of intuitively appealing variants, which we choose for our endeavor, is the so-called *positive* free logic with *dual-domain semantics.* Positive free logics allow atomic formulas with non-denoting terms to be true, false, or indeterminate (truth-valueless). In the dual-domain semantics, a model is equipped with two domains: the *inner domain* D_1 (which is the extension of the *existence predicate,* traditionally denoted by E!) that collects existing objects, and the outer domain $D_0 \supseteq D_1$ containing all objects, both existing and non-existing. Then, the universal and existential quantifiers over existing objects can be defined in such a way that the range of variables is limited to the inner domain [8]; then, however, the objects from $D_0 \setminus D_1$ that are not associated with a logical constant are unreachable by formulas of the given logic. A more flexible possibility is to use the so-called *inner* and *outer quantifiers.* The outer quantifiers range over D_0 and are simply the standard quantifiers of the given logic. The inner quantifiers ranging over D_1 can be defined by relativization of the outer ones.

Since positive free logic allows formulas containing non-denoting terms to be truth-valueless, it is expedient to evaluate formulas in a logic which accommodates *truth-value gaps,* such as three-valued strong Kleene logic K_3 [11] that has the truth values *true, false,* and *neither.* An option less often considered is using a logic that also accommodates truth-value *gluts,* or the truth values for propositions that are *both* true and false. The need for truth-value gluts in positive free logic follows from the fact that nonexistent objects can have contradictory properties: e.g., a *square circle* is both round (being a circle) and not round (being square). A recent system of free logic by Carnielli and Antunes [8] does consider truth-value gluts (though not truth-value gaps), using the three-valued logic LFI1 with the truth values *true, false,* and *both.* In this paper we propose a four-valued free logic that accommodates truth-value gaps as well as gluts, using the four values *true, false, neither,* and *both.* To this end, we employ the first-order logic BDΔ, which was introduced in [15]. We recall the logic BDΔ in Sect. 2 and introduce the (positive dual-domain) free logic over BDΔ in Sect. 3.

A fuzzy variant of a positive free logic with dual-domain semantics has been outlined in [2]. The need for fuzzified free logic is rather natural, as non-denoting terms and terms denoting nonexistent objects can be encountered in fuzzy contexts just like in crisp contexts (e.g., when a fuzzy property is predicated of a nonexistent individual, as in "Sherlock Holmes is clever"). In [2], free fuzzy logic is built over partial fuzzy logic [3], where the additional truth value $*$ represents a truth-value gap. This contribution extends the approach outlined above in such a way that it incorporates also truth-value gluts, and moreover admits the underdeterminacy or contradictoriness of truth values to be graded, i.e., just partially indeterminate (like in interval-valued fuzzy logic) or partially inconsistent (when the degrees of truth and falsity sum up to more than the value 1). We sketch the fuzzification of the four-valued logic BDΔ via Łukasiewicz fuzzy logic Ł_Δ and present the dual-domain semantics over the resulting logic ŁBDΔ in Sect. 4. Some topics for future work are mentioned in Sect. 5.

2   Four-Valued Bilattice Logic

The paradigmatic four-valued logic for reasoning about propositions that can be true, false, both, or neither, is the Belnap–Dunn logic BD [4,9,14]. However, since the logic BD is expressively rather poor, it is expedient to use a suitable extension thereof. For our purposes, the logic BDΔ of [15] is a suitable choice, as it contains (as definable connectives) the normality indicator as well as an implication needed for relativized quantifiers in Sect. 3. Furthermore, the logic BDΔ is defined as a first-order four-valued logic with equality and has a sound and complete natural-deduction axiomatic system [15].

The set of truth values of the logic BDΔ is $\{0,1\}^2$, i.e., the set of pairs of the classical truth values 0 and 1. The first component α of a truth value $\langle \alpha, \beta \rangle \in \{0,1\}^2$ indicates whether the proposition is true or not and the second component β indicates whether it is false or not; the truth and falsity of a proposition are evaluated independently, so a proposition can be both true and false as well as neither true nor false. We can thus define the following four truth values of BDΔ:

$$\mathbf{t} = \langle 1,0 \rangle \quad \text{true (only)}$$
$$\mathbf{f} = \langle 0,1 \rangle \quad \text{false (only)}$$
$$\mathbf{n} = \langle 0,0 \rangle \quad \text{neither true nor false}$$
$$\mathbf{b} = \langle 1,1 \rangle \quad \text{both true and false}$$

It is customary to define two lattice orders on the four truth values (see Fig. 1):

- The *truth order* $\leq_t$, where $\mathbf{f}$ is the least, $\mathbf{t}$ the largest, and $\mathbf{n}$ and $\mathbf{b}$ are intermediate and mutually incomparable; and
- The *information order* $\leq_i$, where $\mathbf{n}$ is the least, $\mathbf{b}$ the largest, and $\mathbf{f}$ and $\mathbf{t}$ are intermediate and mutually incomparable.

The *designated* truth values (i.e., those considered "true" in the definition of entailment) are those which are "at least true" in the information order, i.e., $\mathbf{t}$ and $\mathbf{b}$. Furthermore, we call the truth values $\mathbf{t}$ and $\mathbf{f}$ *normal*.

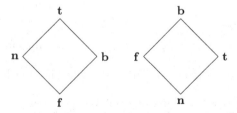

Fig. 1. The Hasse diagrams of the truth order (left) and the information order (right) on the four truth values.

Accordingly, the *valuation* of propositional atoms in BDΔ is a function $v \colon At \to \{0,1\}^2$. It can be decomposed into a pair of functions, $v^+, v^- \colon At \to$

$\{0,1\}$, called the *positive* and *negative valuation*, where $v(p) = \langle v^+(p), v^-(p)\rangle$ for any $p \in At$.

The propositional language of BDΔ consists of the symbols for propositional connectives $\{\sim, \wedge, \vee, \delta\}$ with the following meanings:

$\wedge$	(truth-order lattice) conjunction
$\vee$	(truth-order lattice) disjunction
$\sim$	negation
δ	indicator of designated values

Note that we use the symbol δ for the indicator of designated values (whereas [15] used Δ), to avoid confusion with the connective $\triangle$ of fuzzy logic.

The semantics of these connectives in BDΔ is given by the following Tarski conditions (in the form suitable for easy fuzzification):

$$v^+(\varphi \wedge \psi) = \min(v^+(\varphi), v^+(\psi)) \tag{1}$$
$$v^-(\varphi \wedge \psi) = \max(v^-(\varphi), v^-(\psi)) \tag{2}$$
$$v^+(\varphi \vee \psi) = \max(v^+(\varphi), v^+(\psi)) \tag{3}$$
$$v^-(\varphi \vee \psi) = \min(v^-(\varphi), v^-(\psi)) \tag{4}$$
$$v^+(\sim\varphi) = v^-(\varphi) \tag{5}$$
$$v^-(\sim\varphi) = v^+(\varphi) \tag{6}$$
$$v^+(\delta\varphi) = v^+(\varphi) \tag{7}$$
$$v^-(\delta\varphi) = 1 - v^+(\varphi) \tag{8}$$

Thus, e.g., conjunction is (at least) true if both conjuncts are (at least) true and is (at least) false if either conjunct is (at least) false, and similarly for the other connectives (where "at least" means the information order $\leq_i$). Furthermore we define the following derived connectives of BDΔ (cf. [1,14,15]):

$$\neg\varphi \equiv \sim\delta\varphi \qquad\qquad \text{bivalent negation} \tag{9}$$
$$\varphi \rightarrow \psi \equiv \sim\varphi \vee \psi \qquad\qquad \text{(material) implication} \tag{10}$$
$$\circ\varphi \equiv (\delta\varphi \vee \delta\sim\varphi) \wedge (\sim\delta\varphi \vee \sim\delta\sim\varphi) \qquad \text{normality indicator} \tag{11}$$

The definitions can be summarized by the following truth tables:

$\varphi \wedge \psi$	t	b	n	f
t	t	b	n	f
b	b	b	f	f
n	n	f	n	f
f	f	f	f	f

$\varphi \vee \psi$	t	b	n	f
t	t	t	t	t
b	t	b	t	b
n	t	t	n	n
f	t	b	n	f

φ	$\sim\varphi$
t	f
b	b
n	n
f	t

φ	$\delta\varphi$
t	t
b	t
n	f
f	f

$\varphi \to \psi$	t	b	n	f
t	t	b	n	f
b	t	b	t	b
n	t	t	n	n
f	t	t	t	t

φ	$\neg\varphi$
t	f
b	f
n	t
f	t

φ	$\circ\varphi$
t	t
b	f
n	f
f	t

The first-order language of BDΔ consists of the propositional connectives, the universal and existential quantifiers ∀ and ∃, a countable set of individual variables $Var = \{x_1, x_2, \ldots\}$, a countable set of individual constants $Const = \{c_1, c_2, \ldots\}$, and countable set of predicate symbols (each of a given finite arity) $Pred = \{P_1, P_2, \ldots\}$; in this paper, we omit function symbols for the sake of simplicity. The set of formulas is defined as follows (in the Backus–Naur form):

$$\varphi ::= P(t_1, \ldots, t_n) \mid {\sim}\varphi \mid (\varphi \wedge \varphi) \mid (\varphi \vee \varphi) \mid \delta\varphi \mid (\forall x)\varphi \mid (\exists x)\varphi,$$

where $t_1, \ldots, t_n$ are terms, i.e., individual constants or variables. We employ the usual conventions for omitting parentheses and using the derived connectives. The notions of free and bound variable and sentence are defined as usual.

A model for the (first-order) logic BDΔ is a tuple

$$M = \langle D_M, (P_M)_{P \in Pred}, (c_M)_{c \in Const} \rangle,$$

where D_M is a non-empty set (the *domain* of the model) and $c_M \in D_M$ for all $c \in Const$. Each predicate $P_M \in Pred$ of arity n is interpreted in M by a function

$$P_M : (D_M)^n \to \{0, 1\}^2.$$

It can be decomposed into a pair of functions,

$$P_M^+, P_M^- : (D_M)^n \to \{0, 1\},$$

called the *positive* and *negative extension* (or the *extension* and *anti-extension*) of P in M, so that for all $a_1, \ldots, a_n \in D_M$,

$$P_M(a_1, \ldots, a_n) = \langle P_M^+(a_1, \ldots, a_n), P_M^-(a_1, \ldots, a_n) \rangle.$$

For instance, if $S_M(a) = \mathbf{b} = \langle 1, 1 \rangle$, then $S_M^+(a) = S_M^-(a) = 1$, for a unary predicate $S \in Pred$ and $a \in D_M$ (see Fig. 2).

An *evaluation* of individual variables in M is a function $e\colon Var \to D_M$. By $e[x \mapsto a]$, where $x \in Var$ and $a \in D_M$, we denote the evaluation e' such that $e'(x) = a$ and $e'(y) = e(y)$ for each $y \in Var$ different from x.

The semantic value $\|t\|_{M,e} \in D_M$ of a term t in a model M under an evaluation e is defined as follows:

$$\|x\|_{M,e} = e(x) \qquad \text{for each } x \in Var$$
$$\|c\|_{M,e} = c_M \qquad \text{for each } c \in Const$$

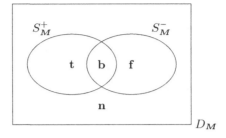

Fig. 2. The positive and negative extensions of a unary predicate S in a model M.

The truth value $\|\varphi\|_{M,e} = \langle \|\varphi\|_{M,e}^+, \|\varphi\|_{M,e}^- \rangle \in \{0,1\}^2$ of a formula φ in a model M under an evaluation e is given by the following Tarski conditions:

$$\|P(t_1,\ldots,t_n)\|_{M,e} = P_M(\|t_1\|_{M,e},\ldots,\|t_n\|_{M,e}) \tag{12}$$

$$\|\heartsuit(\varphi_1,\ldots,\varphi_m)\|_{M,e} = F_\heartsuit(\|\varphi_1\|_{M,e},\ldots,\|\varphi_m\|_{M,e}) \tag{13}$$

$$\|(\forall x)\varphi\|_{M,e}^+ = \inf_{a \in D_M} \|\varphi\|_{M,e[x\mapsto a]}^+ \tag{14}$$

$$\|(\forall x)\varphi\|_{M,e}^- = \sup_{a \in D_M} \|\varphi\|_{M,e[x\mapsto a]}^- \tag{15}$$

$$\|(\exists x)\varphi\|_{M,e}^+ = \sup_{a \in D_M} \|\varphi\|_{M,e[x\mapsto a]}^+ \tag{16}$$

$$\|(\exists x)\varphi\|_{M,e}^- = \inf_{a \in D_M} \|\varphi\|_{M,e[x\mapsto a]}^-, \tag{17}$$

for each n-ary predicate $P \in \textit{Pred}$ and each m-ary connective $\heartsuit$ of BDΔ (where the function $F_\heartsuit$ is given by the truth tables above).

The notion of logical consequence is defined as usual, i.e., as preservation of designated truth values from the premises to the conclusion. In more detail, for a given class K of models, we say that a set of BDΔ-formulas Γ (positively) *entails* a BDΔ-formula φ in K, written $\Gamma \models_K \varphi$, if for all models $M \in K$, the following condition holds: if $\|\psi\|_{M,e}^+ = 1$ for all $\psi \in \Gamma$ and all evaluations e in M, then also $\|\varphi\|_{M,e}^+ = 1$ for all evaluations e in M. We say that K is the class of models of a *theory* Γ (i.e., a set of formulas called the *axioms* of Γ) if $\|\varphi\|_{M,e}^+ = 1$ for all $M \in K$, all evaluations e in M, and all $\varphi \in \Gamma$.

By convention, we can drop the subscript in $\models_K$ if K is the class of all BDΔ-models. We may also omit the subscripts in $\|\varphi\|_{M,e}$ if they are clear from the context.

3   Four-Valued Free Quantification over the Logic BDΔ

As discussed in Sect. 1, we define free quantification over BDΔ in a similar manner as has been done over the three-valued strong Kleene logic K_3 (with truth-value gaps, cf. [2]) and the logic LFI1 (with truth-value gluts, [8]), i.e., by introducing a primitive predicate of existence and restricting quantifiers to it.

The language of the (positive dual-domain) free logic over BDΔ thus extends the language of BDΔ just by the unary *predicate of existence,* traditionally denoted by E!. Intuitively, E! is intended to delimit the objects that exist. For simplicity, in this paper we assume that E! is bivalent, i.e., that all objects either do or do not exist, excluding the possibility of objects that both have and do not have existence, as well as objects that neither exist nor don't exist. This is ensured by assuming the following axiom for E!:

$$\circ E!x \tag{18}$$

This setting leads to a dual-domain semantics that considers two bivalent domains in a given model M:

– The *outer domain* D_0 is the non-empty domain of the model, $(D_0)_M = D_M$.
– The *inner domain* D_1 is the positive extension of E!, i.e., $(D_1)_M = E!_M^+$.

In the following, we fix a model M and can omit the subscripts specifying it. In the inner domain $D_1 \subseteq D_0$ of the model, existent objects are collected. The elements of $E!^- = D_0 \setminus D_1$, i.e., the anti-extension of E! in M, serve as the absent referents of non-denoting terms, and thus represent nonexistent objects. The existence predicate makes it possible to restrict quantification to existing objects only, which leads to two kinds of quantifiers:

– The so-called *outer quantifiers,* which range over the outer domain D_0. These are the standard quantifiers of BDΔ, as they range over the whole domain $D_0 = D$ of the model. In the dual-domain free logic, they are denoted by Π, Σ instead of $\forall, \exists$, as the latter symbols are reserved for the (more commonly used) inner quantifiers over the existing objects.
– The *inner quantifiers* $\forall, \exists$, which range over the inner domain D_1, i.e., over the existing objects only.

The inner quantifiers can be defined from the standard (i.e., outer) BDΔ-quantifiers Π, Σ by restricting them to the inner domain D_1 of existing objects, via relativization by the (bivalent) existence predicate E!:

$$(\forall x)\varphi \equiv_{\mathrm{df}} (\Pi x)(E!x \rightarrow \varphi) \tag{19}$$
$$(\exists x)\varphi \equiv_{\mathrm{df}} (\Sigma x)(E!x \land \varphi) \tag{20}$$

It can be observed that the semantics of the BDΔ-connectives (given by the truth tables in Sect. 2) ensures the intended semantics of the inner quantifiers, i.e., that they range over the inner domain $D_1 = E!^+$ only. Indeed, the resulting Tarski conditions for the inner quantifiers read as follows:

$$\|(\forall x)\varphi\|_{M,e}^+ = \inf_{a \in (D_1)_M} \|\varphi\|_{M,e[x \mapsto a]}^+$$

$$\|(\forall x)\varphi\|_{M,e}^- = \sup_{a \in (D_1)_M} \|\varphi\|_{M,e[x \mapsto a]}^-$$

$$\|(\exists x)\varphi\|_{M,e}^+ = \sup_{a \in (D_1)_M} \|\varphi\|_{M,e[x \mapsto a]}^+$$

$$\|(\exists x)\varphi\|_{M,e}^- = \inf_{a \in (D_1)_M} \|\varphi\|_{M,e[x \mapsto a]}^-$$

Remark 1. In practice, the most commonly used quantifiers are the inner ones, as the intended range of quantification is usually just over existing individuals: for instance, the proposition "some horses fly" is normally considered false, disregarding Pegasus and other fictitious flying horses; thus its adequate formalization is $(\exists x)(Hx \wedge Fx)$. The outer quantifiers are needed for formalization of such propositions as "some things do not exist" (formalized as $(\Sigma x)\sim E!x$), which take into account nonexistent objects as well.

Remark 2. Besides axiom (18) that ensures the bivalence of the existence predicate, further optional axioms can be adopted in free BDΔ. For example, it is reasonable to assume that although non-existing objects such as the round square can be inconsistent, the existing objects in the inner domain D_1 can only have non-contradictory (or even normal, i.e., $\{\mathbf{t},\mathbf{f}\}$-valued) properties. The normality on the inner domain can be ensured by the axioms

$$(\forall x_1)\dots(\forall x_n)\circ P(x_1,\dots,x_n) \tag{21}$$

for all predicates P in the language and the non-contradictoriness by the analogous schema

$$(\forall x_1)\dots(\forall x_n)\neg\big(P(x_1,\dots,x_n)\wedge\sim P(x_1,\dots,x_n)\big) \tag{22}$$

for all $P \in Pred$ (where n is the arity of P).

Remark 3. The described dual-domain free logic over BDΔ can be understood as a formal theory over BDΔ with the special predicate E! and special axiom (18) (stronger variants of the theory can also use axiom (21) or (22)), with the notational convention by which the quantifiers are written as Π and Σ, whereas the inner quantfiers and the derived connectives are regarded as abbreviations that can be expanded according to their definitions (19)–(20) and (9)–(11). The known strong completeness theorem for BDΔ [15, Th. 4.9] then provides a strongly complete (natural deduction style) axiomatic system for the described variants of free logic over BDΔ.

Example 1. As expected in free logic, the classical law of specification and its existential dual do not hold in free BDΔ for the inner quantifiers:

$$(\forall x)\varphi(x) \not\models \varphi(t), \qquad\qquad \varphi(t) \not\models (\exists x)\varphi(x).$$

However, they do hold if the existence of t is explicitly assumed:

$$E!t, (\forall x)\varphi(x) \models \varphi(t), \qquad\qquad E!t, \varphi(t) \models (\exists x)\varphi(x).$$

Moreover, while $\models (\Sigma x)(\varphi(x)\vee\neg\varphi(x))$, the same does not hold for the inner quantifier, $\not\models (\exists x)(\varphi(x)\vee\neg\varphi(x))$; rather, the non-emptiness of the inner domain need be explicitly assumed: $(\exists x)E!x \models (\exists x)(\varphi(x)\vee\neg\varphi(x))$.

4   Free Fuzzy Logic with Graded Gaps and Gluts

We have intentionally formulated the definitions of the logic BD$\triangle$ and its dual-domain free variant in a format that can be easily fuzzified. In the fuzzification, the main change is replacing the bilattice $\{0,1\}^2$ of truth values with a bilattice L^2, for a suitable residuated lattice L; most definitions then stand as stated in the previous sections or require just minor adjustments. For simplicity, in this paper we will only consider the standard MV$_\triangle$-algebra $[0,1]_{Ł_\triangle}$ of Łukasiewicz fuzzy logic (with the operator $\triangle$) as the underlying residuated lattice. As the resulting logic uses the set $[0,1]^2$ of truth values, it is a bilattice-valued *square fuzzy logic* (cf. [10,16]); we will denote the logic by ŁBD$\triangle$.

The set of truth values of the logic ŁBD$\triangle$ is $[0,1]^2$. The first component α of a truth value $\langle \alpha, \beta \rangle \in [0,1]^2$ indicates the degree to which the proposition is true and the second component β indicates the degree to which it is false. Similarly to BD$\triangle$, the degrees of truth and falsity of a proposition are evaluated independently, so unlike in standard fuzzy logics, they need not sum up to 1. We keep the definitions of the truth constants $\mathbf{t} = \langle 1, 0 \rangle$, $\mathbf{f} = \langle 0, 1 \rangle$, $\mathbf{n} = \langle 0, 0 \rangle$, and $\mathbf{b} = \langle 1, 1 \rangle$, which now refer to the corners of the square $[0,1]^2$. The original truth values of fuzzy logic are embedded in $[0,1]^2$ as the pairs $\langle \alpha, 1 - \alpha \rangle$ for $\alpha \in [0,1]$; we will call these values *(fully) normal*. The truth values $\langle \alpha, \beta \rangle$ where $\alpha + \beta < 1$ are *gappy* and those where $\alpha + \beta > 1$ *glutty*; the former can be viewed as partial (or underdetermined) and the latter as contradictory (or overdetermined). The *designated* truth values of the logic ŁBD$\triangle$ are those true to the full degree, i.e., the pairs of the form $\langle 1, \beta \rangle$ for all $\beta \in [0,1]$. The information order $\leq_i$ and the truth order $\leq_t$ are defined on $[0,1]^2$ as follows:

- $\langle \alpha_1, \beta_1 \rangle \leq_i \langle \alpha_2, \beta_2 \rangle$ iff $\alpha_1 \leq \alpha_2$ and $\beta_1 \leq \beta_2$;
- $\langle \alpha_1, \beta_1 \rangle \leq_t \langle \alpha_2, \beta_2 \rangle$ iff $\alpha_1 \leq \alpha_2$ and $\beta_1 \geq \beta_2$.

The notion of valuation is defined as in propositional BD$\triangle$ (see Sect. 2). The propositional language of ŁBD$\triangle$ contains the connectives of BD$\triangle$ with the addition of the connectives specific to the fuzzy logic Ł$_\triangle$, namely, strong conjunction $\&$, strong disjunction $\oplus$, and the Delta operator $\triangle$. The Tarski conditions for the primitive connectives of BD$\triangle$ are defined exactly as in (1)–(8), only evaluated in $[0,1]$. The Tarski conditions for the additional connectives are as follows:

$$v^+(\varphi \,\&\, \psi) = \max(v^+(\varphi) + v^+(\psi) - 1, 0)$$
$$v^-(\varphi \,\&\, \psi) = \min(v^-(\varphi) + v^-(\psi), 1)$$

$$v^+(\varphi \oplus \psi) = \min(v^+(\varphi) + v^+(\psi), 1)$$
$$v^-(\varphi \oplus \psi) = \max(v^-(\varphi) + v^-(\psi) - 1, 0)$$

$$v^+(\triangle\varphi) = 1 - \mathrm{sgn}(1 - v^+(\varphi))$$
$$v^-(\triangle\varphi) = \mathrm{sgn}(1 - v^+(\varphi))$$

In the ŁBD$\triangle$-definitions of the derived connectives $\to$ and $\circ$, we must choose between the weak $(\wedge, \vee)$ and strong $(\&, \oplus)$ connectives of ŁBD$\triangle$ in place of the single conjunction and disjunction $(\wedge, \vee)$ of BD$\triangle$. In order for the definitions to conform with the intended semantics based on Łukasiewicz logic, the strong connectives $(\&, \oplus)$ are the appropriate choice. The ŁBD$\triangle$-definitions of the derived connectives thus read as follows:

$$\neg\varphi \equiv {\sim}\boldsymbol{\delta}\varphi$$
$$\varphi \to \psi \equiv {\sim}\varphi \oplus \psi$$
$$\circ\varphi \equiv (\boldsymbol{\delta}\varphi \oplus \boldsymbol{\delta}{\sim}\varphi) \mathbin{\&} ({\sim}\boldsymbol{\delta}\varphi \oplus {\sim}\boldsymbol{\delta}{\sim}\varphi)$$

For the derived connectives of ŁBD$\triangle$, we obtain the following Tarski conditions:

$$v^+(\neg\varphi) = 1 - v^+(\varphi)$$
$$v^-(\neg\varphi) = v^+(\varphi)$$

$$v^+(\varphi \to \psi) = \min(v^-(\varphi) + v^+(\psi), 1)$$
$$v^-(\varphi \to \psi) = \max(v^+(\varphi) + v^-(\psi) - 1, 0)$$

$$v^+(\circ\varphi) = 1 - \left|v^+(\varphi) + v^-(\varphi) - 1\right|$$
$$v^-(\circ\varphi) = 1 - v^+(\circ\varphi)$$

The first-order language of ŁBD$\triangle$ is the same as that of BD$\triangle$, modulo the added connectives $(\&, \oplus, \triangle)$ of Ł$_\triangle$. The definition of a model for ŁBD$\triangle$ is also the same as for BD$\triangle$, with the only difference that predicate symbols are evaluated in $[0,1]^2$ instead of $\{0,1\}^2$, i.e.,

$$P_M : (D_M)^n \to [0,1]^2.$$

Just like in BD$\triangle$, P_M can be decomposed into its positive and negative extensions $P_M^+, P_M^- : (D_M)^n \to [0,1]$, which are ordinary $[0,1]$-valued fuzzy sets or n-ary fuzzy relations on D_M (see Fig. 3), so that

$$P_M(a_1, \ldots, a_n) = \langle P_M^+(a_1, \ldots, a_n), P_M^-(a_1, \ldots, a_n)\rangle \in [0,1]^2.$$

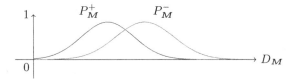

Fig. 3. The positive and negative extensions of a unary predicate P in a model M of ŁBD$\triangle$.

The notion of evaluation and the semantic value of a term in a model are defined in exactly the same way as in BD$\triangle$. Also the truth value of a formula,

$\|\varphi\|_{M,e} = \langle \|\varphi\|^+_{M,e}, \|\varphi\|^-_{M,e} \rangle$, is defined by the same Tarski conditions (12)–(17), only now valued in $[0,1]^2$ instead of $\{0,1\}^2$. Finally, the (positive) consequence relation and the class of models of a theory in LBDΔ are also defined by the same conditions as given in Sect. 2 for BDΔ.

The notions of the dual-domain free logic over LBDΔ, i.e., the existence predicate E!, the outer domain $(D_0) = D_M$, the inner domain $(D_1)_M = \mathrm{E!}^+_M$, and the outer quantifiers Π, Σ are also defined as in Sect. 3 over BDΔ. In LBDΔ the role of axiom (18) is different, as it only ensures the full normality of the existence predicate but allows the inner domain $(D_1)_M = \mathrm{E!}^+_M$ to be fuzzy; in order to ensure the bivalence of the inner domain, the axiom needs to be adjusted as follows:

$$\mathrm{o}\mathrm{E!}x \wedge (\mathrm{E!}x \vee {\sim}\mathrm{E!}x) \tag{23}$$

Assuming (23), the definitions of the inner quantifiers can use the same formulas (19)–(20) as in BDΔ. Like in the dual-domain semantics over BDΔ, it is reasonable to assume the full normality of all predicates on the inner domain, which is ensured by the axiom (21).

Remark 4. It can be observed that adding the following axiom schema to LBDΔ reduces it (including the dual-domain semantics) to the four values of BDΔ:

$$\big(\mathrm{o}\varphi \wedge (\varphi \vee {\sim}\varphi)\big) \vee {\sim}\mathrm{o}\varphi$$

Similarly, the axiom schema $\mathrm{o}\varphi \vee ({\sim}\boldsymbol{\delta}\varphi \wedge {\sim}\boldsymbol{\delta}{\sim}\varphi)$ reduces the setting to that of partial fuzzy logic from [2], which considers only the value **n** (denoted there by $*$) besides the degrees from $[0,1]$. Using the axiom schema $\mathrm{o}\varphi \vee {\sim}\mathrm{o}\varphi$ in LBDΔ adds the value **b** to the setting of [2], giving rise to a logic with the truth values $[0,1] \cup \{\mathbf{n}, \mathbf{b}\}$.

5   Conclusion

In this paper, we have presented a four-valued system of dual-domain positive free logic with truth-value gaps and gluts over the logic BDΔ, sketched a fuzzification of BDΔ via Łukasiewicz fuzzy logic Ł$_\Delta$, and defined the dual-domain semantics over the resulting logic LBDΔ. While the free logic over BDΔ is axiomatized by the strong completeness theorem for BDΔ due to Sano and Omori [15], we have only presented the (first-order and dual-domain) semantics of LBDΔ; a sound and complete axiomatization of LBDΔ is left for future work.

It can be observed that free logics based on truth-functional underlying logics are susceptible to lottery-style paradoxes, where disjuncts are evaluated by **n**, while their exhaustive disjunction should have the value **t**, contrary to the semantics of K$_3$ or (L)BDΔ. A remedy can take the form of using a non-truth-functional modality to handle such cases, similar to the two-layered modalities introduced in [6,7] over related logics. This option is also left for future work.

Acknowledgment. L. Běhounek and A. Dvořák were supported by project No. 22-01137S of the Czech Science Foundation.

References

1. Arieli, O., Avron, A.: Reasoning with logical bilattices. J. Logic Lang. Inform. **5**, 25–63 (1996)
2. Běhounek, L., Dvořák, A.: Non-denoting terms in fuzzy logic: an initial exploration. In: Kacprzyk, J., Szmidt, E., Zadrożny, S., Atanassov, K.T., Krawczak, M. (eds.) IWIFSGN/EUSFLAT -2017. AISC, vol. 641, pp. 148–158. Springer, Cham (2018). https://doi.org/10.1007/978-3-319-66830-7_14
3. Běhounek, L., Novák, V.: Towards fuzzy partial logic. In: Proceedings of the IEEE 45th International Symposium on Multiple-Valued Logics (ISMVL 2015), pp. 139–144. IEEE (2015)
4. Belnap, N.: A useful four-valued logic. In: Dunn, J.M., Epstein, G. (eds.) Modern Uses of Multiple-Valued Logic, pp. 5–37. D. Reidel Publishing Company, Dordrecht (1977)
5. Bencivenga, E.: Free logics. In: Gabbay, D.M., Guenthner, F. (eds.) Handbook of Philosophical Logic, 2nd edn., vol. 5, pp. 147–196. Kluwer (2002)
6. Bílková, M., Frittella, S., Kozhemiachenko, D., Majer, O.: Qualitative reasoning in a two-layered framework. Int. J. Approximate Reasoning **154**, 84–108 (2023)
7. Bílková, M., Frittella, S., Kozhemiachenko, D., Majer, O.: Two-layered logics for paraconsistent probabilities (2023). https://doi.org/10.48550/arXiv.2303.04565
8. Carnielli, W., Antunes, H.: An objectual semantics for first-order LFI1 with an application to free logics. In: Haeusler, E.H., Pereira, L.C.P.D., Viana, J.P. (eds.) A Question is More Illuminating than an Answer. A Festschrift for Paulo A.S. Veloso, pp. 58–81. College Publications, London (2020)
9. Dunn, J.M.: Intuitive semantics for first-degree entailment and "coupled trees.". Philos. Stud. **29**, 149–168 (1976)
10. Genito, D., Gerla, G.: Connecting bilattice theory with multivalued logic. Logic Log. Philos. **23**, 15–45 (2014)
11. Kleene, S.C.: On notation for ordinal numbers. J. Symb. Log. **3**, 150–155 (1938)
12. Klein, D., Majer, O., Rafiee Rad, S.: Probabilities with gaps and gluts. J. Philos. Log. **50**, 1107–1141 (2021)
13. Nolt, J.: Free logic. In: Zalta, E.N. (ed.) The Stanford Encyclopedia of Philosophy. Stanford University, Fall 2021 edn. (2021). https://plato.stanford.edu/archives/fall2021/entries/logic-free/
14. Omori, H., Wansing, H.: 40 years of FDE: an introductory overview. Stud. Logica. **105**(6), 1021–1049 (2017)
15. Sano, K., Omori, H.: An expansion of first-order Belnap-Dunn logic. Logic J. IGPL **22**(3), 458–481 (2013)
16. Turunen, E., Öztürk, M., Tsoukiás, A.: Paraconsistent semantics for Pavelka style fuzzy sentential logic. Fuzzy Sets Syst. **161**, 1926–1940 (2010)

Weighted Fuzzy Rules Based on Implicational Quantifiers

Martina Daňková[(✉)](iD)

University of Ostrava, CE IT4Innovations, 30. dubna 22, 701 03 Ostrava 1,
Czech Republic
martina.dankova@osu.cz

Abstract. In this paper, we explore the use of General Unary Hypotheses Automaton (GUHA) quantifiers, explicitly implicational quantifiers, for analyzing specific relational dependencies. We discuss their suitability in fuzzy modeling and demonstrate their integration with appropriate fuzzy rules to create a new class of weighted fuzzy rules. This study contributes to the advancement of fuzzy modeling and offers a framework for further research and practical applications.

Keywords: Implicational Quantifiers · IF–THEN Rules · Fuzzy Logic · Weighted Fuzzy Rules

1 Introduction

There exist various approaches to modeling dependencies between input and output domains of interest that are applicable, e.g., in the process of gaining knowledge in databases or for confirmation of assertions about patterns in an analyzed database. These assertions can often be expressed using a logical calculus, and items in a database serve as basic observations that allow us to support or reject them. Certain patterns of interest with fuzzy attributes can be analyzed involving a four-fold table, which gathers information from the database about the number of objects that satisfy both the antecedent "A" and the consequent "B", only "A", only "B" or neither, where "A" and "B" can be of a vague nature. This is a key component of both the fuzzy association rules [1,12,13] and the fuzzy GUHA method [5,7,15,17]. Note that fuzzy association rule mining is part of the GUHA method, so we will only report on this broader method below.

Both of the above methods test automatically generated hypotheses, and these hypotheses can take the form of a single fuzzy rule [8,16]. Testing is carried out based on a suitably chosen quantifier [10,11]. In practical applications that generate fuzzy rules using the GUHA quantifier [18], mainly bivalent quantifiers have been used. However, GUHA quantifiers are defined using statistics and can be identified with functions having values in $[0,1]$.

In this paper, we use GUHA quantifiers that are suitable to analyze the dependence of the form

$$\text{"If antecedent then consequent"}.$$

This research was supported by the Czech Science Foundation project No. 23-06280S.

V.-N. Huynh et al. (Eds.): IUKM 2023, LNAI 14375, pp. 27–36, 2023.
https://doi.org/10.1007/978-3-031-46775-2_3

These quantifiers are referred to as implicational [6,19]. Next, we show their suitability in fuzzy modeling. We combine the values of this implicational quantifier with the appropriate fuzzy rules to obtain a new class of weighted fuzzy rules. This expansion provides a promising avenue for diverse applications in various fields; for example, the integration of weighted fuzzy rules can contribute to more precise and robust data mining processes, enabling the discovery of intricate relationships within complex datasets; due to assigning different weights to individual fuzzy rules, the classification system can establish finer decision boundaries, which enables more precise classification of data points that fall within ambiguous or overlapping regions of the feature space.

2   A Four-Fold Table and Implicational Quantifiers

In the sequel, we will use the following symbols:

$$
\begin{array}{ll}
\& & \text{left continuous t-norm} \\
\rightarrow & \text{residuum of \&} \\
\neg & \text{involutive negation} \\
\wedge & \text{minimum} \\
\vee & \text{maximum}
\end{array}
\tag{1}
$$

For simplicity of exposition, consider the following *data matrix*

$$\mathcal{D} = \{(x_i, f(x_i)\}_{i \in I},$$

where $x_i \in X$, $f(x_i) \in Y$, $I = \{1, 2, \ldots n\}$, $X, Y \neq \emptyset$ and $f \colon X \to Y$. This $\mathcal{D}$ can be visualized as follows:

$$
\mathcal{D} = \begin{bmatrix}
x_1 & y_1 = f(x_1) \\
x_2 & y_2 = f(x_2) \\
\vdots & \vdots \\
x_n & y_n = f(x_n)
\end{bmatrix}.
\tag{2}
$$

Definition 1 (4ft-table). *Let A, B be fuzzy sets on $X, Y \neq \emptyset$, respectively, and $\mathcal{D}$ be a data matrix. We define a four-fold table for A, B w.r.t. $\mathcal{D}$ as a matrix 2×2*

$$
4ft(A, B) = \begin{bmatrix} a & b \\ c & d \end{bmatrix},
\tag{3}
$$

where

$$a = \sum_{i \in I} (A(x_i) \,\&\, B(y_i)), \tag{4}$$

$$b = \sum_{i \in I} (A(x_i) \,\&\, \neg B(y_i)), \tag{5}$$

$$c = \sum_{i \in I} (\neg A(x_i) \,\&\, B(y_i)), \tag{6}$$

$$d = \sum_{i \in I} (\neg A(x_i) \,\&\, \neg B(y_i)). \tag{7}$$

The values of the matrix $4ft(A, B)$ are connected to the fuzzy cardinalities of the data matrix within the corresponding fuzzy Cartesian product. For example, the value a is computed as the fuzzy cardinality of $\mathcal{D}$ over the fuzzy Cartesian product of A and B.

The following definition (taken from [11]) of implicational quantifiers was designed to provide a versatile tool for expressing and quantifying various degrees of dependency and causality between (fuzzy) sets based on observations from the data matrix.

Definition 2 (Implicational quantifier). *Let q be a function valued in the interval $[0, 1]$ defined for all pairs (a, b) of real numbers such that $a + b > 0$.*

We say that q is an implicational quantifier *if it satisfies the following property:*

$$\text{if } a \leq a' \text{ and } b' \leq b \text{ then } q(a, b) \leq q(a', b'), \tag{8}$$

is valid for all $a, b, a', b' \in \mathbb{R}$.

For a particular $4ft(A, B)$ of the form (3), we often write $q(A, B)$ instead of $q(a, b)$.

It has been established [11] that there is a direct relationship between implicational quantifiers and fuzzy implications, so that for every fuzzy implication, there is a corresponding way to construct an implicational quantifier.

Example 1. The following are examples of implicational quantifiers:

$$q_1(a, b) = a/(a + b), \tag{9}$$

$$q_2(a, b) = (0.9^{a+1}) \rightarrow_L (0.6^{b+1}), \tag{10}$$

$$q_3(a, b) = (0.8^{a+1}) \rightarrow_P (0.8^{b+1}), \tag{11}$$

$$q_4(a, b) = (b/(a + b)) \rightarrow_L (a/(a + b)), \tag{12}$$

$$q_5(a, b) = (b/(a + b)) \rightarrow_P (a/(a + b)), \tag{13}$$

for all a, b being positive reals, where $\rightarrow_L$ is Łukasiewicz residuum and $\rightarrow_P$ is the product residuum defined as

$$x \rightarrow_L y = \min(1, 1 - x + y), \tag{14}$$

$$x \rightarrow_P y = \min(1, y/x), \tag{15}$$

for all $x, y \in [0, 1]$.

Example 2. – Consider fuzzy sets A from Fig. 2(a), B, C from Fig. 2(c), and input data $\mathcal{D}$ from Fig. 1. Suppose the data from Fig. 1 illustrates commodity sales over time. In this context, the fuzzy set A represents a time segment, while the fuzzy sets B and C represent commodity sales, all characterized by imprecise boundaries. The values $\{a, b\}$ of the four-fold table for A, B w.r.t. $\mathcal{D}$ are $\{a, b\} = \{2.76, 10.64\}$, and for A, C w.r.t. $\mathcal{D}$, we obtain $\{a, b\} = \{0, 13.4\}$.

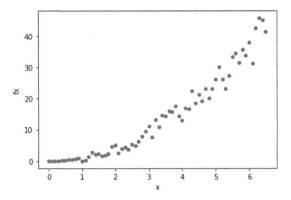

Fig. 1. Input data.

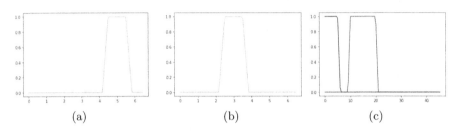

    (a)   (b)   (c)

Fig. 2. Fuzzy sets A (Figs. 2(a) and 2(b)) related to Example 2, Fuzzy sets C (blue line) and B (green line). (Color figure online)

The values of quantifiers $q_1, q_2 \ldots, q_5$ defined by (9)–(13), respectively, are the following:

i	1	2	3	4	5
$q_i(A, B)$	0.21	0.33	0.17	0.41	0.26
$q_i(A, C)$	0.0	0.1	0.05	0.0	0.0

– Consider fuzzy sets A from Fig. 2(b), B, C from Fig. 2(c), and input data $\mathcal{D}$ from Fig. 1. The values $\{a, b\}$ of the four-fold table for A, B w.r.t. $\mathcal{D}$, we obtain $\{a, b\} = \{6.61, 6.79\}$, and for A, C w.r.t. $\mathcal{D}$, we obtain $\{a, b\} = \{2.95, 10.45\}$. The values of quantifiers $q_1, q_2 \ldots, q_5$ defined by (9)–(13), respectively, are the following:

i	1	2	3	4	5
$q_i(A, B)$	0.49	0.57	0.96	0.99	0.97
$q_i(A, C)$	0.22	0.34	0.19	0.44	0.28

Let us recall from [11] that we have two ways of generating implicational quantifier using fuzzy implication, that is,

– Let $p, r \in (0, 1)$ be weights. Then

$$q_{p,r}(a, b) = (p^{a+1}) \rightarrow (r^{b+1}), \tag{16}$$

is the implicational quantifier.

For quantifiers obtained by this construction, we have the following property: If $\frac{a+1}{b+1} \geq \frac{\log r}{\log p}$ then $q_{p,r}(a,b) = 1$. For equal weights $p = r$ the threshold $\frac{\log r}{\log p} = 1$, so, in this case, we obtain $q_{p,r}(a,b) = 1$ whenever $a = b$.

- The following is an implicational quantifier:

$$q(a,b) = \left(\frac{b}{a+b}\right) \rightarrow \left(\frac{a}{a+b}\right). \tag{17}$$

These quantifiers are not as interesting for the residual implication $\rightarrow$ because $q(a,b) = 1$ whenever $a \geq b$. It is more suitable for non-residual implications, such as the Kleene-Dienes implication for which $q(a,b) = 1$ iff $b = 0$ (for more details, see [11]).

By observing the above special constructions of the implicational quantifier, we found that there is a large class of implicational quantifiers that are based on some order-reversing mapping.

Proposition 1. *Let $\mathcal{D}$ be a data matrix and $g\colon \mathbb{R} \mapsto [0,1]$ be a decreasing function. Then*

$$q_g(a,b) = g(a) \rightarrow g(b), \tag{18}$$

is the implicational quantifier.

Proof. It follows from the monotony of $\rightarrow$ in the second argument and the antitony in the second. If $a \leq a'$ and $b' \leq b$ then $g(a') \leq g(a)$ and $g(b) \leq g(b')$, and consequently

$$g(a) \rightarrow g(b) \leq g(a') \rightarrow g(b),$$
$$g(a') \rightarrow g(b) \leq g(a') \rightarrow g(b').$$

Hence,

$$g(a) \rightarrow g(b) \leq g(a') \rightarrow g(b'),$$

which shows that $q_g(a,b) \leq q_g(a',b')$, which means that q_g is an implicational quantifier.

Example 3. Let $n = |\mathcal{D}|$, where $\mathcal{D}$ is a data matrix. For example, we can use to construct (18) the following strictly monotone order-reversing functions:

$$g_1(x) = 1 - (x/n)^2, \tag{19}$$
$$g_2(x) = (n - x)/n, \tag{20}$$
$$g_3(x) = \exp(-x), \tag{21}$$
$$g_4(x) = \sqrt{(1 - (n - x)^2/n^2)}, \tag{22}$$
$$g_5(x) = -\ln(n - x + 1)/\ln(n + 1). \tag{23}$$

3   Fuzzy Rules with Weights Given by an Implicational Quantifier

Weighted fuzzy rules are often used in fuzzy logic systems to improve the accuracy and reliability of the system output [2,9,14]. By adjusting the weighting factors of the fuzzy rules, it is possible to fine-tune the behavior of the system and to adjust its sensitivity to different input conditions. This can be done at several levels, the antecedent level, the subsequent level or the whole rule [4]. The last level will be considered later.

Provided we know the dependency to be modeled, the fuzzy rules can be set without any additional computational effort, as, for example, in the case of monotonic dependency depicted in Fig. 3(a).

In reality, this situation appears rare. Therefore, a number of methods have been developed during the last decades to create a fuzzy rule base (including a

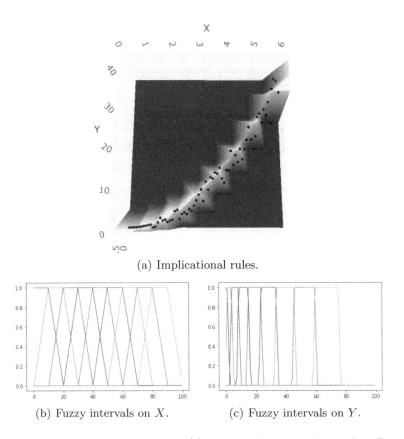

(a) Implicational rules.

(b) Fuzzy intervals on X.   (c) Fuzzy intervals on Y.

Fig. 3. An implicational model in Fig. 3(a) utilizing fuzzy sets depicted in Figs. 3(b) and 3(c) for monotonic dependency of the form $y = x^2$ together with the noisy data $\{x_j, x_j^2 + \text{RandBetween}(-x_j, x_j)\}_{j=1}^{65}$ (scatter plot).

weighted one) based on the input data. One of such models is shown in Fig. 4, where we construct neighborhoods $S(x,p)$ and $S'(y,q)$ using some preset similarity relations S, S' for each data entry (p,q) for all $x \in X, y \in Y$, we join them with the implicative rule $S(x,p) \rightarrow S'(y,q)$, and finally the minimum is taken over all $(p,q) \in \mathcal{D}$. For $\mathcal{D}$ as in Fig. 1, we obtain the implicative model as in Fig. 4. We can observe that the more data in the data matrix, the smaller the degrees of the final implicative model. Moreover, we lose simplicity of the resulting rule base and interpretability.

In some cases it is advantageous to use a fixed number of fuzzy rules in a rule base or to use fuzzy sets with preset linguistic interpretation. Therefore, in the following, we propose a new model that allows one to combine arbitrarily fuzzy sets from the input and output domains, and additionally, there are weights attached that tell us how much the rules suit the input data.

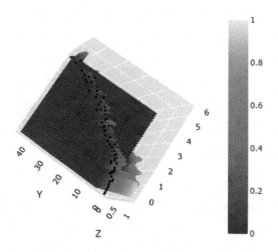

Fig. 4. Example of implicational model based on similarity and input data in Fig. 1.

This new model opens up a new approach to weighted fuzzy rules. Note that basically we have two main interpretations of fuzzy IF–THEN rules. One uses & to join the inner parts of a particular rule and then $\bigvee$ to glow the outer parts. Here, we focus only on the model that uses $\rightarrow$ within the rules and $\bigwedge$ to join the rules together, and therefore we call it the implicational model. A generalization of this model to the weighted implicational model was provided, e.g., in [3]. In the following, we introduce a weighted implicational data model based on implicational quantifier values.

Definition 3. *Let $\mathcal{D}$ be a data matrix of the form* (2), A_i *and* B_i *be fuzzy sets in* X *and* Y, *respectively, for all* $i \in I$, *where* I *is a finite set of indexes. Moreover, let* q *be an implicational quantifier. Then*

$$\text{GRules}_q^{\mathcal{D}}(x,y) = \bigwedge_{i \in I} \left(q(A_i, B_i) \rightarrow [A_i(x) \rightarrow B_i(y)]\right), \tag{24}$$

for all $x \in X, y \in Y$.

We call GRules$_q$ *the* weighted implicative model *w.r.t.* q *and* $\mathcal{D}$.

Since the implicational quantifier is a measure of dependence between two predicates based on observations, it works as a "switch" of the rule in the graded implicative data model. Observe the following (Fig. 5): if $q(A, B) = 1$ then the weighted fuzzy rule $q(A, B) \to (A(x) \to B(y))$ becomes the standard fuzzy rule $A(x) \to B(y)$, while if $q(A, B) = 0$ then the weighted fuzzy rule is evaluated at 1 everywhere, which corresponds to the fact that no implicational dependency was observed between A and B in the given data. In general, we can state that more data supporting the dependency we have higher the weight of the fuzzy rule, and consequently, closer we are to the nonweighted fuzzy rule as stated in the following proposition.

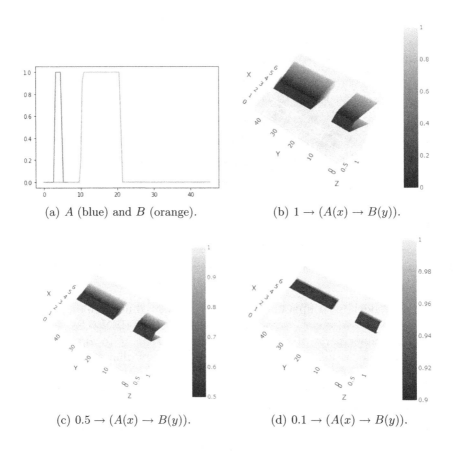

(a) A (blue) and B (orange). (b) $1 \to (A(x) \to B(y))$.

(c) $0.5 \to (A(x) \to B(y))$. (d) $0.1 \to (A(x) \to B(y))$.

Fig. 5. Example of weighted fuzzy rules. (Color figure online)

Proposition 2. *Let $\mathcal{D}, \mathcal{D}'$ be data matrices of the form* (2), A_i *and* B_i *be fuzzy sets on* X *and* Y, *respectively, for all* $i \in I$, *where* I *is a finite set of indexes. Furthermore, let* q *be an implicational quantifier and* a_i $(a'_i), b_i$ (b'_i) *be values of a four-fold table for* A_i, B_i *w.r.t.* $\mathcal{D}$ $(\mathcal{D}')$ *given by* (3) *for all* $i \in I$.
If $a_i \leq a'_i$ *and* $b'_i \leq b_i$ *for all* $i \in I$ *then*

$$\mathrm{GRules}_q^{\mathcal{D}'}(x, y) \leq \mathrm{GRules}_q^{\mathcal{D}}(x, y), \tag{25}$$

is valid for all $x \in X, y \in Y$, *where* $\mathrm{GRules}^{\mathcal{D}(\mathcal{D}')}$ *is the weighted implicative model w.r.t.* q *and* $\mathcal{D}$ $(\mathcal{D}')$ *given by* (24).

Proof. Due to the antitony of $\rightarrow$ in the first argument.

4   Conclusions

We showed a new construction of a subclass of implicational quantifiers using residual operators (see Proposition 1). This method is within the framework of standard fuzzy logic (the truth values are from $[0, 1]$) and is based on a construction introduced in [11]. Additionally, we proposed a well-suited fuzzy relational model (see Definition 3) that utilizes implicational quantifiers as weights. We have provided a justification for this model (see Proposition 2) to establish a new well-founded class of weighted fuzzy rules that precisely align with intuitive expectations for their behavior.

References

1. Agrawal, R., Imieliński, T., Swami, A.: Mining association rules between sets of items in large databases. In: Proceedings of the 1993 ACM SIGMOD International Conference on Management of Data, SIGMOD 1993, pp. 207–216. Association for Computing Machinery, New York, NY, USA (1993). https://doi.org/10.1145/170035.170072
2. Alcalá, R., Ramón Cano, J., Cordón, O., Herrera, F., Villar, P., Zwir, I.: Linguistic modeling with hierarchical systems of weighted linguistic rules. Int. J. Approx. Reasoning **32**(2), 187–215 (2003). https://doi.org/10.1016/S0888-613X(02)00083-X, Soft Computing in Information Mining
3. Daňková, M.: Approximation of extensional fuzzy relations over residuated lattices. Fuzzy Sets Syst. **161**(14), 1973–1991 (2010)
4. delaOssa, L., Gámez, J.A., Puerta, J.M.: Learning weighted linguistic fuzzy rules by using specifically-tailored hybrid estimation of distribution algorithms. Int. J. Approx. Reasoning **50**(3), 541–560 (2009). https://doi.org/10.1016/j.ijar.2008.11.003, Special Section on Bayesian Modelling
5. Hájek, P., Havránek, T.: Mechanizing Hypothesis Formation: Mathematical Foundations for a General Theory. Springer, Heidelberg (1978). https://doi.org/10.1007/978-3-642-66943-9
6. Hájek, P.: Metamathematics of Fuzzy Logic. Kluwer, Dordrecht (1998)
7. Hájek, P., Holeňa, M., Rauch, J.: The GUHA method and its meaning for data mining. J. Comput. Syst. Sci. **76**(1), 34–48 (2010). https://doi.org/10.1016/j.jcss.2009.05.004, Special Issue on Intelligent Data Analysis

8. Holeňa, M.: Fuzzy hypotheses for GUHA implications. Fuzzy Sets Syst. **98**(1), 101–125 (1998). https://doi.org/10.1016/S0165-0114(96)00369-7
9. Ishibuchi, H., Nakashima, T.: Effect of rule weights in fuzzy rule-based classification systems. In: Ninth IEEE International Conference on Fuzzy Systems. FUZZ-IEEE 2000 (Cat. No. 00CH37063), vol. 1, pp. 59–64 (2000). https://doi.org/10.1109/FUZZY.2000.838634
10. Ivánek, J.: On the correspondence between classes of implicational and equivalence quantifiers. In: Żytkow, J.M., Rauch, J. (eds.) Principles of Data Mining and Knowledge Discovery, pp. 116–124. Springer, Heidelberg (1999). https://doi.org/10.1007/978-3-540-48247-5_13
11. Ivánek, J.: Construction of implicational quantifiers from fuzzy implications. Fuzzy Sets Syst. **151**(2), 381–391 (2005). https://doi.org/10.1016/j.fss.2004.07.002
12. Lee, Y.S., Yen, S.J.: Mining utility association rules. In: Proceedings of the 2018 10th International Conference on Computer and Automation Engineering, ICCAE 2018, pp. 6–10. Association for Computing Machinery, New York, NY, USA (2018). https://doi.org/10.1145/3192975.3192987
13. Nanavati, A.A., Chitrapura, K.P., Joshi, S., Krishnapuram, R.: Mining generalised disjunctive association rules. In: Proceedings of the Tenth International Conference on Information and Knowledge Management, CIKM 2001, pp. 482–489. Association for Computing Machinery, New York, NY, USA (2001). https://doi.org/10.1145/502585.502666
14. Nauck, D.D.: Adaptive rule weights in neuro-fuzzy systems. Neural Comput. Appl. **9**, 60–70 (2000)
15. Ralbovský, M.: Fuzzy GUHA. Ph.D. thesis, Prague University of Economics and Business (2009)
16. Rauch, J.: Implicational rules. In: Observational Calculi and Association Rules, vol. 469, pp. 81–97. Springer, Heidelberg (2013). https://doi.org/10.1007/978-3-642-11737-4_7
17. Turunen, E.: GUHA-method in data mining, Pavelka style fuzzy logic, many-valued similarity and their applications in real world problems. In: MTISD 2008 - Methods, Models and Information Technologies for Decision Support Systems, pp. 49–50 (2008)
18. Turunen, E.: Mathematics Behind Fuzzy Logic. Advances in Soft Computing. Springer, Heidelberg (1999)
19. Turunen, E., Coufal, D.: Short term prediction of highway travel time using GUHA data mining method. Neural Network World **3–4**, 221–231 (2004)

Differentially Private Probabilistic Social Choice in the Shuffle Model

Qingyuan Ding⑩, Keke Sun⑩, Lixin Jiang⑩, Haibo Zhou⑩, and Chunlai Zhou⁽⊠⁾⑩

School of Information, Renmin University of China, Beijing 100872, China
czhou@ruc.edu.cn

Abstract. Given a profile of ranking lists over a finite set of alternatives, probabilistic social choice seeks to select a probability function over the alternatives on the basis of the pairwise comparison voting data. In this paper, we establish a differentially private formalism for probabilistic social choice in the shuffle model. In the shuffle model, individual voters submit their locally randomized data *anonymously* through a trusted shuffler which randomly permutes these individual data. Anonymity here plays a double role as a fairness condition and privacy amplification. The crucial step in our construction is to employ spectral clustering to find *data-independent* cluster centers and then to approximately round each input ranking order to these centers. We proceed to define a local differentially private randomizer plus the shuffler and then implement standard probabilistic social choice protocols such as maximal lottery and random dictatorship. Moreover, we analyze both privacy and utility of the proposed shuffle model and run some experiments to show the effectiveness of our formalism. In the last section, we discuss some related works and future directions.

Keywords: Probabilistic social choice · differential privacy · shuffle model

1 Introduction

Computational social choice theory is an important AI field that studies computational paradigms and techniques for the aggregation of individual preferences into a *deterministic* collective choice which is *usually* either an alternative or a ranking order over alternatives [8]. But, in certain situations with cyclic majorities or/and tied votes, there is no deterministic way to choose a clear majority winner from a finite set of alternatives or candidates on the basis of paired-comparison voting data without violating one of the two basic fairness conditions known as anonymity and neutrality [7]. Anonymity requires that the collective choice ought to be invariant with respect to permutations of individual preferences whereas neutrality requires impartiality towards the alternatives. Allowing *probabilistic* social outcomes hence seems like a necessity for *fair* collective decision. *Probabilistic social choice functions* (PSCFs) map a profile of individual preference relations over alternatives to a probability function (or lottery) over alternatives. In this paper, we focus on two main probabilistic social choice rules: one is random dictatorship (RD) and the other is maximal lottery (ML). *Random dictatorship* randomly picks an individual preference order and chooses the most-preferred alternative as the winner. It is the only strategy-proof and ex post efficient

ⓒ The Author(s), under exclusive license to Springer Nature Switzerland AG 2023
V.-N. Huynh et al. (Eds.): IUKM 2023, LNAI 14375, pp. 37–48, 2023.
https://doi.org/10.1007/978-3-031-46775-2_4

probabilistic social choice functions [16]. The *maximal lottery* method uses the *pair-wise* alternative-comparison voting data to find a lottery over the alternatives that are weakly preferred to any other lottery. It is characterized by two important consistency properties: population-consistency and composition-consistency [5]. Population consistency requires that, whenever two disjoint electorates agree on a lottery, this lottery should also be chosen by the union of both electorates. Composition consistency prescribes that the probability of an alternative within a component where the alternatives bear the same relationship to all other alternatives should be directly proportional to the probability that the alternative receives when the component is considered in isolation.

In this paper we study *differentially private(DP)* probabilistic social choice. In social choice theory, the anonymity property is usually treated as a fairness condition that the social outcome is independent of individual identities. So one might consider a social choice algorithm that has access only to the histogram of individual ranking orders [23]. However, in many cases for example, in a secret ballot, we don't have access to the voter identities and treat anonymity as *privacy protection*. But a *linkage attack* can match anonymized records with other non-anonymized records in a different dataset to recover these anonymized data [26]. Data cannot be fully anonymized and remain useful [12]. *Differential privacy* neutralizes such a linkage attack and addresses the trade-off between data privacy and utility. An algorithm is differentially private if, after seeing its output summary about the input database, an adversary cannot tell if a particular individual's data were used in the computation. In particular, the framework in this paper is just a differentially private mechanism with the adversary represented by a probabilistic social choice function.

Our main contribution is to establish a differentially private formalism for probabilistic social choice in the *shuffle model*. Unlike in the local model where the adversary collecting locally randomized data *directly* from individuals can track back an input to a specific individual, individuals in the shuffle model submit their randomized inputs to an adversary *anonymously* instead through a *shuffler* which randomly permutes individual randomized data. This setup yields a trust model which sits in between the classical central (or curator) and local models for differential privacy. Anonymity plays a *double* role in our framework: in probabilistic social choice, it is a fairness condition and in the shuffle model, it acts as *privacy amplifier*. It is this role as privacy amplification that reduces the high level of noise required in local randomizer and hence improves the accuracy of the model [2, 13]. Our model here is similar to the one-message shuffle model for the problem of summation of real numbers. The *crucial step* to construct the shuffle model is to partition the set of all $k!$ permutations over k candidates to l clusters of *equal size* and locate the l cluster centers for a fixed l. Since the Kemeny consensus problem associated in the clustering is intractable, we choose to encode all linear orders $\succ$ over k candidates as elements $\Phi(\succ)$ in $\mathbb{R}^{\binom{k}{2}}$, which faithfully represents the paired-comparison information in these rankings, and then employ spectral clustering technique to partition the encodings $\Phi(\succ)$ of all $k!$ rankings instead into l same-size clusters. The cluster centers $C_1, C_2, \cdots, C_l$ are chosen to be the barycenters of these clusters. It is important to note that these centers are independent of the input voting profile $P = \{(1, \succ_1), (2, \succ_2), \cdots, (n, \succ_n)\}$. Next we round the encoding $\Phi(\succ_i)$ of each individual voting data $(i, \succ_i)$ to its nearest cluster center $C_{r(\succ_i)}$ and then implement an

l-ary randomized response mechanism from $\{C_1, \cdots, C_l\}$ to itself. After shuffling, we obtain a histogram of $\{C_1, \cdots, C_l\}$. For maximal lottery, we debias the histogram and then apply the standard procedure. For random dictatorship, we debias the shuffled message, use KwikSort algorithm to turn those barycenters $C_1, \cdots, C_l$ into ranking orders and then apply the standard procedure. We illustrate our protocol for maximal lottery in Fig. 1. The privacy and accuracy analysis proves the effectiveness of our framework (Theorems 1). The rest of the paper is organized as follows. In Sect. 2, we fix some conventions and briefly introduce probabilistic social choice theory, differential privacy and shuffle model. Section 3 is our main contribution and it formulates our DP framework for probabilistic social choice in the shuffle model. In Sect. 4, we run some comparison experiments to show the effectiveness of our method and conclude in Sect. 5. To save space, we relegate the proofs and some relevant algorithms to the Supplementary Materials.

2   Preliminary Background

In this section, we provide a background about probabilistic social choice, differential privacy and the shuffle model. In this paper, we use calligraphic letters such as $\mathcal{R}, \mathcal{S}, \mathcal{Q}$ to denote algorithms or operators, capital letters such as X, Y, C to denote different domains, lower-case letters such as x, y, c, c' for elements in those domains. As usual, $\mathbb{N}$ and $\mathbb{R}$ denote the set of natural and real numbers, respectively. The set $\mathbb{N}_n^Y$ denotes the collection of all multisets or histograms over Y with cardinality n.

2.1   Probabilistic Social Choice

Let C be a finite set of alternatives or candidates $c_1, c_2, \cdots, c_{|C|}$ and $N :=$ $\{1, 2, \cdots, n\}$ be a set of n voters. For simplicity, k denotes $|C|$, the size of C, and the i-th voter is denoted as i. Each voter i contributes a *ranking* or a linear preference relation $\succ_i$ over all candidates, which is a complete, transitive and anti-symmetric binary relation over C. A *permutation* on $[k] := \{1, 2, \cdots, k\}$ is a one-to-one mapping from the finite set $[k]$ of natural numbers to itself. Let $\mathcal{L}(C)$ denote the set of all ranking orders on C and $\mathbb{S}_k$ denote the set of all permutations on $[k]$. There is a one-to-one correspondence between ranking orders $\succ$ on C and permutations τ on $[k]$. For example, if $k = 5$, the ranking $c_1 \succ c_4 \succ c_2 \succ c_5 \succ c_3$ is associated with the permutation

$$\tau = \begin{pmatrix} 1\,2\,3\,4\,5 \\ 1\,4\,2\,5\,3 \end{pmatrix}$$

which means that $\tau(1) = 1, \tau(2) = 4, \tau(3) = 2, \tau(4) = 5$ and $\tau(5) = 3$. So, for each ranking $\succ$ on C, its corresponding permutation on $[k]$ is denoted as $\tau^\succ$ and, for each permutation τ on $[k]$, its corresponding ranking on C is denoted as $\succ^\tau$. In the paper, we need the following important notions. Given two permutations $\tau, \tau' \in \mathbb{S}_k$, the numbers of pairwise agreements and disagreements between these two permutations are defined respectively as

$$n_a(\tau, \tau') = |\{(i, j) : i < j, \tau(i) < \tau(j) \text{ and } \tau'(i) < \tau'(j)\}|$$
$$+ |\{(i, j) : i < j, \tau(i) > \tau(j) \text{ and } \tau'(i) > \tau'(j)\}|$$
$$n_d(\tau, \tau') = |\{(i, j) : i < j, \tau(i) < \tau(j) \text{ and } \tau'(i) > \tau'(j)\}|$$
$$+ |\{(i, j) : i < j, \tau(i) > \tau(j) \text{ and } \tau'(i) < \tau'(j)\}|$$

The number $n_d(\tau, \tau')$ is usually called the *Kendall-tau* distance between τ and τ'.

A (ranking) *profile* P is a tuple $(\succ_1, \succ_2, \cdots, \succ_n)$ which consists of a ranking order for each voter. Let $\mathcal{L}(C)^N$ be the set of all such profiles and $\Delta(C)$ denotes the set of all probability functions on C. A mapping $f : \mathcal{L}(C)^N \to \Delta(C)$ is called a *probabilistic social choice function*. Perhaps the most-studied probabilistic social choice rules are *random dictatorship* and *maximal lottery*. According to random dictatorship, a voter is uniformly picked at random and this voter's most preferred candidate is selected. Let f_{RD} denotes the probabilistic social choice function according to random dictatorship. Formally, for any profile $P = (\succ_1, \cdots, \succ_n)$ and any candidate $c_j (1 \leq j \leq k)$, $f_{RD}(P)(c_j) = \frac{|\{i : \tau \succ_i (1) = c_j\}|}{n}$. In order to introduce maximal lottery method, we need to prepare some technical notions. For a profile $P = (\succ_1, \succ_2, \cdots, \succ_n)$ and $c, c' \in C$, we denote by $n_{cc'} = \{i \in N : c \succ_i c'\}$ the number of voters who prefers c to c'. The *majority margin* $m_{cc'}$ of c over c' is defined as $n_{cc'} - n_{c'c}$. A candidate c is called a *Condorcet winner* if $m_{cc'} > 0$ for all other candidate $c' \in C$. It is well-known that Condorcet winners do not exist in general. When the context is clear, we sometimes write $m_{c_i c_j}$ as m_{ij} for simplicity. So, we obtain a *majority margin matrix* M whose (i, j) entry is m_{ij}. It is easy to see that M is integer-valued and skew-symmetric, i.e., $m_{ij} = -m_{ji}$ for all $1 \leq i, j \leq k$. According to von Neumann's Minimax Theorem, there always exists a probability (row) vector $\boldsymbol{p} = (p_1, \cdots, p_k)$ such that $\boldsymbol{p}M \geq 0$. It follows that, for any probability vector $\boldsymbol{q} = (q_1, \cdots, q_k)$, $\boldsymbol{p}M\boldsymbol{q}^T \geq 0$. In other words, according to the profile P, the lottery $\boldsymbol{p}$ is socially preferred to or indifferent to any lottery $\boldsymbol{q}$. Such a probability vector $\boldsymbol{p}$ is called a *maximal lottery* on C. It can be obtained *efficiently* via linear programming. And it defines a probabilistic social choice function f_{ML} as $f_{ML}(P) = \boldsymbol{p}$. Note that maximal lotteries are not necessarily unique. However, most profiles admits a unique maximal lottery, for example, when the number of voters is odd. The maximal lottery method will always picks a Condorcet winner whenever one exists. For more details to probabilistic social choice, one may refer further to [6, 7, 14, 24].

2.2   Differential Privacy and the Shuffle Model

There are two well-known models for differential privacy: one is the local model and the other the central model. Let X denote the domain of individual data and Y the outcome set. The *local (DP) model* assumes that each individual uses randomized algorithm $\mathcal{R}$ locally to protect her data and then submits to a (possibly untrusted) analyzer. The randomized algorithm $\mathcal{R} : X \to \Delta(Y)$ here is called (ϵ, δ)-*locally differentially private* (LDP for short) if, for all data value $x, x' \in X$ and any $E \subseteq Y$, $Pr[\mathcal{R}(x) \in E] \leq e^\epsilon Pr[\mathcal{R}(x') \in E] + \delta$. Note that both $\mathcal{R}(x)$ and $\mathcal{R}(x')$ are probability functions on Y. In particular, $(\epsilon, 0)$-LDP is simply called ϵ-*LDP* or *pure LDP*. Randomized response techniques are the prototypical examples of local differential privacy. In this paper, we will employ a special optimal k-ary randomized response mechanism (k-ary RR for short) in which we assume that $X = Y$ and their sizes are k: for any x in X,

$$Pr[\mathcal{R}(x) = y)] = \begin{cases} \frac{e^x}{e^x + k - 1} & \text{if } x = y, \\ \frac{1}{e^x + k - 1} & \text{if } x \neq y. \end{cases}$$

It is easy to check that it is ϵ-LDP. The well-known Warner's randomized response technique is just the 2-ary optimal RR [30]. To the contrary, the central model assumes the availability of a trusted curator who has access to raw individual data in a central-ized location. Two datasets $x = (x_1, x_2, \cdots, x_n), x' = (x'_1, x'_2, \cdots, x'_n) \in X^n$ are called *neighboring datasets* if they differ in one element, denoted by $x \simeq x'$. Both x_i and x'_i are contributed by individual (or voter in the social-choice scenario) i. Let $\epsilon \in [0, 1]$ and $\delta \in [0, 1]$. A randomized algorithm $\mathcal{Q} : X^n \to \Delta(Y)$ is called (ϵ, δ)-*differentially private* if, for all $E \subseteq Y$ and for all $x, x' \in X^n$ such that $x \simeq x'$: $Pr[\mathcal{Q}(x) \in E] \le e^{\epsilon} Pr[\mathcal{Q}(x') \in E] + \delta$. The above (ϵ, δ)-differential privacy is also called (ϵ, δ)-*central differential privacy* and the associated model is called *the central (DP) model*. Whenever the context is clear, we sometimes omit the quantifier "central". An important property for DP algorithms is *postprocessing*. It says, if an algorithm pro-tects an individual's privacy, then a data analyst cannot increase privacy loss simply by "sitting in a corner and thinking about the output of the algorithm" [12]. Formally, an (ϵ, δ)-DP algorithm post-processed with a *data-independent* mapping is also (ϵ, δ) differentially private. In order to perform the privacy analysis, we also need a "prepro-cessing" lemma, which states that any (*not necessarily data-independent*) preprocessing procedure composed with an (ϵ, δ)-DP as a new randomized algorithm is still (ϵ, δ)-DP.

The shuffle model sits between the local and central models. Actually the above two kinds of models are related and the product of local models can also be regarded as a central model. The mechanism $\mathcal{R}^n : X^n \to Y^n$ given by $\mathcal{R}^n(x_1, x_2, \cdots, x_n) := (\mathcal{R}(x_1), \mathcal{R}(x_2), \cdots, \mathcal{R}(x_n))$ is (ϵ', δ')-differentially private in the central sense for some ϵ' and δ'. The *shuffle model* is the randomized algorithm $\mathcal{R}^n$ plus an addi-tional *trusted* random shuffler $\mathcal{S}$ in place to provide anonymity to the locally ran-domized $\mathcal{R}(x_i)$'s so that the data analyzer is unable to associate those messages to individuals or voters. Formally, a (single-message) protocol in the shuffle model is $\mathcal{S} \circ \mathcal{R}^n : X^n \to \Delta(\mathbb{N}_n^Y)$ where $\mathcal{R} : X \to \Delta(Y)$ is a locally differential pri-vate algorithm, and $\mathcal{S}$ is a permutation on the set of locally randomized messages $\{\mathcal{R}(x_1), \mathcal{R}(x_2), \cdots, \mathcal{R}(x_n)\}$. The protocol operates as follows: each voter employs $\mathcal{R}$ to locally randomize her data x_i and then send the randomized message $\mathcal{R}(x_i)$ to the shuffler, which is unknown to the analyzer. The shuffled messages are then col-lected by the analyzer $\mathcal{A}$. Note that the shuffled messages are multisets or histograms of elements in Y. The structure of the shuffle model is illustrated in Fig. 1 (a). For a further introduction to differential privacy and the shuffle model, one may refer to [2, 9, 10, 12, 13, 20, 29].

3   Probabilistic Social Choice in the Shuffle Model

This section is our main contribution, which is illustrated in Fig. 1 (b). Without further notice, n and k denote the number of voters (or voting data) and candidates, respec-tively. Let $\mathbb{S}_k$ denote the set of all permutations over $[k] = \{1, 2, \cdots, k\}$ and is called the *symmetric group*. The crux in this section is to find l cluster centers in $\mathbb{S}_k$ where l is much smaller than $k!$ and *approximately round* each permutation to these l fixed centers.

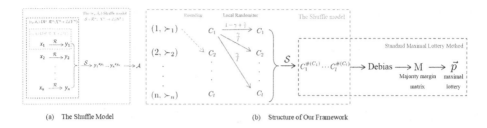

(a)   The Shuffle Model   (b)   Structure of Our Framework

Fig. 1. Structure of Our Framework

3.1 Spectral Clustering and Approximate Rounding

A natural choice to find the l cluster centers would be k-means algorithm which partitions $\mathbb{S}_k$ into l clusters $\{S_i\}_{i=1}^l$ so as to minimize the within-cluster sum of squares. Formally, the objective is to find

$$\arg\min_{S_j, \pi_j \in S_j} \sum_{j=1}^{l} \sum_{\tau \in S_j} n_d(\tau, \pi_j) \tag{1}$$

The standard k-means algorithm includes two steps: the *assignment step* assigns each permutation to the cluster with the nearest center; and the *update step* recalculate each center for permutations assigned to each cluster; the algorithm repeats until centers remain unchanged in an iteration. Note that, when $l = 1$, the above k-means problem in Eq. (1) is reduced to the commonly known Kemeny consensus problem, which is proved to be NP-hard [4]. The above update step is also equivalent to solve the Kemeny consensus problem for each cluster. So with the above standard k-means algorithm it is *computationally* difficult to solve the clustering for $\mathbb{S}_k$ of permutations.

Instead we choose to implement spectral clustering algorithm [21,22,28] to find the l cluster centers for approximately rounding. In order to bypass the Kemeny consensus problem in the k-means-clustering step of the algorithm and to preserve the pairwise-comparison voting data, we need the following embedding mapping Φ for permutations [18]. Let $\mathbb{R}^{\binom{k}{2}}$ be the Euclidean space of dimension $\binom{k}{2}$ and each point is regarded as a column vector. Now we embed $\mathbb{S}_k$ into $\mathbb{R}^{\binom{k}{2}}$ and the embedding function $\Phi : \mathbb{S}_k \to \mathbb{R}^{\binom{k}{2}}$ is defined as: for any $\tau \in \mathbb{S}_k$,

$$\Phi(\tau) = \left(\frac{1}{\sqrt{\binom{k}{2}}} (\chi_{\{\tau(i) > \tau(j)\}} - \chi_{\{\tau(i) < \tau(j)\}}) \right)_{1 \leq i < j \leq k}$$

In other words, for any $i < j$, if $\tau(i) > \tau(j)$, then the (i,j)-th coordinate of $\Phi(\tau)$ is $1/\sqrt{\binom{k}{2}}$; otherwise, the (i,j)-th coordinate of $\Phi(\tau)$ is $-1/\sqrt{\binom{k}{2}}$. The factor $1/\sqrt{\binom{k}{2}}$ is for normalization. It follows immediately that $\|\Phi(\tau) - \Phi(\tau')\|^2 = \frac{4}{\binom{k}{2}} n_d(\tau, \tau')$. In some sense, the embedding into the Euclidean space preserves the Kendall-tau distance. In order to make the cluster centers "evenly distributed", we need a special k-means algorithm for clusters of equal size (see in Supplementary Materials).

For any $1 \le j \le l$, C_j as the barycenters of $\Phi(S_j) := \{\Phi(\tau) : \tau \in S_j\}$ is formally computed as $C_j = \frac{\sum_{\tau \in S_j} \Phi(\tau)}{|S_j|}$. The set $\mathbb{C}_l := \{C_1, \cdots, C_l\}$ of l cluster centers for $\Phi(S_k)$ is the rounding basis that we are looking for. It is also the input database that we define local differentially private mechanism $\mathcal{R}_l$ on. More importantly, these cluster centers $C_1, C_2, \cdots, C_l$ are independent of the input voting data $P = (\succ_1, \cdots, \succ_n)$. This *data-independence* property is required for a local randomizer in the shuffle model.

For the dataset S_k of permutations over k alternatives, spectral clustering algorithm needs to form an $k! \times k!$ similarity matrix and compute the eigenvectors of this matrix, which usually has a computational complexity of $O((k!)^3)$ in general. And the k-means step can be solved in time $O((k!)^{l\binom{k}{2}+1})$. Recall that l is the number of clusters. All other operations are at most linear. For applications with $k!$ on the order of thousands (or the number of candidates $k \ge 7$), spectral clustering methods begin to become infeasible and then we need to use the so-called Fast Approximate Spectral Clustering [31].

There are two obvious *approximately rounding* approaches. The first one is *deterministic rounding* which rounds the encoded permutation to its closest center. Mathematically, for any $1 \le j \le k!$, the encoded permutation $\Phi(\tau_j)$ will be approximately rounded to the nearest cluster center $C_{r(\Phi(\tau_j))} := \arg\min_{C_{j'} \in \{C_1, \cdots, C_l\}} \|\Phi(\tau_j) - C_{j'}\|$. The second method is called *stochastic rounding* which rounds $\Phi(\tau_j)$ to all cluster centers according to the probabilities proportional to their similarities. In other words, $\Phi(\tau_j)$ is rounded to $C_{j'} (1 \le j' \le l)$ with probability $p_{j'} = \frac{A(j,j')}{\sum_{r=1}^l A(j,r)}$. It is important to note that both rounding methods are biased. For example, in stochastic rounding, $\Phi(\tau_j)$ is not necessarily equal to $p_1 C_1 + p_2 C_2 + \cdots + p_l C_l$.

3.2 Locally DP Randomization and Shuffling

After rounding, we apply local randomized response mechanism $\mathcal{R}_l$ to the l cluster centers $C_1, C_2 \cdots$, and C_l, which are independent of the voting data. The main idea is similar to the "privacy blanket" approach in [2]. Each center $C_i \in \mathbb{C}_l$ is locally randomized truthfully to itself with probability $1 - \gamma$, and is uniformly randomized to $\mathbb{C}_l$ with probability γ. This implies that, for all $1 \le i, j \le l$,

$$\mathcal{R}_l(C_i) = \begin{cases} C_i & \text{with probability} (1 - \gamma) + \frac{\gamma}{l}, \\ C_j (j \ne i) & \text{with probability} \frac{\gamma}{l}. \end{cases}$$

We may regard both deterministic and stochastic rounding as preprocessing. Note that the (ϵ, δ)-local randomizer preprocessed with rounding is still an (ϵ, δ)-LDP. After local randomization, voting messages are shuffled by a random permutation unknown to the adversary. In other words, the adversary is unable to associate messages with voters. So we may assume that the output of the shuffler $\mathcal{S}$ is a histogram of $C_1, C_2, \cdots, C_l$: $C_1^{\#(C_1)} C_2^{\#(C_2)} \cdots C_l^{\#(C_l)}$ where $\sum_i \#(C_i) = n$.

Theorem 1. *The above shuffle model $\mathcal{S} \circ \mathcal{R}_l \circ Rounding$ is (ϵ, δ)-DP for any $l, n \in \mathbb{N}, \epsilon \le 1$ and $\delta \in (0,1]$ such that $\gamma = \max\{\frac{27l}{(n-1)\epsilon}, \frac{14 \times l \times log(2/\delta)}{(n-1)\epsilon^2}\} < 1$.*

One may refer to Corollary 5.3.1 in [2] for a more general result about privacy analysis of the shuffle model with other more general LDP mechanisms than randomized response techniques here.

3.3 Debiasing

In the shuffle model, there are two steps where bias is introduced. The first step is the approximate rounding. Here we consider only the deterministic nearest center rounding. For each rounding of a voting data $(j, \succ_j)$ to its closest center $C_{r(\succ_j)}$, a bias of $C_{r(\succ_j)} - \Phi(\succ_j)$ is added. In the extreme case when all voting data is the same as $\succ_j$, the largest possible total bias is $n(C_{r(\succ_j)} - \Phi(\succ_j))$. When all rounded voting data are evenly distributed among $C_1, C_2, \cdots, C_l$, the total bias should be very small and even equals 0. Any possible debiasing approach to the rounding step would depend on the *original* voting data and destroy the privacy-preserving mechanism from the local randomizer and hence in the whole shuffle model. We still don't know how to debias the rounded data without increasing the privacy loss. The second step for bias is the local randomizer $\mathcal{R}_l$. It is straightforward to see that the bias in this step is equal to $\frac{\gamma}{l} \sum_{i=1}^{l} C_i - \gamma C_j$ when the true input is C_j. We can debias the local randomization by

$$\mathcal{D}(C_j) = \frac{C_j - \frac{\gamma}{l} \sum_{i=1}^{l} C_i}{1 - \gamma} \tag{2}$$

for any C_j as an output message from the shuffle model. The debiasing in this step is *data-independent*, can be regarded as postprocessing and hence the privacy-preserving property in the shuffle model is maintained.

3.4 Probabilistic Social Choice

After debiasing, we obtain a histogram $\mathcal{D}(C_1)^{\#(C_1)} \mathcal{D}(C_2)^{\#(C_2)} \cdots \mathcal{D}(C_l)^{\#(C_l)}$. The social voting information is actually contained in a single vector $C = \sum_{i=1}^{l} (\#(C_i)) \mathcal{D}(C_i) \in \mathbb{R}^{\binom{k}{2}}$. Here we describe how to apply the maximal lottery method to C. Each debiased message $\mathcal{D}(C_j)$ can be regarded as a noisy ranking order coming from some original voting data $(i, \succ_i)$. But it is important to note that the vector $\mathcal{D}(C_j)$ does not necessarily represent a true ranking order over the k candidates. This is because we employed the kernel k-means algorithm to bypass the computationally expensive Kemeny consensus problem and the cluster centers are obtained by computing the barycenter of each cluster. However, these vectors $\mathcal{D}(C_j)$ approximate the true ranking order $(i, \succ_i)$ through the paired-comparison voting data. And the maximal lottery method needs only the data about all these pairwise comparisons of candidates from all voters. So we can still use maximal lottery method to approximate the true maximal lottery. One difference from the standard maximal lottery is that the majority margin matrix is real-valued rather than integer-valued. For any vector $\mathcal{D}(C_j) \in \mathbb{R}^{\binom{k}{2}}$ and $1 \leq i_1 < i_2 \leq k$, let the (i_1, i_2)-th coordinate denoted as $\mathcal{D}(C_j)(i_1, i_2)$ and it represents the relative preference of the candidate c_{i_1} to c_{i_2}. Now we turn the debiased vector $\mathcal{D}(C)$ into a majority margin matrix M whose (i, j)-entry $M(i, j)$ is just the (i, j)-coordinate of $\mathcal{D}(C)$ for $1 \leq i < j \leq k$. Note that k is the number of candidates and $M(j, i) = -M(i, j)$. Now we describe an algorithm to find a maximal lottery p such that $pM \geq 0$. The existence of this maximal lottery is guaranteed by von Neumann's Minimax Theorem. Moreover maximal lotteries are almost always unique, and can be efficiently computed using linear programming [5]. Here we adapt the algorithm from [24] which has a motivating two-person zero-sum symmetric game-theoretic interpretation.

For real-world voting data, the number k of candidates is usually within 10 and hence the dimension is usually not large. For reasonable privacy requirement, if we want the number l of cluster centers to be reasonable (say within 50), then we have to restrict the size of the original dataset P to be within tens of thousands (see our following experiments).

4   Experiments

In this section, we perform relevant experiments to evaluate the effectiveness of the proposed DP probabilistic social choice algorithms. We use data generated randomly as used in [24], and work with 5,000 voters and 5 candidates. In the experiments, we use A, B, C, D, E to represent candidates. Each voter submits his preference order over the candidates. We use *mean square error* for the measure of accuracy. We repeat the algorithm 1,000 times and take the average as final results to reduce random error.

Firstly, we demonstrate the accuracy of our algorithm by comparing it with the standard Maximal Lottery method. We implement the algorithm shown in Fig. 1 to clarify the impact of adding rounding and local randomizer to Maximal Lottery algorithm on the voting results. We set different privacy parameters ϵ and cluster parameter l when $\delta = 10^{-3}$. We also compare the effect of shuffle models with fixed ϵ but with different numbers of clusters. The results are shown in Table 1, where column represents the probability vector p^*. The values in the first column are those from the traditional maximal lottery method. We view it as true value and reference point to evaluate the performance of our algorithm with different ϵ and l. We can see that when only perform *Rounding*, the results are closest to the real value. The greater the value of ϵ, the closer the results are to the real values. Secondly, we illustrate the effect of different l on voting results when ϵ and δ are fixed. When $\epsilon = 1.3$, $\delta = 10^{-3}$, $k = 5$, $n = 5000$, our result supports that $l = 8$ is the approximately optimal cluster number. Tables 2 show the results of experiments carried out with Maximal Lottery method. The results show that when $l = 8$, the difference between the results and the Ground Truth is indeed the smallest. The results for RD are in Supplementary Materials.

Table 1. Compared the algorithm described in Fig. 1 with traditional Maximal Lottery method under different l and ϵ. Each column in the table represents the output, which is probability vector p^*. The last row shows the MSE between the corresponding p^* and the Truth.

	Truth	Only Rounding			$\epsilon = 0.1$			$\epsilon = 0.5$			$\epsilon = 1$		
		$l=1$	$l=5$	$l=10$	$l=1$	$l=5$	$l=10$	$l=1$	$l=5$	$l=10$	$l=1$	$l=5$	$l=10$
A	0.0	0.0	0.0	0.01	0.0	0.0	0.0	0.0	0.0	0.0	0.0	0.0	0.04
B	0.46	0.51	0.51	0.68	0.53	0.21	0.33	0.53	0.28	0.22	0.52	0.38	0.83
C	0.05	0.0	0.0	0.0	0.0	0.1	0.02	0.0	0.07	0.66	0.0	0.09	0.0
D	0.49	0.49	0.49	0.31	0.47	0.69	0.65	0.47	0.64	0.12	0.48	0.53	0.13
E	0.0	0.0	0.0	0.0	0.0	0.0	0.0	0.0	0.0	0.0	0.0	0.0	0.0
MSE	0	0.0009	0.0009	0.0167	0.0015	0.0210	0.0090	0.0016	0.0115	0.0394	0.0012	0.0021	0.0538

5 Related Works and Conclusion

In this paper we investigate differentially private probabilistic social choice in the shuffle model. It looks quite natural to apply differential privacy to social choice theory because both fields deals with the issue of individual data and population summary. But it is surprising that there is not much literature about this combination [1,17,25,32]. All of them consider *deterministic* social choice functions and all of those voting rules are anonymous. But none of them regards anonymity as privacy amplifier. Since a shuffler can be implicitly assumed to locate between local randomizer and aggregation function in those voting rules, the privacy guarantee there could be much improved according to a similar privacy analysis in this paper (Theorem 1). To the best of our knowledge, there is only one paper about this connection of probabilistic social choice with DP [27]. The paper considers only random dictatorship and the motivation there is just inverse to ours: the author studied the use of probabilistic social choice to achieve differential privacy. Our streamline is similar to the one-message shuffle model for the summation of real numbers [2,10]. But, since they worked with real numbers, they adopted a much simpler stochastic rounding, which can be easily seen to be unbiased. To deal with the rounding problem for permutations i.e., the crux of our paper, we have to develop a much more advanced machinery which first encodes all permutations as elements in a high-dimensional space and then use spectral clustering to partition these encodings into a fixed number of clusters of equal size. Our work in this aspect can be regarded as an advance to a recent research on kernel-k-means algorithm for permutations [18].

Table 2. The results of different l of Maximal Lottery method when $\epsilon = 1.3$, $\delta = 10^{-3}$, $k = 5$, $n = 5000$.

	Truth	ML Method				
		$l=2$	$l=4$	$l=8$	$l=10$	$l=15$
A	0.0	0.02	0.0	0.0	0.03	0.0
B	0.46	0.97	0.64	0.36	0.94	0.0
C	0.05	0.0	0.12	0.26	0.0	0.02
D	0.49	0.01	0.24	0.38	0.03	0.98
E	0.0	0.0	0.0	0.0	0.0	0.0
Diff		0.49	0.1	**0.07**	0.45	0.47

Our framework is basic in the sense that, for more complex scenarios, we may adapt our algorithm to accommodate those complex elements. If the number of candidates is more than 7, our spectral clustering algorithm becomes infeasible. We will develop an approximate spectral clustering algorithm for more candidates based on [31]. For permutations, we need to understand better about the discrete geometry of $\mathbb{S}_k$ especially in high-dimensional space and machine learning techniques about permutations [19]. Also we are looking for a better rounding technique than those in this paper. Here we consider only the basic one-message shuffle model. We will develop other many-message shuffle models for probabilistic social choice based on recent advances in this area [3,11,15].

References

1. Ao, L., Lu, Y., Xia, L., Zikas, V.: How private are commonly-used voting rules? In: UAI2020, pp. 629–638. PMLR (2020)
2. Balle, B., Bell, J., Gascón, A., Nissim, K.: The privacy blanket of the shuffle model. In: Boldyreva, A., Micciancio, D. (eds.) CRYPTO 2019. LNCS, vol. 11693, pp. 638–667. Springer, Cham (2019). https://doi.org/10.1007/978-3-030-26951-7_22
3. Balle, B., Bell, J., Gascón, A., Nissim, K.: Private summation in the multi-message shuffle model. In: CCS 2020, pp. 657–676 (2020)
4. Bartholdi, J., Tovey, C.A., Trick, M.A.: Voting schemes for which it can be difficult to tell who won the election. Soc. Choice Welfare 6(2), 157–165 (1989)
5. Brandl, F., Brandt, F., Seedig, H.G.: Consistent probabilistic social choice. Econometrica 84(5), 1839–1880 (2016)
6. Brandl, F., Brandt, F., Stricker, C.: An analytical and experimental comparison of maximal lottery schemes. Soc. Choice Welfare 58(1), 5–38 (2021). https://doi.org/10.1007/s00355-021-01326-x
7. Brandt, F.: Rolling the dice: recent results in probabilistic social choice. In: Trends in Computational Social Choice, pp. 3–26 (2017)
8. Brandt, F., Conitzer, V., Endriss, U., Lang, J., Procaccia, A.D.: Handbook of Computational Social Choice. Cambridge University Press, Cambridge (2016)
9. Cheu, A.: Differential privacy in the shuffle model. Ph.D. thesis, Northeastern University (2021)
10. Cheu, A., Smith, A., Ullman, J., Zeber, D., Zhilyaev, M.: Distributed differential privacy via shuffling. In: Ishai, Y., Rijmen, V. (eds.) EUROCRYPT 2019. LNCS, vol. 11476, pp. 375–403. Springer, Cham (2019). https://doi.org/10.1007/978-3-030-17653-2_13
11. Cheu, A., Ullman, J.: The limits of pan privacy and shuffle privacy for learning and estimation. In: STOC 2021, pp. 1081–1094 (2021)
12. Dwork, C., Roth, A.: The algorithmic foundations of differential privacy. Found. Trends Theor. Comput. Sci. 9(3–4), 211–407 (2014)
13. Erlingsson, Ú., Feldman, V., Mironov, I., Raghunathan, A., Talwar, K., Thakurta, A.: Amplification by shuffling: from local to central differential privacy via anonymity. In: SODA 2019, pp. 2468–2479. SIAM (2019)
14. Fishburn, P.C.: Probabilistic social choice based on simple voting comparisons. Rev. Econ. Stud. 51(4), 683–692 (1984)
15. Ghazi, B., Kumar, R., Manurangsi, P., Pagh, R., Sinha, A.: Differentially private aggregation in the shuffle model: almost central accuracy in almost a single message. In: ICML 2021, pp. 3692–3701. PMLR (2021)

16. Gibbard, A.: Manipulation of schemes that mix voting with chance. Econometrica **45**(3), 665–81 (1977)
17. Hay, M., Elagina, L., Miklau, G.: Differentially private rank aggregation. In: ICDM 2017, pp. 669–677. SIAM (2017)
18. Jiao, Y., Vert, J.P.: The Kendall and Mallows Kernels for permutations. IEEE Trans. Pattern Anal. Mach. Intell. **40**(7), 1755–1769 (2018)
19. Jiao, Y., Vert, J.P.: The weighted Kendall and high-order Kernels for permutations. In: ICML 2018, pp. 2314–2322. PMLR (2018)
20. Kairouz, P., Bonawitz, K., Ramage, D.: Discrete distribution estimation under local privacy. In: Balcan, M., Weinberger, K.Q. (eds.) ICML 2016, New York City, NY, USA, 19–24 June 2016, vol. 48, pp. 2436–2444. JMLR.org (2016)
21. Liu, J., Han, J.: Spectral clustering. In: Data Clustering, pp. 177–200. Chapman and Hall/CRC (2018)
22. Ng, A.Y., Jordan, M.I., Weiss, Y.: On spectral clustering: analysis and an algorithm. In: NIPS 2002, pp. 849–856 (2002)
23. Prasad, A., Pareek, H., Ravikumar, P.: Distributional rank aggregation, and an axiomatic analysis. In: ICML 2015, pp. 2104–2112. PMLR (2015)
24. Rivest, R.L., Shen, E.: An optimal single-winner preferential voting system based on game theory. In: Proceedings of 3rd International Workshop on Computational Social Choice, pp. 399–410 (2010)
25. Shang, S., Wang, T., Cuff, P., Kulkarni, S.: The application of differential privacy for rank aggregation: privacy and accuracy. In: FUSION 2014, pp. 1–7. IEEE (2014)
26. Sweeney, L.: Weaving technology and policy together to maintain confidentiality. J. Law Med. Ethics **25**(2–3), 98–110 (1997)
27. Torra, V.: Random dictatorship for privacy-preserving social choice. Int. J. Inf. Secur. **19**(5), 537–545 (2020)
28. Von Luxburg, U.: A tutorial on spectral clustering. Stat. Comput. **17**(4), 395–416 (2007)
29. Wang, T., Blocki, J., Li, N., Jha, S.: Locally differentially private protocols for frequency estimation. In: Kirda, E., Ristenpart, T. (eds.) 26th USENIX Security Symposium, USENIX Security 2017, Vancouver, BC, Canada, 16–18 August 2017, pp. 729–745. USENIX Association (2017)
30. Warner, S.: Randomized response: a survey technique for eliminating evasive answer bias. J. Am. Stat. Assoc. **60**(309), 63–69 (1965)
31. Yan, D., Huang, L., Jordan, M.I.: Fast approximate spectral clustering. In: KDD 2009, pp. 907–916 (2009)
32. Yan, Z., Li, G., Liu, J.: Private rank aggregation under local differential privacy. Int. J. Intell. Syst. **35**(10), 1492–1519 (2020)

Reasoning About Games
with Possibilistic Uncertainty

Churn-Jung Liau$^{(\boxtimes)}$

Institute of Information Science, Academia Sinica, Taipei 115, Taiwan
liaucj@iis.sinica.edu.tw

Abstract. Game theory is the mathematical study of strategic interaction among rational agents and has found many applications in economics, politics, logic, AI, and computer science. Because of pervasive
uncertainty in game-playing environments, how to deal with uncertainty
is also a key issue in game theory. While probability calculus has been
the standard theory for uncertainty management in games, information
of exact probability assessment may be not available in many realistic situations. In such cases, possibility theory is an alternative tool for
modeling uncertainty in games. In past decades, the cross-fertilization
between logic and game theory has proved to be very successful. Therefore, the paper is aimed at the integration of possibilistic uncertainty into
modal logics for reasoning about games including game logic and coalition logic. We will study syntax and semantics of the integrated logics
as well as their reasoning problems.

Keywords: Game logic · Coalition logic · Fuzzy logic · Possibilistic
logic

1 Introduction

In past decades, the research of intelligent agents has received much attention
in the field of AI and continues to play an important role in the development of
modern autonomous AI systems [21,24]. As the goal of intelligent agent research
is to design rational agents that can choose optimal actions given their perception of the surrounding environment, it shares many common interests and
notions with the study of rational decision-making and reasoning. Consequently,
economics and logic, which are concerned with studies of agent's rational behaviors and valid reasoning patterns respectively, have become the most prominent
paradigms in agent theory [5,16]. In particular, game theory and the logic of
agency have been extensively applied to modeling and verification of agent systems. Since mid twentieth century, several logicians have noticed that game
theory can provide a conceptual tool for the analysis of key notions in logic.
The most remarkable examples are Lorenzen's dialogical logic and Hintikka's

Supported by NSTC of Taiwan under Grant No. 110-2221-E-001-022-MY3.

game-theoretical semantics [13]. On the other hand, a lot of logics to model game-playing situations have grown at a fast pace in the past twenty years, mostly in terms of logics for rational or strategic interactions [1,2,4,14].

As classical logic focuses on bivalent reasoning, most game-theoretical logics do not involve modeling of uncertain information (or at most qualitative uncertainty). However, uncertainty is pervasive in almost all game-playing environments. For example, a player may be not certain of the current states, the outcomes of some action, and other players' strategies. Hence, dealing with uncertain information is inevitable in game-theoretic models. As probability calculus is the standard formalism for uncertainty management in game theory, in recent years, there have been several attempts to extend different logical formulations of games with the capability of probabilistic reasoning [7,8,15,22]. Nevertheless, there exist various types of uncertainties other than probability regarding information in a more realistic environment. Indeed, in many real-world applications, exact probabilistic information is rarely available so that uncertainty cannot be represented by probability. For example, in some cases, a player may consider a state more possible than another one but cannot assess exact probability values of these states. This kind of ordinal uncertainty can be better modeled by possibility theory than the probability one [26]. Therefore, to fulfill such needs, we explore logical formalisms for possibilistic reasoning in game-theoretical situations.

The remainder of the paper is organized as follows. In Sect. 2, we review some basic notions about logic, fuzzy sets, and games. Next, we propose the extensions of game logic and coalition logic to deal with possibilistic uncertainty in Sects. 3 and 4 respectively. Finally, in Sect. 5, we present the conclusion and indicate main directions of future work on the proposed logics.

2 Preliminaries

2.1 Strategic Games

One of the most basic models in game theory is that of a strategic game. In such a game, each player chooses one of alternative actions/strategies available to it, and together, these joint actions determine the outcome of the game. Mathematically, a *strategic game form* is defined as a tuple $(S, P, (\Sigma_i)_{i\in P}, f)$, where S is a nonempty set of states (i.e. the outcome space), P is a set of players, Σ_i is a nonempty set of actions/strategies for each $i \in P$, and $f : \Pi_{i\in P}\Sigma_i \to S$ is the outcome function. Then, a *strategic game* is a strategic game form with an additional component expressing different players' preferences over the outcome space. There are at least two ways to express a player's preference. One is to use a binary relation $\succeq_i \subseteq S \times S$. Thus, $s \succeq_i t$ means that player i prefers state s at least as much as state t. The other way is to define the payoff function $u_i : S \to \Re$ so that player i prefers state s at least as much as state t iff $u_i(s) \geq u_i(t)$.

2.2 Fuzzy Set and Possibility Theory

Fuzzy set theory is invented by Zadeh [25] to model vagueness in set membership. A fuzzy set is a set without a precise boundary between its elements and non-elements. Given a universe W, a fuzzy set A over W is defined as a membership function $A : W \to [0,1]$. For any $x \in W$, $A(x)$ is called the *membership degree* of x in A. Hence, a classical set (aka. crisp set) A is regarded as a special case of fuzzy set such that $A(x) = 0$ or 1 for any $x \in W$. Let W and U be two universes. Then, a fuzzy subset of $W \times U$ is also called a (binary) fuzzy relation. As in classical set theory, where $\mathcal{P}(W)$ or 2^W is used to denote the power set of W, we also denote the class of all fuzzy subsets over W by $\mathcal{F}(W)$ or $[0,1]^W$.

Classical set-theoretic operations can be generalized to fuzzy sets by using operations on the value domain $[0,1]$ such as t-norm, t-conorm, and residuated implication [9]. A *t-norm* is a binary operation $\otimes$ on $[0,1]$ that satisfies commutativity, associativity, and $1 \otimes c = c$ and $0 \otimes c = 0$ for all $c \in [0,1]$; and is non-decreasing in both arguments. Dually, a *t-conorm* (or s-norm) is a binary operation $\oplus : [0,1]^2 \to [0,1]$ that satisfies commutativity, associativity, and $0 \oplus c = c$ and $1 \oplus c = 1$ for all $c \in [0,1]$; and is non-decreasing in both arguments. For each t-norm $\otimes$, there is a corresponding t-conorm $\oplus$ defined by $\oplus(a,b) = 1 - \otimes(1-a, 1-b)$ and vice versa. The *residuum* of a t-norm $\otimes$ is a binary operation $\Rightarrow: [0,1]^2 \to [0,1]$ defined as $a \Rightarrow b = \sup\{c \mid a \otimes c \le b\}$ for all $a,b \in [0,1]$. We also define the unary negation function $\neg : [0,1] \to [0,1]$ by $\neg a = a \Rightarrow 0$ for any $a \in [0,1]$. Several well-known examples of t-norms include Gödel t-norm $a \otimes b = \min(a,b)$, product t-norm $a \otimes b = a \cdot b$ and Łukasiewicz t-norm $a \otimes b = \max(0, a+b-1)$.

In [9], two additional binary operations $\wedge$ and $\vee : [0,1]^2 \to [0,1]$ are defined by

$$a \wedge b := a \otimes (a \Rightarrow b)$$

$$a \vee b := ((a \Rightarrow b) \Rightarrow b) \wedge ((b \Rightarrow a) \Rightarrow a)$$

and it is shown that, for any continuous t-norm $\otimes$, $\wedge$ and $\vee$ are exactly the Gödel t-norm and t-conorm respectively.

Let A and B be two fuzzy sets over W. Then, their general intersection and union are defined as follows:

$$(A \otimes B)(x) = A(x) \otimes B(x),$$

$$(A \oplus B)(x) = A(x) \oplus B(x)$$

for any $x \in W$. In particular, when $\otimes$ is the Gödel t-norm, we write the general intersection and union as $A \cap B$ and $A \cup B$ respectively. In addition, we can also define the implication between A and B as $A \Rightarrow B$ with the membership function

$$(A \Rightarrow B)(x) = A(x) \Rightarrow B(x),$$

and the negated set $\neg A$ has the membership function $\neg A(x) = A(x) \Rightarrow 0$. When the negation is involutive (corresponding to the Łukasiewicz t-norm), then $\neg A$ is called its complemented set and denoted by $\overline{A}$. In this case, its membership function is $\overline{A}(x) = 1 - A(x)$.

Generally, most applications of fuzzy set theory to game-playing and decision-making scenarios employ its twofold interpretation. On one hand, a fuzzy subset of the outcome space can usually represent a fuzzy event that has only vague description, like "The payoff is high". On the other hand, a fuzzy subset of states can specify the possibility degree of each state. In this regard, a fuzzy set is called a possibility distribution and the membership degree of a state is actually its degree of possibility [26].

2.3 Modal Logic: Kripke and Neighborhood Semantics

Next, we briefly review the syntax and two semantics of modal logic. More detailed account can be found in well-known textbooks [3,6].

For the syntax of propositional modal logic, its alphabet consists of the logical constant $\perp$ (*falsum* or *falsehood*), a set of propositional symbols $\Phi_0 = \{p, q, r, \ldots\}$, the Boolean connective $\rightarrow$ (material implication), and the modal operator $\Box$ (necessity). The formation rules of well-formed formulas (wffs) are

$$\varphi ::= p \mid \perp \mid \varphi \rightarrow \varphi \mid \Box\varphi,$$

$p \in \Phi_0$. As usual, we take other logical connectives and modality as abbreviations: $\neg\varphi = \varphi \rightarrow \perp$, $\top = \neg\perp$, $\varphi \wedge \psi = \neg(\varphi \rightarrow \neg\psi)$, $\varphi \vee \psi = \neg\varphi \rightarrow \psi$, $\varphi \equiv \psi = (\varphi \rightarrow \psi) \wedge (\psi \rightarrow \varphi)$, $\Diamond\varphi = \neg\Box\neg\varphi$.

The standard semantics for modal logic is based on the Kripke model (possible world model), which is defined as a triple $\mathfrak{M} = \langle W, R, V \rangle$, where W is a set of possible worlds, $R \subseteq W \times W$ is a binary relation on W, called *accessibility relation*, and $V : \Phi_0 \rightarrow 2^W$ assigns to each propositional symbol in Φ_0 a subset of W. Given a model $\mathfrak{M}$, we can define a satisfaction relation $\models_{\mathfrak{M}}$ between W and modal logic formulas by the following rules (we omit the subscript $\mathfrak{M}$ for simplicity):

- $w \models p$ iff $w \in V(p)$ for any $p \in \Phi_0$;
- $w \not\models \perp$ for all $w \in W$;
- $w \models \varphi \rightarrow \psi$ iff $w \models \varphi$ implies $w \models \psi$;
- $w \models \Box\varphi$ iff for any u such that $(w, u) \in R$, $u \models \varphi$.

For a given Kripke model $\mathfrak{M} = \langle W, R, V \rangle$ and a wff φ, we can define the truth set of φ with respect to $\mathfrak{M}$ by

$$|\varphi|_{\mathfrak{M}} = \{w \mid w \in W, w \models \varphi\}.$$

We usually drop the subscript and simply write $|\varphi|$ for the truth set of φ when the model $\mathfrak{M}$ is clear from the context. Let Σ be a set of wffs. Then, we use $w \models_{\mathfrak{M}} \Sigma$ to denote that $w \models_{\mathfrak{M}} \psi$ for all $\psi \in \Sigma$. Furthermore, a wff φ is a logical consequence of Σ, denoted by $\Sigma \models \varphi$, if, for any $\mathfrak{M}$ and w, $w \models_{\mathfrak{M}} \Sigma$ implies $w \models_{\mathfrak{M}} \varphi$. A wff φ is valid, denoted by $\models \varphi$, if it is a logical consequence of the empty set.

A modal logic characterized by Kripke models is called a *normal modal logic*. The smallest (i.e., logically weakest) normal modal logic is the system K, whose main axiom is $\Box(\varphi \rightarrow \psi) \rightarrow (\Box\varphi \rightarrow \Box\psi)$. However, because, in many applications, the system K is still too strong, weaker systems based on alternative semantics are needed. The neighborhood semantics, which is independently invented by Dana Scott and Richard Montague in 1970 and largely ignored outside pure logicians, has revived in recent years [17]. A neighborhood model (aka. Scott-Montague model) is a triple $\mathfrak{M} = \langle W, N, V \rangle$, where W and V are the same as in the Kripke model and $N : W \to \mathcal{P}(\mathcal{P}(W))$ is called a *neighborhood function*. For each $w \in W$, $N(w) \subseteq \mathcal{P}(W)$ is called the neighborhood system of w and each $X \in N(w)$ is called a neighborhood of w. The satisfaction relation between modal formulas and a world in the neighborhood model $\mathfrak{M} = \langle W, N, V \rangle$ is defined by

$- \ w \models \Box\varphi$ iff $|\varphi| \in N(w)$,

and the satisfaction of other formulas remains the same as in the Kripke model. We will see that neighborhood models naturally arises when studying game theory.

2.4 Dynamic Logic

In reasoning about games, we can understand players' behaviors according to their belief, action, and preference. While modal logic can be easily used to represent and reason about agents' beliefs and knowledge if we adopt an epistemic reading of modal operators, reasoning about action requires the extension of modalities. Dynamic logic is one of the earliest attempt along this direction [11,12]. The original motivation of dynamic logic is to reason about program. However, it can be applied to any structural set of actions. Here, we introduce the propositional dynamic logic (PDL) for reasoning about regular program. The language of regular PDL has two sorts of expressions: well-formed formulas $\varphi, \psi, \cdots$ and programs $\alpha, \beta, \cdots$. The alphabet of PDL consists of

1. a set of propositional symbols, $\Phi_0 = \{p, q, \ldots\}$,
2. a set of atomic actions, $\Pi_0 = \{a, b, c, \ldots\}$, and
3. logical symbols $\bot$, $\to$, $[$, $]$, $;$, *, $\cup$, and $?$.

The formation rules of wffs and programs are as follows:

$$\varphi ::= p \mid \bot \mid \varphi \to \varphi \mid [\alpha]\varphi,$$

$$\alpha ::= a \mid \alpha; \alpha \mid \alpha \cup \alpha \mid \alpha^* \mid ?\varphi,$$

where $p \in \Phi_0$ and $a \in \Pi_0$.

Intuitively, the formula $[\alpha]\varphi$ means that φ is necessarily true after executing α. The formation rules of programs stipulate how more complicated programs are composed from simpler ones. The simplest program is an atomic action and the other constructs have the following intuitive meanings:

- $\alpha; \beta$: the sequential composition of α and β,
- $\alpha \cup \beta$: execute either α or β nondeterministically,

- α^*: execute α nondeterministically chosen finite number of times (including zero),
- $?\varphi$: test φ, i.e., proceed if φ is true and fail otherwise.

Based on such intuition, PDL is interpreted in generalized Kripke semantics, where each atomic action is associated with a binary transition relation on the state space. Hence, a PDL model is a tuple $\mathfrak{M} = \langle W, (R_a)_{a \in \Pi_0}, V \rangle$, where W is a set of states, $R_a \subseteq W \times W$ is a binary relation on W, called a *transition relation*, for each $a \in \Pi_0$, and $V : \Phi_0 \to 2^W$ assigns to each propositional symbol in Φ_0 a subset of W. The transition relations can be extended to all programs by using the following rules inductively:

- $R_{\alpha;\beta} = R_\alpha \circ R_\beta = \{(u,v) \mid \exists w \in W, (u,w) \in R_\alpha \wedge (w,v) \in R_\beta\}$
- $R_{\alpha \cup \beta} = R_\alpha \cup R_\beta$
- $R_{\alpha^*} = R_\alpha^* = \bigcup_{n \geq 0} R_\alpha^n$, where $R_\alpha^0 = \{(w,w) \mid w \in W\}$ and $R_\alpha^n = R_\alpha^{n-1} \circ R_\alpha$
- $R_{?\varphi} = \{(w,w) \mid w \models \varphi\}$

In addition, the satisfaction of dynamic formulas in a state is defined by

- $w \models [\alpha]\varphi$ iff for any u such that $(w,u) \in R_\alpha$, $u \models \varphi$.

3   Fuzzy Game Logic

The game logic (GL) is one of the earliest system for reasoning about determined 2-player games [18,20]. The syntax of GL is an extension of PDL and comprises of games and formulas. The games of GL involve two players, player 1 (Angel) and player 2 (Demon), and its formulas describe the propositions that hold in a state. In GL, compound games are composed from simpler games by some basic construction such as sequential composition, choice, and role exchanging. The semantics of GL is given with a generalized form of *neighborhood model* in modal logic [6,17]. In the model, a neighborhood system E_g for each primitive game g is used to represent the Angel's effectivity function, from which we can derive a set transformation that transform a goal event (i.e. an arbitrary set of states) to the precondition (also represented by a set of states) on which the Angel has a strategy playing game g to achieve the event. The transformation can be inductively extended to any compound games. Then, a state satisfying the precondition will ensure the existence of a strategy for the Angel to achieve the goal.

The syntax of fuzzy game logic (FGL) is the same as that of GL which consists of two sorts, formulas and games. Let Φ_0 and Γ_0 denote the set of atomic propositions and atomic games respectively. Then, games γ and formulas φ of FGL can be define by simultaneous induction as follows:

$$\gamma ::= g \mid \varphi? \mid \gamma; \gamma \mid \gamma \cup \gamma \mid \gamma^* \mid \gamma^d$$
$$\varphi ::= \bot \mid p \mid \varphi \otimes \varphi \mid \varphi \to \varphi \mid \langle \gamma \rangle \varphi.$$

We abbreviate $\varphi \to \bot$ as $\neg\varphi$, $\neg\bot$ as $\top$, $\varphi \otimes (\varphi \to \psi)$ as $\varphi \wedge \psi$, $((\varphi \to \psi) \to \psi) \wedge ((\psi \to \varphi) \to \varphi)$ as $\varphi \vee \psi$, and $(\varphi \to \psi) \otimes (\psi \to \varphi)$ as $(\varphi \leftrightarrow \psi)$.

The formula $\langle\gamma\rangle\varphi$ means that Angel has a strategy by playing game γ to achieve the goal φ. The intuition behind the constructions of games is as follows. The test game $\varphi?$ consists of checking whether φ holds at a state. The sequential composition $\gamma_1;\gamma_2$ consists of first playing γ_1 and then γ_2, and the choice $\gamma_1\cup\gamma_2$ denotes the game where Angel chooses either of the two subgames to continue playing. For the iterated game γ^*, Angel can choose how many times to play γ (possibly not at all); each time she has played γ, she can decide whether to play it again or not. Finally, the dual game γ^d is the same as playing γ with two players exchanging their roles.

Although the syntax of GL and FGL is a straightforward extension of PDL syntax, the PDL semantics cannot be easily applied to them. The main reason is that in PDL, $\alpha\cup\beta$ means a nondeterministic choice, whereas in GL or FGL, it means the Angel's decision. Hence, while $\alpha;(\beta\cup\gamma)=\alpha;\beta\cup\alpha;\gamma$ is intuitively reasonable in PDL, it is no longer true for the game interpretation. After all, the Angel's choice between $\alpha;\beta$ and $\alpha;\gamma$ is not necessarily equivalent to her choice between β and γ depending on the outcomes of playing α. Therefore, as in the case of GL, we have to use the more general neighborhood semantics. As a consequence, a *fuzzy game model* is a triple $\mathfrak{M}=(S,(E_g)_{g\in\Gamma_0},V)$, where S is a set of states, $V:\Phi_0\to\mathcal{F}(S)$ is a fuzzy valuation for the propositional symbols, and for each atomic game $g\in\Gamma_0$, $E_g:S\times\mathcal{F}(S)\to[0,1]$ is a fuzzy effectivity relation such that $E_g(s,X)$ specifies the possibility degree to which the Angel playing g has a strategy at s to achieve a fuzzy event X. Hence, in the semantics, we use both interpretations of fuzzy sets to model fuzzy events and their possibility degrees respectively. We assume that E_g is monotonic with respect to the second argument, i.e. $E_g(s,X)\le E_g(s,X')$ for any $s\in S$ and $X\subseteq X'$. From the fuzzy effectivity relation E_g, we can define a fuzzy effectivity transformation $\widetilde{E_g}:\mathcal{F}(S)\to\mathcal{F}(S)$ such that $\widetilde{E_g}(X)(s)=E_g(s,X)$ for any $X\in\mathcal{F}(S)$ and $s\in S$.

The fuzzy effectivity transformation can be extended to all games in the following way:

1. $\widetilde{E_{\varphi?}}(X)=V(\varphi)\otimes X$,
2. $\widetilde{E_{\alpha;\beta}}(X)=\widetilde{E_\alpha}(\widetilde{E_\beta}(X))$
3. $\widetilde{E_{\alpha\cup\beta}}(X)=\widetilde{E_\alpha}(X)\cup\widetilde{E_\beta}(X)$
4. $\widetilde{E_{\alpha^*}}(X)=\mu Y.(X\cup\widetilde{E_\alpha}(Y))$, i.e., the least fixed point of the function $F(Y)=X\cup\widetilde{E_\alpha}(Y)$
5. $\widetilde{E_{\alpha^d}}(X)=\neg\widetilde{E_\alpha}(\neg X)$

By the monotonicity of the fuzzy effectivity relation E_g for any primitive game g and the inductive construction of compound games, $\widetilde{E_\alpha}$ is also monotonic for any game α. As a result, the function $F:\mathcal{F}(S)\to\mathcal{F}(S)$ defined by $Y\mapsto X\cup\widetilde{E_\alpha}(Y)$ where $X\in\mathcal{F}(S)$ is also monotonic. Hence, by the well-known Knaster-Tarski theorem [23], the least fixed point of F exists and the definition of $\widetilde{E_{\alpha^*}}$ is well-defined.

The fuzzy valuation can also be extended to any formula φ inductively:

- $V(\bot) = \emptyset$
- $V(\varphi_1 \otimes \varphi_2) = V(\varphi_1) \otimes V(\varphi_2)$
- $V(\varphi_1 \rightarrow \varphi_2) = V(\varphi_1) \Rightarrow V(\varphi_2)$
- $V(\langle\gamma\rangle\varphi) = \widetilde{E_\gamma}(V(\varphi))$

By using the (extended) fuzzy valuation, the *degree of truth or satisfaction* of a formula φ in state s is simply $V(\varphi)(s)$. A formula φ is true in s if its degree of truth is 1 and is valid in $\mathfrak{M}$ if it is true in all states of $\mathfrak{M}$. We can then extend the definitions of truth and validity to a set of formulas. Let $\Sigma \cup \{\varphi\}$ be a set of FGL formulas. Then, φ is an FGL-consequence of Σ, denoted by $\Sigma \models_{FGL} \varphi$ (or simply $\Sigma \models \varphi$ if it is clear from the context), if for every fuzzy game model $\mathfrak{M}$ and every state s in $\mathfrak{M}$, all formulas in Σ being true in s implies that φ being true in s. When Σ is empty, we write it as $\models_{FGL} \varphi$ and say that φ is valid. To characterize the FGL-validity, a Hilbert style axiomatic system FGL is presented in Fig. 1[1].

---

1. **Axioms:**
   (a) The standard set of axioms for the basic many-valued logic BL
   (b) $\langle\alpha \cup \beta\rangle\varphi \leftrightarrow \langle\alpha\rangle\varphi \vee \langle\beta\rangle\varphi$
   (c) $\langle\alpha; \beta\rangle\varphi \leftrightarrow \langle\alpha\rangle\langle\beta\rangle\varphi$
   (d) $\langle\psi?\rangle\varphi \leftrightarrow \psi \otimes \varphi$
   (e) $(\varphi \vee \langle\gamma\rangle\langle\gamma^*\rangle\varphi) \rightarrow \langle\gamma^*\rangle\varphi$
   (f) $\langle\gamma^d\rangle\varphi \leftrightarrow \neg\langle\gamma\rangle\neg\varphi$
2. **Rules of Inference:**
   (a) $\dfrac{\varphi, \varphi \rightarrow \psi}{\psi}$
   (b) $\dfrac{\varphi \rightarrow \psi}{\langle\gamma\rangle\varphi \rightarrow \langle\gamma\rangle\psi}$
   (c) $\dfrac{(\varphi \vee \langle\gamma\rangle\psi) \rightarrow \psi}{\langle\gamma^*\rangle\varphi \rightarrow \psi}$

Fig. 1. The axiomatic system FGL

An FGL formula φ is *derivable* or *provable* in the system FGL if there exists a finite sequence of formulas $\varphi_0, \cdots, \varphi_n$ such that $\varphi_n = \varphi$ and for any $0 \leq i \leq n$, each φ_i is an instance of some axiom in FGL or follows from one or more φ_j's for $j < i$ by the application of a rule of inference. We use $\vdash_{FGL} \varphi$ (or simply $\vdash \varphi$ in case of no confusion) to denote that φ is derivable in the system FGL.

As usual, the most important property of an axiomatic system is its soundness and completeness. For the system FGL, we can easily verify its soundness.

Theorem 1. *Let φ be an FGL formula. Then $\vdash \varphi$ implies $\models \varphi$.*

---

[1] See [9] for the axiomatization of the basic many-valued logic BL.

PROOF. The proof of soundness is quite standard. We can verify the validity of each axiom and the validity-preserving property of each inference rule in a routine way. As examples, we show that axiom (e) is valid and rule (c) is validity-preserving. For the validity of axiom (e), let $\mathfrak{M} = (S, (E_g)_{g \in \Gamma_0}, V)$ be a fuzzy game model. Then, by the definition of fuzzy valuation,

$$V(\varphi \vee \langle \gamma \rangle \langle \gamma^* \rangle \varphi) = V(\varphi) \cup \widetilde{E_\gamma}(\widetilde{E_{\gamma^*}}(V(\varphi)))$$
$$= \widetilde{E_{\gamma^*}}(V(\varphi))$$
$$= V(\langle \gamma^* \rangle \varphi)$$

as $\widetilde{E_{\gamma^*}}(V(\varphi))$ is a fixed point of $V(\varphi) \cup \widetilde{E_\gamma}(X)$. Thus, $V((\varphi \vee \langle \gamma \rangle \langle \gamma^* \rangle \varphi) \to \langle \gamma^* \rangle \varphi) = V(\varphi \vee \langle \gamma \rangle \langle \gamma^* \rangle \varphi) \Rightarrow V(\langle \gamma^* \rangle \varphi) = \mathbf{1}$, where $\mathbf{1}(s) = 1$ for any $s \in S$. This shows that axiom (e) is valid in any model.

To show that rule (c) is validity-preserving, assume that $(\varphi \vee \langle \gamma \rangle \psi) \to \psi$ is valid. Then, for any model $\mathfrak{M} = (S, (E_g)_{g \in \Gamma_0}, V)$, we have $V(\varphi) \cup \widetilde{E_\gamma}(V(\varphi)) \subseteq V(\psi)$ by the definition of fuzzy valuation. As the least fixed point of a monotonic function F is the least element X such that $F(X) \leq X$ [23], we have $\widetilde{E_{\gamma^*}}(V(\varphi)) \subseteq V(\psi)$. Hence, $V(\langle \gamma^* \rangle \varphi \to \psi) = V(\langle \gamma^* \rangle \varphi) \Rightarrow V(\psi) = \mathbf{1}$, i.e., $\langle \gamma^* \rangle \varphi \to \psi$ is also valid.   □

On the other hand, while we conjecture the completeness of the system, i.e. $\models \varphi$ implies $\vdash \varphi$ for any FGL formula φ, its proof is still an open question.

4   Possibilistic Coalition Logic

While GL models 2-player games, coalition logic (CL) aims at modeling multi-agent system by focusing on the coalitional ability of multiple players [10,19]. Unlike GL, the games in CL are not decomposed into simpler games. Thus each game is regarded as atomic in CL. Consequently, the modalities in CL represent coalitions (i.e. subsets of agents) instead of games. Semantically, a neighborhood system is associated with each coalition to denote the α-effectivity function of the coalition. An effectivity function is α-effective for a coalition C iff there exists a joint strategy for C that can achieve a goal no matter what strategies the players outside C choose. Then, a formula $[C]\varphi$ expressing that coalition C is α-effective for the goal φ can be interpreted in the semantic model of CL. In this section, we consider a mild generalization of CL, called possibilistic coalition logic (PCL).

The syntax of PCL is a graded version of that for CL and its semantics is thus two-valued. Given a set of agents or players Ag and a set of propositional symbols Φ_0, the definition of PCL formulas is as follows:

$$\varphi ::= \bot \mid p \mid \varphi \to \varphi \mid [C]^{\sim c}\varphi,$$

where $C \subseteq Ag$ is a coalition, $\sim \in \{>, \geq\}$ and $c \in [0,1]$ is a rational number. The semantics of PCL formulas can still be given by using a fuzzy effectivity relation like in the case of FGL. Nevertheless, there are two main differences. On one hand, unlike the case of FGL, we do not have to consider the composition (or decomposition) of games in PCL. On the other hand, the relation must be indexed with respect to each coalition. Hence, a possibilistic coalition model is a triple $\mathfrak{M} = (S, E, V)$, where S is a set of states, $E : \mathcal{P}(Ag) \times S \times \mathcal{P}(S) \to [0, 1]$ is an indexed fuzzy effectivity relation, and $V : \Phi_0 \to 2^S$ is the truth assignment of propositional symbols. Essentially, $E(C, s, X)$ denotes the possibility degree to which the coalition C at s has a strategy to achieve the goal event X. Unlike in the case of FGL, we only use the possibility theory-based interpretation of fuzzy sets in the semantics of PCL as events in the models are crisp. In addition, we impose several reasonableness conditions on E, which are also generalizations of corresponding conditions on α-effectivity function for CL. These conditions are

1. liveness: $E(C, s, \emptyset) = 0$;
2. safety: $E(C, s, S) = 1$;
3. monotonicity: $E(C, s, X) \leq E(C, s, Y)$ for any $X \subseteq Y$;
4. Ag-maximal: $1 - E(\emptyset, s, \overline{X}) \leq E(Ag, s, X)$;
5. super-additivity: $E(C, s, X) \otimes E(C', s, Y) \leq E(C \cup C', s, X \cap Y)$ for any $C \cap C' = \emptyset$;

where $C, C' \subseteq Ag$, $s \in S$, and $X, Y \subseteq S$. Then, the satisfaction of PCL formulas φ in a model $\mathfrak{M} = (S, E, V)$ at a state $s \in S$, denoted by $\mathfrak{M}, s \models \varphi$, is defined as in the case of modal logic except that:

– $\mathfrak{M}, s \models [C]^{\sim c}\varphi$ iff $E(C, s, |\varphi|) \sim c$.

We can then present an axiomatization of PCL in Fig. 2. As in the case of FGL, we can also define the validity and provability of PCL formulas, denoted by $\models_{PCL}$ and $\vdash_{PCL}$ respectively, and relate them by soundness and completeness.

1. Axioms:
   (a) The standard set of axioms for the classical propositional logic
   (b) $\neg[C]^{>0}\bot$
   (c) $[C]^{\geq 1}\top$
   (d) $[C]^{\sim c}(\varphi \wedge \psi) \to [C]^{\sim c}\varphi$
   (e) $\neg[\emptyset]^{>c}\neg\varphi \to [Ag]^{\geq 1-c}\varphi$
   (f) $\neg[\emptyset]^{\geq c}\neg\varphi \to [Ag]^{>1-c}\varphi$
   (g) $[C_1]^{\sim c}\varphi \wedge [C_2]^{\sim d}\psi \to [C_1 \cup C_2]^{\sim c \otimes d}(\varphi \wedge \psi)$ if $C_1 \cap C_2 = \emptyset$
   (h) $[C]^{\geq 0}\varphi$
   (i) $\neg[C]^{>1}\varphi$
   (j) $[C]^{>c}\varphi \to [C]^{\geq c}\varphi$
   (k) $[C]^{\geq c}\varphi \to [C]^{>d}\varphi$ if $c > d$
2. Rules of Inference:
   (a) $\dfrac{\varphi, \varphi \to \psi}{\psi}$
   (b) $\dfrac{\varphi \leftrightarrow \psi}{[C]^{\sim c}\varphi \leftrightarrow [C]^{\sim c}\psi}$

Fig. 2. The axiomatic system PCL

Theorem 2. *Let φ be an PCL formula. Then $\vdash \varphi$ implies $\models \varphi$.*

PROOF. To verify validity of axioms, we simply note that axioms (b)-(g) correspond to conditions on effectivity function and axioms (h)-(k) easily follow from the property of numerical inequality. In addition, the validity-preserving of the rule (b) depends on the fact that equivalent formulas have the same truth set.   □

As in the case of FGL, we also conjecture the completeness of PCL without proof yet.

5   Concluding Remarks

In this paper, we propose the extensions of modal logics for reasoning about games to accommodate possibilistic uncertainty. The main contribution of the work is to integrate possibility theory into semantic models of FGL and PCL. While we need to model the possibility degrees of achieving some events in both logics, in FGL, we also have to additionally consider vaguely-described events as fuzzy subsets.

As we are mainly concerned with the semantic aspects of the proposed logics, there remain many logical and computational issues not addressed, including proving open conjectures about the completeness of FGL and PCL, explicitly expressing degrees of truth and uncertainty in the language of FGL, dealing with vague information in PCL with fuzzy events like in the case of FGL, and studying decidability and complexity of satisfiability and model checking problems of the proposed logics. These will be possible directions for future work.

References

1. van Benthem, J.: Logical dynamics of information and interaction. Cambridge University Press (2011)
2. van Benthem, J.: Logic in games. The MIT Press (2014)
3. Blackburn, P., de Rijke, M., Venema, Y.: Modal logic. Cambridge University Press (2001)
4. Bonanno, G., Dégremont, C.: Logic and game theory. In: Baltag, A., Smets, S. (eds.) Johan van Benthem on Logic and Information Dynamics. OCL, vol. 5, pp. 421–449. Springer, Cham (2014). https://doi.org/10.1007/978-3-319-06025-5_15
5. Chaudhuri, S., Kannan, S., Majumdar, R., Wooldridge, M.: Game theory in AI, logic, and algorithms (dagstuhl seminar 17111). Dagstuhl Reports **7**(3), 27–32 (2017)
6. Chellas, B.: Modal logic : an introduction. Cambridge University Press (1980)
7. Chen, T., Lu, J.: Probabilistic alternating-time temporal logic and model checking algorithm. In: Proceedings of the Fourth International Conference on Fuzzy Systems and Knowledge Discovery (FSKD), vol. 2, pp. 35–39 (2007)
8. Doberkat, E.-E.: Towards a probabilistic interpretation of game logic. In: Kahl, W., Winter, M., Oliveira, J.N. (eds.) RAMICS 2015. LNCS, vol. 9348, pp. 43–47. Springer, Cham (2015). https://doi.org/10.1007/978-3-319-24704-5_3
9. Hájek, P.: Metamathematics of fuzzy logic. Kluwer Academic Publisher (1998)

10. Hansen, H.H., Pauly, M.: Axiomatising nash-consistent coalition logic. In: Flesca, S., Greco, S., Ianni, G., Leone, N. (eds.) JELIA 2002. LNCS (LNAI), vol. 2424, pp. 394–406. Springer, Heidelberg (2002). https://doi.org/10.1007/3-540-45757-7_33
11. Harel, D.: Dynamic logic. In: Gabbay, D., Guenthner, F. (eds.) Handbook of Philosophical Logic, Vol. II: Extensions of Classical Logic, pp. 497–604. D. Reidel Publishing Company (1984)
12. Harel, D., Kozen, D., Tiuryn, J.: Dynamic logic. MIT Press (2000)
13. Hodges, W.: Logic and games. In: Zalta, E. (ed.) The Stanford Encyclopedia of Philosophy. Metaphysics Research Lab, Stanford University, spring 2013 edn. (2013)
14. van der Hoek, W., Pauly, M.: Modal logic for games and information. In: Blackburn, P., Benthem, J.V., Wolter, F. (eds.) Handbook of Modal Logic, pp. 1077–1148. Elsevier (2007)
15. Huang, X., Su, K., Zhang, C.: Probabilistic alternating-time temporal logic of incomplete information and synchronous perfect recall. In: Proceedings of the Twenty-Sixth AAAI Conference on Artificial Intelligence (2012)
16. Michalak, T., Wooldridge, M.: AI and economics. IEEE Intell. Syst. **32**(1), 5–7 (2017)
17. Pacuit, E.: Neighborhood Semantics for Modal Logic. Springer, Cham (2017). https://doi.org/10.1007/978-3-319-67149-9
18. Parikh, R.: Propositional game logic. In: Proceedings of the 24th Annual Symposium on Foundations of Computer Science (FOCS), pp. 195–200. IEEE Computer Society (1983)
19. Pauly, M.: A modal logic for coalitional power in games. J. Log. Comput. **12**(1), 149–166 (2002)
20. Pauly, M., Parikh, R.: Game logic - an overview. Stud. Logica. **75**(2), 165–182 (2003)
21. Russell, S., Norvig, P.: Artificial intelligence: a modern approach (3rd ed.). Pearson (2010)
22. Sack, J.: Logics for dynamic epistemic behavioral strategies. In: Yang, S.C.-M., Deng, D.-M., Lin, H. (eds.) Structural Analysis of Non-Classical Logics. LASLL, pp. 159–182. Springer, Heidelberg (2016). https://doi.org/10.1007/978-3-662-48357-2_8
23. Tarski, A.: A lattice-theoretical fixpoint theorem and its applications. Pac. J. Math. **5**(2), 285–309 (1955)
24. Wooldridge, M., Jennings, N.: Intelligent agents: theory and practice. Knowl. Eng. Rev. **10**(2), 115–152 (1995)
25. Zadeh, L.: Fuzzy sets. Inf. Control **8**(3), 338–353 (1965)
26. Zadeh, L.: Fuzzy sets as a basis for a theory of possibility. Fuzzy Sets Syst. **1**(1), 3–28 (1978)

Application of Interval Valued Fuzzy Sets in Attribute Ranking

Bich Khue Vo[1] and Hung Son Nguyen[2]($\boxtimes$)

[1] University of Finance and Marketing, Ho Chi Minh City, Vietnam
vokhue@ufm.edu.vn
[2] Institute of Computer Science, University of Warsaw, ul. Banacha 2,
Warsaw 02-097, Poland
son@mimuw.edu.pl

Abstract. In this paper, a new methodology for ranking the attributes of a given decision table is proposed. It is a combination of discernibility relations in rough set theory and decision-making methods based on interval-valued fuzzy sets. Several acceleration methods based on randomized techniques are also presented to reduce the time complexity of the proposed methodology. The experiment results shows that the proposed methods are very up-and-coming.

Keywords: Interval Valued Fuzzy Sets · Rough Sets · Decision making · Attribute importance

1 Introduction

In machine learning (ML), input data usually consists of a set of input attributes or variables which are individual properties generated from a dataset, used as input to ML models and represented as numerical or string columns in datasets. When the ML model outputs prediction, it relies more on some attributes than others. Fundamentally, model predictions directly depend on the quality of attributes. Attribute importance is defined as the problem of looking for attribute ranking according to their importance in classification or regression tasks.

Attribute importance can be measured using many different techniques, but one of the most popular techniques is the *Random Forest Classifier* [4]. Using the Random forest algorithm, the feature importance can be measured as the average impurity decrease computed from all decision trees in the forest.

Another simple and commonly used technique is *Permutation Feature Importance*. The method focuses on observing how predictions of the ML model change when we change the values of a single variable.

In [12], we propose a method for feature ranking called RAFAR (Rough-fuzzy Algorithm For Attribute Ranking). The main idea of RAFAR is to assign a weight w_i to the i-th alternative so that the higher weight means the preferable

---

Supported by University of Finance and Marketing.

V.-N. Huynh et al. (Eds.): IUKM 2023, LNAI 14375, pp. 61–72, 2023.
https://doi.org/10.1007/978-3-031-46775-2_6

choice. Additionally, we assume that the weight vector can be determined in the form of *a probability vector*, and our feature ranking algorithm is based on solving some optimization problems, which are constructed from the given decision table.

The experiment's results are very promising. However, it exposes certain shortcomings that we want to improve in this paper. We propose two improvements including (1) replacing real weight vectors by interval value vectors and (2) applying a randomized technique to accelerate the process of building optimization problems from data.

The paper is organized as follows: Sect. 2 contains the basic notions about Interval-Valued Fuzzy sets (IVFS), fuzzy preference relations, and decision-making methods based on IVFS. The RAFAR methodology and its improved components are presented in Sect. 3. In Sect. 4, the results of the experiment on the performance of the RAFAR methodology are presented. The concluding remarks are described in Sect. 5.

2 Preliminaries

In many real life applications, membership value in the fuzzy set [15] cannot be exactly defined. A type-2 fuzzy set is a set in which the values of the membership function are also uncertain.

Below are brief reminders about the Interval-valued fuzzy set (IVFS) [1] defined over a finite universal, i.e. $X = \{x_1, \cdots, x_n\}$.

Each traditional subset $A \subset X$ can be identified with a function $f_A : X \to \{0,1\}$ and inversely $A = \{x : f_A(x) = 1\}$. For any mapping $\mu_A : X \to [0,1]$, the value $\mu_A(x)$ is called *fuzzy membership function* of x to A. Let

$$\Gamma([0,1]) = \{(x,y) : 0 \leq x \leq y \leq 1\} = \{[x,y] : [x,y] \subseteq [0,1]\}$$

Any mapping $m : X \to \Gamma([0,1])$ is called *interval-valued fuzzy subset* (IVFS). The family of all IVFS of X is denoted by $\mathcal{IVFS}(X)$. Any $A \in \mathcal{IVFS}(X)$ is a mapping $m_A = [l_A, r_A] : X \to \Gamma([0,1])$ and it can be interpreted as a family of fuzzy sets S, such that $l_A(x_i) \leq \mu_S(x_i) \leq r_A(x_i)$ for each $x_i \in X$. The mapping $l_A : X \to [0,1]$ and $r_A : X \to [0,1]$ are called *the lower and the upper bound* of the membership interval $m_A(x)$, respectively.

2.1 Fuzzy Relations

In traditional set theory, a binary relation (over a set X) is defined to be a subset of the Cartesian product $X \times X$. It is quite natural to define fuzzy relations over X as fuzzy subset of $X \times X$. If $X = \{x_1, \cdots, x_n\}$ is a finite set, then any fuzzy relation over X can be represented by an $n \times n$ matrix $R = (\rho_{ij})_{n \times n}$, where ρ_{ij} is a fuzzy value describing the membership degree of (x_i, x_j) to this relation. If ρ_{ij} are intuitionistic fuzzy numbers (IFNs) or interval valued numbers (IVNs), then the matrix R is called intuitionistic fuzzy matrix (IFM) or interval valued fuzzy matrix (IVFM), correspondingly.

Recall that any set $R \subset X \times X$ is called *a relation* on X. The relation $R \subset X \times X$ is called *a (partial) order on* X if it is reflexive, antisymmetric and transitive. In addition, R is *a linear order* if for any $x, y \in X$, either $(x, y) \in R$ or $(y, x) \in R$. We can extend the concept of linear order into *fuzzy preference relation* and *intuitionistic fuzzy preference relation* as follows:

Definition 1 (Fuzzy Preference Relation - FPR). A fuzzy preference relation on X is a fuzzy set on $X \times X$, which is characterized by a membership function $\mu_P : X \times X \to [0, 1]$, satisfying the additive reciprocal conditions, i.e.:

$$\mu_P(x_i, x_j) + \mu_P(x_j, x_i) = 1 \text{ and } \mu_P(x_i, x_i) = 1/2 \quad \text{for all } i, j = 1, \cdots, n.$$

If we denote $p_{ij} = \mu_P(x_i, x_j)$, then the fuzzy preference relation can be represented by the $n \times n$ matrix $P = (p_{ij})_{i,j=\overline{1,n}}$. The value $p_{ij} = \mu_P(x_i, x_j) \in [0, 1]$ is interpreted as the preference degree of x_i over x_j. If $p_{ij} = 1/2$, then we say that there is no difference between x_i and x_j, $p_{ij} = 1$ indicates that x_i is absolutely better than x_j (classical order relation), $p_{ij} > 1/2$ indicates that x_i is preferable to x_j. Moreover, the transitive property of an FPR can be expressed by either additive or multiplicative consistency:

Definition 2 (Additive and Multiplicative Consistent FPR [11]). Let $P = (p_{ij})$ be a fuzzy preference relation. P is called *additive consistent* if

$$p_{ij} + p_{jk} + p_{ki} = \frac{3}{2} \quad \text{for all } i, j, k = 1, \cdots, n.$$

P is called *multiplicative consistent* if

$$p_{ij} \cdot p_{jk} \cdot p_{ki} = p_{ji} \cdot p_{ik} \cdot p_{kj} \quad \text{for all } i, j, k = 1, \cdots, n.$$

Notice that if a fuzzy preference relation $P = (p_{ij})$ is additive consistent and if both $p_{ij} > 1/2$ (x_i is preferable to x_j) and $p_{jk} > 1/2$ (x_j is preferable to x_k) then $p_{ki} < 1/2$, which implies that $p_{ik} > 1/2$ (x_i is preferable to x_k). This means the additive consistent FPRs are also transitive. This fact is also true for multiplicative consistency.

Definition 3 (IVFPR). The matrix $C = (c_{ij})_{n \times n}$ is called interval-valued fuzzy preference relation (IVFPR) if $c_{ij} = [\lambda_{ij}, \rho_{ij}] \subset \mathbf{\Gamma}([0, 1])$ satisfying:

$$c_{ii} = [0.5, 0.5] \quad \text{and} \quad c_{ji} = [\lambda_{ji}, \rho_{ji}] = [1 - \rho_{ij}, 1 - \lambda_{ij}]$$

for all $i, j \in \{1, \cdots, n\}$. In fact, each IVFPR can be represented by a pair $C = (L_C = (l_{ij}), U_C = (u_{ij}))$ of two FPR, where

$$l_{ij} = \begin{cases} \lambda_{ij} & \text{if } i \leq j \\ 1 - \lambda_{ji} & \text{if } i > j \end{cases} \quad \text{and} \quad u_{ij} = \begin{cases} \rho_{ij} & \text{if } i \leq j \\ 1 - \rho_{ji} & \text{if } i > j \end{cases}$$

Matrices L_C and U_C are called lower and upper bounds of IVFPR C.

The concepts of interval-valued fuzzy relation and interval-valued fuzzy matrix (IVFM) have been studied by many authors [2, 7]. IVFM is a generalization of the Fuzzy Matrix and has been useful in dealing with decision-making, clustering analysis, relational equations, etc.

3   Attribute Ranking and RAFAR Method

In machine learning, the supervised learning task can be defined in term of training dataset $D = \{(\mathbf{x}_1, y_1), \cdots, (\mathbf{x}_m, y_m)\}$, where $\mathbf{x}_1, \cdots, \mathbf{x}_m$ are the descriptions of training objects $u_1, \cdots, u_m$ and $y_1, \cdots, y_m$ are the class label of those objects. Usually, the description of objects are represented by attributes or features, namely $a_1, ..., a_n$. Each attribute a_i is in fact a function of form: $a_i : U \to V_{a_i}$, where V_{a_i} is called the domain of a_i. In other words, $\mathbf{x}_i = (a_1(u_i), \cdots, a_n(u_i))$.

Thus, why in rough set theory, the training dataset can be presented in the form of a decision table, which is a tuple $\mathbf{T} = (U, A \cup \{d\})$, where $U = \{u_1, \cdots, u_m\}$ is the set of objects, $A = \{a_1, \cdots a_n\}$ is the set of attributes and d is the decision attribute, i.e. $d(x_i) = y_i$.

The goal of the attributes ranking problem for the decision table $\mathbf{T} = (U, A \cup \{d\})$ is to order the features in a ranking list so that the more important features are at the beginning of the list, while the less important features are located at the end of the ranking list.

In [12], we propose a new method for features ranking called RAFAR (Rough-fuzzy Algorithm for Attributes Ranking). This is a hybrid method that combines discernibility relation of the rough set theory and the ranking method described in the previous section. The RAFAR method consists of two main steps: (1) constructing Intuitionistic Fuzzy Preference Relation (IFPR) for the set of attributes and (2) searching for an optimal *probability vector*, i.e. a vector $\mathbf{w} = (w_1, \ldots, w_n)$ such that $w_i \in [0, 1]$ for $i = 1, \cdots, n$ and $\sum_{i=1}^{n} w_i = 1$, which is consistent with this IFPR. Each value w_i is related to attribute $a_i \in A$ so that the higher weight means the more preferred choice. In [12], four optimization models including additive consistent models **(A1)**, **(A2)** and multiplicative consistent models **(M1)** or **(M2)** are applied to generate the probability weight vector for the set of attributes.

In this paper, we present another ranking method based on IVFPR and interval-valued fuzzy vectors (IVFV) and show that this approach has more advanced properties.

3.1   Ranking Method Based on Interval-Valued Fuzzy Sets

Let $\mathbf{T} = (U, A \cup \{d\})$ be a given decision table, where $U = \{u_1, \cdots, u_m\}$ and $A = \{a_1, \cdots a_n\}$. For any symbolic attribute $a_k \in A$, we define

$$D_k = \{(u_i, u_j) \in U \times U : d(u_i) \neq d(u_j) \text{ and } a_k(u_i) \neq a_k(u_j)\}$$

If $a_k \in A$ is a continuous attribute, we discretize its domain into b equal length intervals, where b is a parameter, and use the discernibility relation for the discretised feature. D_k is in fact the discernibility relation, which is a building block in rough set theory [6].

At the first step we create an IVFPR denoted by $C_{\mathbf{T}} = ([\lambda_{ij}, \rho_{ij}])_{n \times n}$, where

$$\lambda_{ij} = \begin{cases} 1 - \frac{|A_j|}{|A_i \cup A_j|} & \text{if } i \neq j \\ \frac{1}{2} & \text{if } i = j \end{cases} \qquad \rho_{ij} = \begin{cases} \frac{|A_i|}{|A_i \cup A_j|} & \text{if } i \neq j \\ \frac{1}{2} & \text{if } i = j \end{cases} \tag{1}$$

and the problem is searching for an interval-valued fuzzy vector (IVFV):

$$\mathbf{w} = ([l_1, r_1], \cdots, [l_n, r_n]) \in \mathbf{\Gamma}([0, 1])^n$$

where $0 \leq l_i \leq r_i \leq 1$ for $i = 1, \cdots, n$ that is somehow consistent with $C_{\mathbf{T}}$.

Equation 1 is in fact a modification of the simplified IFPR generation method presented in [12]. In order to switch from probability weight vector into IVFV, the normalization condition $\sum_{i=1}^{n} w_i = 1$, the additive and multiplicative consistency conditions as well as the corresponding optimization problems should be also modified. In [10,14], these conditions are redefined as follows:

– **Normalization:** If the vector $\mathbf{w} = ([l_1, r_1], \cdots, [l_n, r_n])$ satisfies:

$$\sum_{j=1,j\neq i}^{n} l_j + r_i \leq 1 \leq l_i + \sum_{j=1,j\neq i}^{n} r_j \quad \text{for each } i \in \{1, \cdots, n\} \qquad (2)$$

then it is called the *normalized IVFV*.
– **Consistency:** For a given vector $\mathbf{w}$ one can define two matrices $\mathbf{A} = (\mathbf{A}^-, \mathbf{A}^+) = ([a_{ij}^-, a_{ij}^+])$ and $\mathbf{P} = (\mathbf{P}^-, \mathbf{P}^+) = ([p_{ij}^-, p_{ij}^+])$, where

$$[a_{ij}^-, a_{ij}^+] = \begin{cases} [0.5, 0.5] & \text{if } i = j \\ \left[\frac{1+l_i-r_j}{2}, \frac{1+r_i-l_j}{2}\right] & \text{if } i \neq j \end{cases} \quad [p_{ij}^-, p_{ij}^+] = \begin{cases} [0.5, 0.5] & \text{if } i = j \\ \left[\frac{l_i}{l_i+r_j}, \frac{r_i}{r_i+l_j}\right] & \text{if } i \neq j \end{cases}$$

- $C = (L_C, U_C)$ is *additive consistent* IVFPR if there exists a normalized IVFV $\mathbf{w}$ such that $(L_C, U_C) = (\mathbf{A}^-(\mathbf{w}), \mathbf{A}^+(\mathbf{w}))$,
- $C = (L_C, U_C)$ is *multiplicative consistent* IVFPR if there exists a normalized IVFV $\mathbf{w}$ such that $(L_C, U_C) = (\mathbf{P}^-(\mathbf{w}), \mathbf{P}^+(\mathbf{w}))$,

Similar to the case of the probability weight vector, not every IVFPR is either additive or multiplicative consistent. The relaxation is based on the minimization of the differences $\|L_C - \mathbf{A}^-(\mathbf{w})\|$ and $\|U_C - \mathbf{A}^+(\mathbf{w})\|$. Thus the problem of searching for the best interval-valued fuzzy vector IVFV can be formulated as the following optimization problem:

$$\min \sum_{i=1}^{n-1} \sum_{j=i+1}^{n} \left(\left|\frac{l_i-r_j+1}{2} - \lambda_{ij}\right| + \left|\frac{r_i-l_j+1}{2} - \upsilon_{ij}\right|\right)$$

$$\text{s.t.} \begin{cases} \sum_{j=1,j\neq i}^{n} l_j + r_i \leq 1 \leq l_i + \sum_{j=1,j\neq i}^{n} r_j, \text{ for } i = 1, ..., n \\ 0 \leq l_i \leq r_i \leq 1, \text{ for } i = 1, ..., n \end{cases}$$

This problem can be transformed into a linear programming model by the fact:

Lemma 1. *For any* $x \in \mathbb{R}$*, if* $\xi^+ = \frac{|x|+x}{2}$ *and* $\xi^- = \frac{|x|-x}{2}$ *then* $\xi^+, \xi^- \geq 0$ *and* $|x| = \xi^+ + \xi^-$*,* $x = \xi^+ - \xi^-$*.*

Applying the above lemma, the modified additive consistent model, and the multiplicative model for the interval weight vector [14] are redefined as follows:

Model (A3):	**Model (M3):**
$\min \mathbf{J} = \sum\limits_{i=1}^{n-1} \sum\limits_{j=i+1}^{n} (\xi_{ij}^{+} + \xi_{ij}^{-} + \eta_{ij}^{+} + \eta_{ij}^{-})$	$\min \mathbf{K} = \sum\limits_{i=1}^{n-1} \sum\limits_{j=i+1}^{n} (\xi_{ij}^{+} + \xi_{ij}^{-} + \eta_{ij}^{+} + \eta_{ij}^{-})$
s.t. $\left\{ \begin{array}{l} \frac{l_i - r_j + 1}{2} - \lambda_{ij} - \xi_{ij}^{+} + \xi_{ij}^{-} = 0 \\ \frac{r_i - l_j + 1}{2} - v_{ij} - \eta_{ij}^{+} + \eta_{ij}^{-} = 0 \\ \xi_{ij}^{+}, \xi_{ij}^{-}, \eta_{ij}^{+}, \eta_{ij}^{-} \geq 0 \\ \sum\limits_{k \neq i} l_k + r_i \leq 1 \leq l_i + \sum\limits_{k \neq i} r_k \\ 0 \leq l_i \leq r_i \leq 1 \end{array} \right\} \begin{array}{l}(*)\\ \\ \\(**)\end{array}$	s.t. $\left\{ \begin{array}{l} l_i - \lambda_{ij}(l_i + r_j) - \xi_{ij}^{+} + \xi_{ij}^{-} = 0 \\ r_i - v_{ij}(r_i + l_j) - \eta_{ij}^{+} + \eta_{ij}^{-} = 0 \\ \xi_{ij}^{+}, \xi_{ij}^{-}, \eta_{ij}^{+}, \eta_{ij}^{-} \geq 0 \\ \sum\limits_{k \neq i} l_k + r_i \leq 1 \leq l_i + \sum\limits_{k \neq i} r_k \\ 0 \leq l_i \leq r_i \leq 1 \end{array} \right\} \begin{array}{l}(*)\\ \\ \\(**)\end{array}$
where (*) holds for all $1 \leq i < j \leq n$ and (**) holds for all $1 \leq i \leq n$.	where (*) holds for all $1 \leq i < j \leq n$ and (**) holds for all $1 \leq i \leq n$.

Using one of the models **(A3)** or **(M3)**, we find out the interval weight vectors $w_i = (l_i, r_i)$ which is assigned to the ith attributes for $i = 1, ..., n$.

3.2   Ordering Interval-Valued Fuzzy Numbers

Many useful methods have been developed to compare two interval-valued fuzzy numbers as well as to arrange a set of IVFN in a linear order. The first idea is to evaluate each interval $a = [l_a, r_a]$ by a real value $S(a)$ called *score*, i.e.:
$a \leq b \iff S(a) \leq S(b)$.

The simplest score is the interval center $S_0(a) = \dfrac{l_a + r_a}{2}$. Based on this function, a parameterized score function was defined in [5]:

$$S_\lambda(a) = (l_a + r_a - 1)(1 + r_a - l_a) + \lambda \cdot (r_a - l_a)^2$$

where $\lambda \in [-1, 1]$ is the risk parameter given by the decision maker in consensus, reflecting a DM's attitude towards risk.

Another score function [16] has a form $S(a) = (L(a), H(a))$, where

$$L(a) = \frac{r_a}{1 + r_a - l_a}; \quad H(a) = l_a - r_a + 1;$$

H and L are called the *Accuracy* and the *Similarity functions*. Using those functions two intervals can be compared as follows:

$$a > b \iff L(a) > L(b) \text{ or } (L(a) = L(b) \text{ and } H(a) > H(b))$$

In [13], the authors introduced a method for comparisons and rankings of intervals based on calculating the degree of preference a and b by:

$$p(a \geq b) = \frac{\max(0, r_a - l_b) - \max(0, l_a - r_b)}{w_a + w_b}$$

The value $p(a \geq b) \in [0, 1]$ can be interpreted as the likelihood of the event that when for two randomly picked numbers $x \in a$ and $y \in b$ the inequality $x \geq y$ holds. This function satisfies the condition: $p(a \geq b) + p(b \geq a) = 1$, therefore if

Table 1. Example of 2 additive consistency ranking methods for WDBC data set

Att.Nr	(A1)	(A3)		Scores			Att.Nr	(A1)	(A3)		Scores		
i	w_i	l_i	r_i	S_0	L	p		w_i	l_i	r_i	S_0	L	p
V11	0	0	0	0	0		V5	0	0	0.056	0.028	0.059	0.516
V12	0	0	0	0	0	0.5	V2	0	0	0.282	0.141	0.393	0.835
V13	0	0	0	0	0	0.5	V24	0	0	0.321	0.161	0.473	0.532
V14	0	0	0	0	0	0.5	V25	0	0	0.328	0.164	0.488	0.505
V15	0	0	0	0	0	0.5	V4	0	0	0.350	0.175	0.538	0.516
V16	0	0	0	0	0	0.5	V26	0	0	0.355	0.178	0.551	0.504
V17	0	0	0	0	0	0.5	V22	0	0	0.410	0.205	0.696	0.536
V18	0	0	0	0	0	0.5	V6	0	0	0.450	0.225	0.817	0.523
V19	0	0	0	0	0	0.5	V1	0.142	0	0.503	0.251	1.012	0.528
V20	0	0	0	0	0	0.5	V3	0.006	0	0.528	0.264	1.118	0.512
V29	0	0	0	0	0	0.5	V27	0	0	0.554	0.277	1.241	0.512
V30	0	0	0	0	0	0.5	V23	0.146	0.004	0.558	0.281	1.250	0.504
V10	0	0	0.028	0.014	0.029	1.000	V21	0.193	0.011	0.565	0.288	1.266	0.506
V9	0	0	0.052	0.026	0.055	0.648	V7	0	0.028	0.581	0.305	1.303	0.515
							V8	0.216	0.141	0.694	0.418	1.556	0.602
							V28	0.296	0.262	0.816	0.539	1.829	0.610

$p(a \geq b) > 0.5$, a is said to be superior to b to the degree of $p(a \geq b)$, and it is denoted by $a \overset{p(a \geq b)}{\geq} b$.

An example of application of models (A1) and (A3) on the WDBC data set [9] is shown in Table 1. The WDBC dataset contains features extracted from a digitized image of a fine needle aspirate of a breast mass which describes the characteristics of the cell nuclei in the image. This dataset consists of 569 instances with 30 attributes and two decision classes. The features are denoted by $V1, V2, \cdots, V30$.

We can notice that it returns a non-zero value to six features and a 0 value to the other 24 features. This fact means these 24 features are not comparable by model (A1). The result of model (A3) is shown in columns l_i and r_i. In this case, only 12 features are not comparable. The next two columns in Table 1 present the values of S_0 and L scores and the features $V1, \cdots, V30$ are ranked with respect to S_0. The column p presents the probability that the current feature is better than the feature in the previous line. One can notice the fact that in this example, all three score functions are consistent. Moreover, features V7 and V27 are ranked very high by model (A3), while they are treated as not important by the model (A1).

3.3 Randomized RAFAR

The main drawback of RAFAR method is the calculation time for the matrix construction step, in which either IFPR or IVFPR is calculated. Equation 1 suggests that the time complexity for matrix calculation is $O(n^2 \cdot m \log m)$, where n is the number of attributes and m is the number of objects. In this section, we propose several modification methods that make RAFAR more scalable, but still maintain the same level of accuracy.

The first method called *randomized RAFAR* is in fact the sampling technique. The idea is based on the observation that RAFAR method is dealing with an uncertain information about the order relation between pairs of attributes as well as the weight of each single attribute. IVFPR contains intervals of possible uncertain values and can be approximately determined. The sketch of the randomized IVFPR construction method is as follows:

For each pair (k, l) such that $1 \le k < l \le n$:
1. Select randomly a sample set of objects $X \subset U$ such that $|X| \approx p \cdot |U|$, where $p \in [0, 1]$ is a parameter.
2. Calculate $|A_k(X)|, |A_l(X)|$ and $|A_k(X) \cup |A_l(X)|$, where

$$A_k(X) = \{(u_i, u_j) \in X \times X : d(u_i) \ne d(u_j) \text{ and } a_k(u_i) \ne a_k(u_j)\}$$

3. Set $\lambda_{kl} = \begin{cases} 1 - \frac{|A_l(X)|}{|A_k(X) \cup A_l(X)|} & \text{if } k \ne l \\ \frac{1}{2} & \text{if } k = l \end{cases}$ $\rho_{kl} = \begin{cases} \frac{|A_k(X)|}{|A_k(X) \cup A_l(X)|} & \text{if } k \ne l \\ \frac{1}{2} & \text{if } k = l \end{cases}$

The time complexity of the first method is $O(p \cdot n^2 \cdot pm)$. The experiment results on benchmark data sets show that the accuracy of RAFAR remains very high for $0.01 \le p \le 0.1$.

For the datasets with large number of attributes, the second method based on the decomposition technique can be applied. Given a decision table $\mathbf{T} = (U, A \cup \{d\})$ with the set of objects $U = \{u_1, \cdots, u_m\}$ the set of attributes $A = \{a_1, \cdots, a_n\}$ and the decision attributes d.

Decomposition (p, q)
1. Sampling n subsets of objects $X_1, \cdots, X_n \subset U$ such that $|X_i| \approx pm$
2. Sampling n subsets of attributes $A_1, \cdots, A_n \subset A$ such that $|A_i| \approx qn$ and $a_i \in A_i$. Create n decision tables $\mathbf{T}_1, \cdots, \mathbf{T}_n$, where $\mathbf{T}_i = (X_i, A_i \cup \{d\})$.
3. Apply model **A3** or **M3** to each of decision sub-tables $\mathbf{T}_1, \cdots, \mathbf{T}_n$ to generate optimal interval-valued fuzzy vectors $\mathbf{w}_1, \cdots, \mathbf{w}_n \in \mathbf{\Gamma}([0, 1])^n$.
4. Combine all IVFV $\mathbf{w}_1, \cdots, \mathbf{w}_n$ to create a final ranking list of attributes.

We propose two combination methods to making the final ranking from n IVFV $\mathbf{w}_1, \cdots, \mathbf{w}_n$. Assume $[l_{1k}, r_{1k}], \cdots, [l_{nk}, r_{nk}]$ are the IVFN corresponding to attribute a_k in $\mathbf{w}_1, \cdots, \mathbf{w}_n$. The first method, called *arithmetic mean of intervals*, the combined IVFN of a_k is defined by

$$\left[\frac{l_{1k} + \cdots + l_{nk}}{n}, \frac{r_{1k} + \cdots + r_{nk}}{n} \right]$$

The correctness of this method is based on the following observations:

– The weighted arithmetic mean of two normalized interval weight vectors is also a normalized interval weight vector
– Any interval weight vector $\mathbf{w} = ([l_1, r_1], \cdots, [l_n, r_n])$ of length $n > 2$ can be normalized, e.g. there exists a monotone transformation T so that

$$T(\mathbf{w}) = (T([l_1, r_1]), \cdots, T([l_n, r_n]))$$

is a normalized interval weight vector.

The proofs of these facts are quite long and technical. We will present them in the extended version of this work.

In the second method, called *fuzzyfication-defuzzyfication*, we associate each attribute a_k with a fuzzy subset μ_k of the interval $[0, 1]$ defined by

$$\mu_k(x) = \frac{\mu_{1k}(x) + \cdots + \mu_{nk}}{n}, \text{ where } \mu_{ik}(x) = \begin{cases} 1 & \text{if } x \in [l_{ik}, r_{ik}] \\ 0 & \text{otherwise.} \end{cases}$$

Let w_k be the defuzzyfication value of μ_k. The attributes from A can be ranked according to $w_1, \cdots, w_n$.

4   Experiment Results

The quality of an attributes ranking list can be represented by the *accuracy curve*. For a given ranking list $\mathbf{RL} = (a[1], a[2], \cdots, a[n])$ of attributes from A (n is the number of attributes from A), *the accuracy curve* of ranking list $\mathbf{RL}$ using classification algorithm $\mathcal{A}$ is the graph of the function $M_{\mathbf{A}} : \{1, \cdots, n\} \to [0, 1]$, where $M_{\mathbf{A}}(m)$ is the accuracy of $\mathcal{A}$ for the dataset restricted to the first m attributes of the ranking list $\mathbf{RL}$. For a fixed value of m we evaluate the accuracy of the classifier using 5-fold-cross-validation technique. Intuitively, we say that the ranking list $\mathbf{RL}_1$ is better than $\mathbf{RL}_2$ if the accuracy curve of $\mathbf{RL}_1$ is above the accuracy curve of $\mathbf{RL}_1$.

4.1   Accuracy Analysis

We present the accuracy comparison of the ranking lists returned by Random Forest Feature Importance (RFFI) [8] and four RAFAR optimisation models **(A1)**, **(A3)**, **(M1)**, **(M3)** on the WDBC dataset [9], which has been mentioned in Sect. 3.2.

We apply 3 classifiers kNN, SVM and Decision Tree and WDBC dataset to perform this comparison analysis and the results of this experiment are presented in Fig. 1.

The first classifier in our experiment is kNN. As it was proved in [12], the optimal value of k for kNN classifier on WDBC dataset equals to 9. Therefore the 9NN classifier is used in our experiments. We can notice the fact that in case of 9NN, the ranking lists generated by models **(A1)** and **(A3)** have the same accuracy curve and both are better than the ranking generated by RFFI. However, when comparing with the multiplicative models, the ranking list generated by RFFI is better than by **(M1)** but is worse than **(M3)**.

In case of a Decision Tree, RFFI should be the best technique because random forest is very close to the tree. However, the ranking list generated by interval value vector approach (models **(A3)** and **(M3)**) has very similar accuracy curve. For the SVM classifier (Fig. 2) all ranking lists generated by different RAFAR models, except **(M1)**, are a little bit better than RFFI.

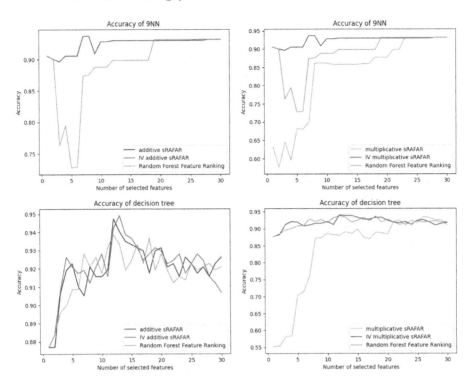

Fig. 1. The accuracy comparison between five attributes ranking methods using 9NN and the Decision Tree classifiers on WDBC dataset.

4.2 Scalability Analysis

To demonstrate the applicability of RAFAR to larger data, we consider the HIGGS dataset [3], which has been produced using Monte Carlo simulations. The data consists of 10.5 million in observations and 21 attributes related to the kinematic properties of the particle in the accelerator. The last seven attributes are high-level features calculated from the 21 original attributes. These seven synthetic attributes have been derived by physicists and have greater discriminating power when classifying between two classes.

The experiment on the HIGGS dataset was performed on all 28 attributes. We compare, with respect to execution time and accuracy, two methods for matrix generation: (1) the simplified method presented in [12] and (2) the randomized technique presented in the previous Session. The execution time for IFPR generation is 1545.78 s, while the randomized algorithm with sample size $p = 10\%$ completes the calculation in 228.15 s only. Figure 3 presents the accuracy curves of SVM and Decision Tree classifiers over 4 ranking lists:

- **No ranking:** the original list of attributes;
- **Additive r-sRAFAR:** the ranking list generated by randomized RAFAR with sample size $p = 10\%$, using model (**A1**)

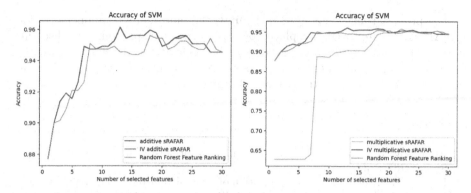

Fig. 2. The accuracy comparison between five attributes ranking methods using SVM classifier on WDBC dataset.

- **Multiplicative r-sRAFAR:** the ranking list generated by randomized RAFAR with sample size $p = 10\%$, using model **(M3)**
- **Multiplicative sRAFAR:** the ranking list generated by RAFAR using model **(M3)**

Since the ranking lists generated by model **(A1)** for two matrices are almost the same, only one accuracy curve is shown. We can notice that randomized RAFAR is 7 times faster than standard RAFAR, but their quality is on a similar level.

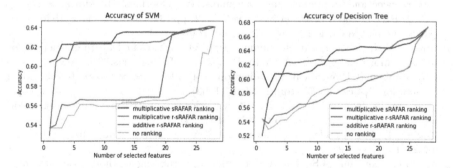

Fig. 3. Accuracy comparison between different strategies using SVM and Decision Tree classifiers on the HIGGS dataset.

5   Conclusions

This paper describes a Rough-Fuzzy hybrid method for attribute ranking problems. The idea of using interval-valued numbers to manage the uncertain information about the final weight of attributes seems very effective.

The proposed methodology is quite universal, and many components inside RAFAR can be improved or replaced by others. In this paper, only the discernibility power of attributes was considered when calculating the final ranking. In the future, we plan to add more aspects and apply multi-criteria group decision making methods to the RAFAR methodology.

References

1. Atanassov, K., Gargov, G.: Interval valued intuitionistic fuzzy sets. Fuzzy Sets Syst. **31**(3), 343–349 (1989)
2. Bakar, S.A., Kamal, S.N.: An application of interval valued fuzzy matrix in modeling clinical waste incineration process. J. Phys. Conf. Ser. **1770**(1), 012057 (2021)
3. Baldi, P., Sadowski, P., Whiteson, D.: Searching for exotic particles in high-energy physics with deep learning. Nat. Commun. **5**(1) (2014)
4. Breiman, L.: Random forests. Mach. Learn. **45**, 5–32 (2001)
5. Lin, Y., Wang, Y.: Group decision making with consistency of intuitionistic fuzzy preference relations under uncertainty. IEEE/CAA J. Automat. Sinica **5**, 741–748 (2018)
6. Nguyen, H.S.: Approximate boolean reasoning: foundations and applications in data mining. In: Peters, J.F., Skowron, A. (eds.) Transactions on Rough Sets V. LNCS, vol. 4100, pp. 334–506. Springer, Heidelberg (2006). https://doi.org/10.1007/11847465_16
7. Pal, M.: Interval-valued fuzzy matrices with interval-valued fuzzy rows and columns. Fuzzy Inf. Eng. **7**, 335–368 (2015)
8. Saeys, Y., Abeel, T., Van de Peer, Y.: Robust feature selection using ensemble feature selection techniques. In: Daelemans, W., Goethals, B., Morik, K. (eds.) ECML PKDD 2008. LNCS (LNAI), vol. 5212, pp. 313–325. Springer, Heidelberg (2008). https://doi.org/10.1007/978-3-540-87481-2_21
9. Street, W.N., Wolberg, W.H., Mangasarian, O.L.: Nuclear feature extraction for breast tumor diagnosis. In: Acharya, R.S., Goldgof, D.B. (eds.) Biomedical Image Processing and Biomedical Visualization, vol. 1905, pp. 861–870. International Society for Optics and Photonics, SPIE (1993)
10. Sugihara, K., Ishii, H., Tanaka, H.: Interval priorities in AHP by interval regression analysis. Eur. J. Oper. Res. **158**, 745–754 (2004)
11. Tanino, T.: Fuzzy preference orderings in group decision making. Fuzzy Sets Syst. **12**(2), 117–131 (1984)
12. Vo, B.K., Nguyen, H.S.: Feature selection and ranking method based on intuitionistic fuzzy matrix and rough sets. In: Proceedings of the 17th Conference on Computer Science and Intelligence Systems. Annals of Computer Science and Information Systems, vol. 30, pp. 279–288 (2022)
13. Wang, Y.M., Yang, J.B., Xu, D.L.: A two-stage logarithmic goal programming method for generating weights from interval comparison matrices. Fuzzy Sets Syst. **152**(3), 475–498 (2005)
14. Wang, Z.J., Li, K.W.: Goal programming approaches to deriving interval weights based on interval fuzzy preference relations. Inf. Sci. **193**, 180–198 (2012)
15. Zadeh, L.A.: Fuzzy sets. Inf. Control **8**, 338–353 (1965)
16. Zhang, X., Xu, Z.: A new method for ranking intuitionistic fuzzy values and its application in multi-attribute decision making. Fuzzy Optim. Decis. Mak. **11**(2), 135–146 (2012)

Utilization of Data Envelopment Analysis for Grid Traffic Accident Data

Daisuke Maruyama[1], Masahiro Inuiguchi[1(✉)], Naoki Hayashi[1], Hirosato Seki[1], Hiromitsu Kurihara[2], and Tomoya Ueura[2]

[1] Graduate School of Engineering Science, Osaka University, Toyonaka, Osaka 560-8531, Japan
maruyama@inulab.sys.es.osaka-u.ac.jp, inuiguti@sys.es.osaka-u.ac.jp
[2] Traffic Planning Division, Hyogo Prefectural Police, Kobe-city, Hyogo 650-8510, Japan

Abstract. In this paper, given data about the population, traffic accidents, roads, and facilities of every grid square of two cities, we propose a method for analyzing traffic accidents. Considering each grid square a decision making unit (DMU) outputs traffic accidents for conceivable risk factors, we apply the Data Envelopment Analysis (DEA). Namely, the efficient DMUs are risky grid squares and the efficiency score of each DMU works as an indicator of the risk. This approach enabled us to evaluate the level of the traffic accident risk of each grid square in a standardized manner. Performing the DEA method in each city under several scenarios, risky grid squares are found. Applying weights that qualify the risky grid squares to all grid squares of the two cities, we compare the traffic accident risk between the two cities. Through this paper, the potentiality of the DEA applications in analyzing traffic accidents is demonstrated.

Keywords: Traffic accident analysis · data envelopment analysis (DEA) · grid division

1 Introduction

In recent years, traffic accidents have decreased due to the development of safety devices for automobiles and road safety improvements. Nevertheless, traffic accidents are still a significant problem close to the people as many lives are lost annually by traffic accidents [1]. Therefore collecting and analyzing traffic accident data are valuable for creating more safe and secure societies by reducing the number of traffic accidents. Indeed, traffic accident data have been recorded and composed of traffic accident statistics [2,3]. The analysis of traffic accident data is active not only in Japan but also all over the world [4,5]. Collecting data through telematics [6] and analyzing it would work very well for reducing traffic accidents in the future. The importance of data analysis is increasing.

Supported by IDS Interdisciplinary Co-creation Project of Osaka University.

In this study, we analyze data about the population, traffic accidents, roads, and facilities of every grid square of two cities. The ultimate goal is to build a method for predicting traffic accidents from the assumption that grid squares sharing the same characters will be similar in traffic accidents. The establishments of the analysis and the procedure for finding characters closely related the traffic accidents are the uneasy tasks for the goal at this stage. As the first stage of this study, we propose an analysis applicable to the given data for providing some useful and meaningful results.

The obtained data includes traffic accident and road data but no traffic volume data. From those data, we find some data regarded as risk factors and the numbers of traffic accidents regarded as results. Then, we consider that each grid square is a decision making unit producing traffic accidents as outputs from risk factors as inputs. Accordingly, we apply the data envelopment analysis (DEA) [7,8] for the given data. In this analysis, efficient grid squares are found and identified as risky grid squares because they produce more traffic accidents from fewer risk factors. Moreover, the efficiency score of each grid square works as a risk measure. Picking out risky grid squares would be valuable because we may find reasons why the grid squares are risky by comparing the grid squares having similar risk factors.

This paper is organized as follows. The data envelopment analysis (DEA) is briefly reviewed in the next section. In Sect. 3, the idea as well as the methods for coping with difficulties in the application of DEA to the grid traffic accident data are described together with data treatments. The results of the applications of the DEA are shown in Sect. 4. In Sect. 5, some concluding remarks are given.

2   Data Envelopment Analysis

In this paper, we use the data envelopment analysis (DEA) [7,8] for evaluating the traffic accident risk of grid squares, i.e., zones of a city. DEA [7,8] was proposed originally for evaluating the efficiency of activities of people or organizations called decision-making units (DMUs) having multiple inputs and outputs. Although many DEA models [8] have been proposed so far, we use the original DEA model called the CCR model [7] proposed by Charnes, Cooper, and Rhodes in 1978. Then we introduce briefly only the CCR model.

Consider n DMUs and let DMU_k indicate the k-th DMU. Let $\boldsymbol{x}_k = (x_{1k}, x_{2k}, ..., x_{pk})^{\text{T}}$ be the vector of values of the p input items of DMU_k and $\boldsymbol{y}_k = (y_{1k}, y_{2k}, ..., y_{qk})^{\text{T}}$ be the vector of values of the q output items of DMU_k ($k = 1, 2, ..., n$). The efficiency score θ_k of the DMU_k is assumed to be given by the following equation, where the weight of the i-th input is $w_i > 0$, $i = 1, 2, ..., p$ and the weight of the j-th output is $v_j > 0$, $j = 1, 2, ..., q$.

$$\theta_k = \frac{\text{Total output}}{\text{Total input}} = \frac{\sum_{j=1}^{q} v_j y_{jk}}{\sum_{i=1}^{p} w_i x_{ik}}. \tag{1}$$

When weights $w_i > 0$, $i = 1, 2, ..., p$ and $v_j > 0$, $j = 1, 2, ..., q$ are specified concretely, we obtain the efficiency score θ_k, $k = 1, 2, ..., n$. However, the determination of those weights would not be an easy task. Consider that each DMU has a different policy so that the standpoints can be different by DMUs. Accordingly their weights of items will be different. To overcome this difficulty, DEA evaluates the efficiency score of a DMU by selecting the most favorable weights for the DMU under the constraints all efficiency scores of DMUs with that weights are not larger than 1. Then the efficiency score of DMU_k is obtained as the optimal value of the following linear fractional programming problem:

$$\text{maximize } \theta_k = \frac{\sum_{j=1}^{q} v_j y_{jk}}{\sum_{i=1}^{p} w_i x_{ik}}$$

$$\text{subject to } \frac{\sum_{j=1}^{q} v_j y_{jl}}{\sum_{i=1}^{p} w_i x_{il}} \leq 1, \ l = 1, 2, ..., n, \tag{2}$$

$$w_i \geq 0, \ i = 1, 2, ..., p, \ v_j \geq 0, \ j = 1, 2, ..., q.$$

This linear fractional programming problem is reduced to the following linear programming problem, which is solved, for example, by the simplex method.

$$\text{maximize } \theta_k = \sum_{j=1}^{q} v_j y_{jk}$$

$$\text{subject to } \sum_{i=1}^{p} w_i x_{ik} = 1,$$

$$\sum_{j=1}^{q} v_j y_{jl} \leq \sum_{i=1}^{p} w_i x_{il}, \ l = 1, 2, ..., n, \tag{3}$$

$$w_i \geq 0, \ i = 1, 2, ..., p, \ v_j \geq 0, \ j = 1, 2, ..., q.$$

DMU_k is called D-efficient when the optimal value of this linear programming problem is 1, i.e., $\theta_k^* = 1$. It is known that D-efficiency is a weaker concept than the perfect efficiency. If $\theta_k^* < 1$, the activity of DMU_k is not efficient. Even DMU_k is not efficient, the optimal value θ_k^* shows the degree of efficiency. θ_k^* is called the D-efficiency score of DMU_k.

When $\theta_k^* = 1$, the perfect efficiency is checked by solving the following linear programming problem:

$$\text{maximize } \omega = \sum_{i=1}^{p} s_i + \sum_{j=1}^{q} t_j$$

$$\text{subject to } \sum_{l=1}^{n} \lambda_l x_{il} + s_i = x_{ik}, \ i = 1, 2, ..., p,$$

$$\sum_{l=1}^{n} \lambda_l y_{jl} - t_j = y_{jk}, \ j = 1, 2, ..., q, \tag{4}$$

$$s_i \geq 0, \ i = 1, 2, ..., p, \ t_j \geq 0, \ j = 1, 2, ..., q,$$

$$\lambda_l \geq 0, \ l = 1, 2, ..., n.$$

If the optimal value of this linear programming problem is zero, i.e., $\omega^* = 0$, the activity of DMU_k is perfectly efficient.

3   Application of DEA to Grid Traffic Accident Data

3.1   Data and Preparations

We analyze the data about the population, traffic accidents, roads, and facilities of every grid square of two cities collected from 2017 to 2021 (five years). The size of the grid square is 300m × 300m. The data consists of environmental data and traffic accident data given for each grid square of two cities. The environmental data includes the population, the number of households, the number of crossings, the number of educational facilities, etc. while traffic accident data includes the frequency distributions over various categorizations of accidents, the frequency distribution over the speeds when the drivers of cars/motorcycles in accidents perceived as dangerous (for short, danger perception speeds), etc. From items of data, we couple the items of risk factors with the items of traffic accidents. Namely, the data about risk factors work as the inputs while the data about traffic accidents work as the outputs. Then we apply the DEA to the grid traffic accident data regarding a grid square as a DMU. Then we find efficient DMUs which are considered risky grid squares in this application.

The first difficulty is that the data includes many zeros because of no populations or no traffic accidents in the grid squares. The original DEA assumes all input values are positive because the denominator of the linear fractional functions of Problem (2) can become zeros. Then we exclude DMUs (grid squares) which take zeros for all items of risk factors. Moreover, we exclude the items of risk factors whose values of DMU_k are zero. Similarly, if the output data of a DMU are all zeros, the D-efficiency score becomes zero. We exclude all DMUs (grid squares) such that all items of traffic accidents take zeros.

The second difficulty is that the power of potential influence of item categories does not reasonably reflected in DEA when we apply the given items directly. For example, traffic accidents are classified into three categories, i.e., fatal accidents, serious injury accidents, and slight injury accidents. If a big positive number is given to the weight of slight injury accidents and zeros are given to the weights of serious and fatal accidents, a strange result is obtained. For example, a grid square with a certain number of slight injury accidents and no fatal and serious injury accident is evaluated as riskier than a grid square with no slight injury accident and big number of fatal and serious injury accidents. To avoid such a strange result, we should take care of the power of potential influence of item categories reasonably when we apply the DEA. Namely, in the example above, the power of potential influence of fatal accidents is more than that of slight injury accidents and that of serious injury accidents. To overcome this difficulty, we use the number of cases in all item categories influences not less than an item category, instead of the number of cases in an item category. In the case of the example, we consider the items, accidents of not less severe than slight injury, accidents of not less severe than serious injury, and fatal accidents.

Table 1. Statistics of the grid traffic accident data of two cities

Number of grid squares	City A	City B
total	450	1517
with no population	74	793
with no traffic accidents	95	942
with no danger perception speed data	109	953
with no danger perception speed not less than 10 km/h	33	50
with no danger perception speed not less than 20 km/h	70	84
with no danger perception speed not less than 30 km/h	72	155
with no danger perception speed not less than 40 km/h	64	133
with no danger perception speed not less than 50 km/h	65	84
with at least one danger perception speed not less than 60 km/h	37	58

3.2 The Input and Output Data

Although we considered several combinations of input and output data, we could not find many good combinations, because (i) the given traffic data does not include very many factors potentially increasing traffic accidents, and (ii) inputs independently increasing traffic accidents can produce strange results that DMUs with normal output values become efficient if a few inputs take small values and other inputs take sufficiently big values. The strange results of (ii) come from discarding inputs with sufficiently big values in Model (1). Therefore, factors conjunctly increasing traffic accidents should be selected as inputs. Among good combinations, we describe the results when we choose the danger perception speed as the input factor, i.e., the risk factor, and the traffic accident as the output factor. The danger perception speed is divided into six categories, i.e., 0 km/h–10 km/h, 10 km/h–20 km/h, ..., 50 km/h–60 km/h, and the number of cases in each category is given. Considering the power of potential influence of item categories, we convert the item categories to 0 km/h or more ($I1$), 10 km/h or more ($I2$), ..., and 50 km/h ($I6$) or more. Therefore, we have six inputs. On the other hand, traffic accidents are classified into three categories, i.e., fatal accidents, serious injury accidents, and slight injury accidents. Similarly, considering the power of the potential influence of item categories, we convert the item categories to accidents of not less severe than slight injury ($O1$), accidents of not less severe than serious injury ($O2$), and fatal accidents ($O3$) in the same way as described in the previous subsection. Then we have three outputs. Substituting the value of Ii and the value of Oj for x_i and y_j ($i = 1, 2, ..., 6$, $j = 1, 2, 3$), respectively, we apply the DEA to data of each city, i.e., City A and City B, separately, and to the joint data of both cities.

4 Results of the Applications of DEA

4.1 Outline of Data

Applying the DEA to data of City A and City B, separately. City A is a small city located in the suburbs of metropolitan areas, while City B is a central city

Fig. 1. The color-coded map of traffic accident risk levels for City A (Map Data by OpenStreetMap, under ODbL (https://www.openstreetmap.org/copyright))

Table 2. Frequency distributions of D-efficiency scores in Cities A and B, respectively.

D-efficiency score	Color	Risk Level	City A	City B
1.0 (perfectly efficient)	red	risky	3	2
1.0 (D-efficient)	purple	almost risky	0	22
0.5–1.0	orange	fairly risky	16	19
0.3–0.5	green	modestly risky	43	35
0–0.3	blue	not very risky	279	466

located between two metropolitan areas including an arterial road. City B is bigger than City A and includes the sea coast and mountain area. Those cities are adjacent. Both cities are divided into many 300m × 300m squares. Then we have grid squares.

The statistics of the grid traffic accident data of Cities A and B are shown in Table 1. In Table 1, the set of grid squares with no traffic accidents is included in the set of grid squares with no danger perception speed data. The difference shows that traffic accidents without cars and motorcycles and exceptional accidents. We note that the set of grid squares with no population is not always included in the set of grid squares with no traffic accidents. However, this inclusion holds in the cases of those two cities. As shown in Table 1, the ratio of grid squares having high danger perception speeds in City A is higher than that of City B. This is because City B has many grid squares with no population as well as with no traffic accidents as it includes a mountain area.

Fig. 2. The color-coded map of traffic accident risk levels for City B (Map Data by OpenStreetMap, under ODbL (https://www.openstreetmap.org/copyright))

4.2 Results of the Analysis

Applying DEA to the data in City A and City B, separately, we obtain D-efficiency scores for grid squares having danger perception speeds. The grid squares are categorized into five classes by the D-efficiency scores. The five classes are perfectly efficient class, D-efficient but not perfectly efficient class, the class of D-efficiency score in $[0.5, 1)$, the class of D-efficiency score in $[0.3, 0.5)$ and the class of D-efficiency score less than 0.3 and they are considered class of risky squares, the class of almost risky squares, the class of fairly risky squares, the class of modestly risky squares and the class of not very risky squares, respectively. Figures 1 and 2 show the results of the analysis of Cities A and B, respectively, where red, purple, orange, green, and blue colors represent risky squares, almost risky squares, fairly risky squares, modestly risky squares, and not very risky squares, respectively. The number of grid squares classified into each class is shown in Table 2. We found three risky grid squares and no almost risky grid squares in City A, and two risky grid squares and 22 almost risky grid squares in City B. We observe that City B has a much higher proportion of grid squares considered risky and almost risky than City A. This may come from the fact that an arterial road skirts City B.

We compare the evaluations of traffic accident risks between City A and City B. To this end, we draw correlation diagrams of efficiency (risk) scores using the weights u_i, $i = 1, 2, ..., p$ and v_j, $j = 1, 2, 3$ of (1) obtained for risky

Table 3. Weights for Risky Grid Squares in Cities A and B

	O1	O2	O3	I1	I2	I3	I4	I5	I6
Square A1	0.142857	0	0	1	–	–	–	–	–
Square A2	0.0714286	0.178571	0.678571	0.5	–	–	–	–	–
Square A3	0.142857	0.357143	0	1	0	0	0	–	–
Square B1	0	1	0	1	–	–	–	–	–
Square B2	0.166667	0	0.833333	1	0	0	0	–	–

Table 4. Input and Output Data of Risky Grid Squares in Cities A and B

	O1	O2	O3	I1	I2	I3	I4	I5	I6
Square A1	7	0	0	1	–	–	–	–	–
Square A2	2	1	1	2	–	–	–	–	–
Square A3	2	2	0	1	1	1	1	–	–
Square B1	6	1	0	1	–	–	–	–	–
Square B2	1	1	1	1	1	1	1	–	–

grid squares (perfect efficient ones) in Cities A and B, where we note that the number of inputs p depends on the risky grid square because we erase inputs taking zero values. The weights u_i, $i = 1, 2, ..., p$ and v_j, $j = 1, 2, 3$ are shown in Table 3. In Table 3, Square A1, Square A2, and Square A3 are risky grid squares in City A while Square B1 and Square B2 are risky grid squares in City B. As for input, only the weight of $I1$ is positive. Namely, danger perception speeds do not influence the risk evaluation but the total number of positive danger perception speeds. The input and output data of those risky grid squares are shown in Table 4. From Table 4, we observe that there are many slight injury accidents with no running cars in Square A1 and Square B1. The weight vector obtained for Square B1 is interesting, the weight of output $O1$ takes zero while its value looks large in proportion to the value of input $I1$.

The correlation diagrams for all 3×2 combinations are shown in Figs. 3 and 4. In Fig. 3, data of City A are plotted while in Fig. 4, data of City B are plotted. In those figures, "Score by Ai" and "Score by Bj" written along the vertical and horizontal axes imply that axes show efficiency scores using weights of Squares Ai and Bj ($i = 1, 2, 3$, $j = 1, 2$), respectively. The red lines in those figures show the points that the two compared evaluations are the same. From those figures, we see that the evaluation by the weights obtained for Square B1 is different from others. Because the weight of $O1$ is zero in Suquare B1 while they are positive in Suquares A1, A2, and A3. Indeed, Square B1 locates in the mountain area while other squares locate in the town. This is why there are many points on the horizontal axis. The evaluation by the weights obtained for Square B2 gives larger values than the evaluation by those obtained for Square A1 so that points in the upper right figures locate over the red lines in Figs. 3 and 4. The

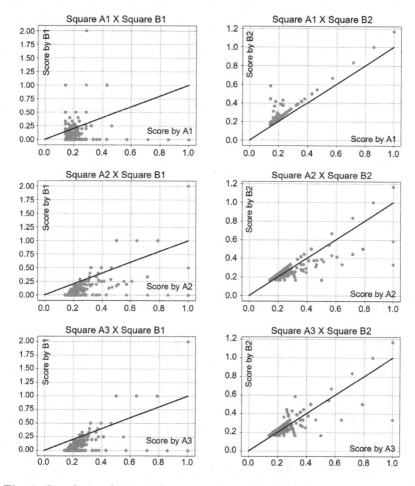

Fig. 3. Correlation diagrams for comparing Cities A and B with City A data

evaluation by the weights obtained for Square B2 is more or less similar to the evaluations by the weights obtained for Squares A2 and A3. The largest score of a grid square in City B exceeds 1 in the evaluations obtained by the weights for Squares A1, A2, and A3 as shown in Fig. 3. Similarly, the largest score of a grid square in City A exceeds 1 in the evaluations obtained by the weights for Squares B1 and B2 as shown in Fig. 4. These imply that all potential traffic accident patterns are not encompassed by the data of a city and the analysis with data from more cities is preferable. Although there are no almost risky grid squares in City A, we observe a few grid squares in City A take 1 in the evaluation by the weights obtained for Squares A2 and A3. This implies that there are multiple weights for evaluating Squares A1, A2, and A3 as D-efficient. Indeed, in Table 3, the weight of $O1$ is essentially same among Squares A1, A2 and A3 as their ratios of the weight $O1$ to the weight $I1$ are same. Therefore,

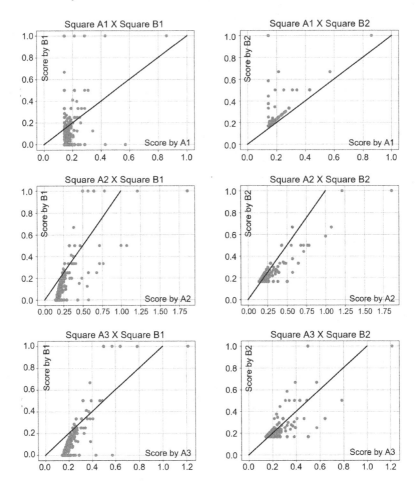

Fig. 4. Correlation diagrams for comparing Cities A and B with City B data

Square A1, a perfectly efficient grid square with only two positive weights on $I1$ and $O1$, takes 1 in all evaluations by weights obtained for Squares A1, A2, and A3.

4.3 Analysis of the Joint Data of Two Cities

As the data of a city was not sufficient for encompassing all potential traffic accident patterns, we apply the DEA for the joint traffic accident data of Cities A and B. Figure 5 shows the color-coded map of traffic accident risk levels as the result of the application of the DEA to the joint data. As shown in Fig. 5, we obtained four risky grid squares, two almost risky grid squares, 35 fairly risky grid squares, 61 modestly risky grid squares, and 803 not very risky grid squares. From those results, we found that one of the risky grid squares obtained by individual

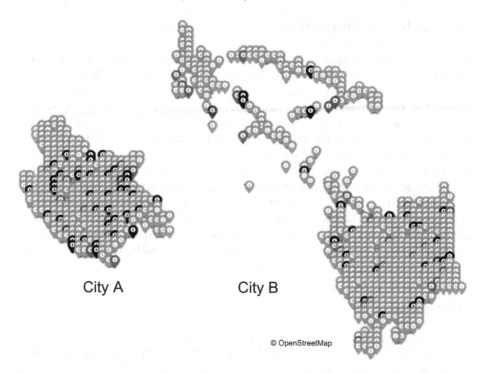

Fig. 5. The color-coded map of traffic accident risk levels in the joint data (Map Data by OpenStreetMap,under ODbL (https://www.openstreetmap.org/copyright))

analysis is degraded and many almost risky grid squares are classified into less risky categories. Indeed, Square A2 is downgraded to a fairly risky grid square. The weights obtained for Squares A1, A3, B1 and B2 are shown in Table 5. The weights are adjusted so that there are no grid squares whose risk scores exceed 1. We note that it is sufficient to use only risky (efficient) grid squares obtained in DEA applications to the individual data of cities for obtaining the weights in DEA applications to the joint Data.

Table 5. Weights for Risky Grid Squares Obtained by the Analysis of the Joint Data

	O1	O2	O3	I1	I2	I3	I4	I5	I6
Square A1	0.142857	0	0	1	–	–	–	–	–
Square A3	0.1	0.4	0	1	0	0	0	–	–
Square B1	0.1	0.4	0	1	–	–	–	–	–
Square B2	0.142857	0.142857	0.714286	1	0	0	0	–	–

5   Concluding Remarks

We have proposed the application of DEA to the analysis of grid traffic accident data. By the DEA, we found risky grid squares and we obtained the risk scores of all grid squares of the analyzed cities. By comparing the risky grid squares obtained by DEA applications to individual cities, we found a difference in risk evaluation. The difference would come from the character of the grid square. Moreover, we found that the data of a city is not always sufficient for covering all possible traffic accident patterns. It would be recommended to analyze the joint data of cities. By surveying the factors for the traffic accident risk at the risky grid squares not appear in the data, we hope that the safety of that square is improved. The selection of input items and output items for application to the DEA is important. We will find more combinations of input and output items by introducing a sufficient volume of telematics data. Finally, we may examine the applications of other models of DEA to the grid traffic accident data for improving the usefulness of the analysis.

References

1. https://www.npa.go.jp/publications/statistics/koutsuu/index_jiko_e.html
2. Noguchi, N., Inuiguchi, M., Hayashi, N., Seki, H., Suyama, T., Hasegawa J.: Factor investigation of serious accidents in traffic accident data. In: Proceedings of 37th Fuzzy System Symposium (SOFT), pp. 675–680 (2021). https://doi.org/10.14864/fss.37.0_675
3. Onishi, H., Fujiu M.: A fundamental analysis about the accident that is walking of the child using traffic accident statistics data. Intell. Inf. Infrast. 2(J2), 848–855 (2021). https://doi.org/10.11532/jsceiii.2.J2_848 https://doi.org/10.11532/jsceiii.2.J2_848
4. Hammas, M., Al-Modayan, A.: Spatial analysis of traffic accidents in the city of Medina using GIS. Geograph. Inf. Syst. 14(5), 462–477 (2022). https://doi.org/10.4236/jgis.2022.145025
5. Yeole, M., Jain, R.K., Menon R.: Prediction of road accident using artificial neural network. Eng. Trends Technol. 70(2), 143–150 (2022). https://doi.org/10.14445/22315381/IJETT-V70I2P217 https://doi.org/10.14445/22315381/IJETT-V70I2P217
6. https://en.wikipedia.org/wiki/Telematics
7. Charnes, A., Cooper, W.W., Rhodes, E.: Measuring the efficiency of decision making units. Eur. J. Oper. Res. 2, 429–444 (1978)
8. Sickles, R., Zelenyuk, V.: Measurement of Productivity and Efficiency - Theory and Practice (PDF). Cambridge University Press (2019)

Different Types of Decision Criteria in a Decision Problem

Tomoe Entani[✉] [ID]

University of Hyogo, Kobe 651-0047, Japan
entani@gsis.u-hyogo.ac.jp

Abstract. In a decision problem, there are various types of decision criteria. Some criteria are quantitative, represented by objective numerical data, and others are qualitative, represented by subjective linguistic data given by a decision maker. Moreover, a decision maker has the ideal and negative ideal in her mind regarding some criteria even if she cannot afford them now. However, she has yet to specialize in some criteria and accepts the alternatives positively. The conventional methods were proposed under the conditions of their indigenous types of criteria. Hence, other types of criteria than the specific ones are ignored or forced to be transformed into the type by some additional tasks. Our approach considers quantitative and qualitative criteria and the ideal and negative alternatives if there are in a decision maker's mind. We propose the models to obtain local evaluations of alternatives under each type of criterion. The processes are built to normalization as a premise since the local evaluations under all criteria are aggregated into overall evaluations.

Keywords: Analytic Hierarchy Process · Qualitative and quantitative · Ideal and negative ideal · Multi-criteria decision analysis

1 Introduction

We make various decisions in our daily lives repeatedly. Decision problems are simple or complex, private issues or concerned public issues with some stakeholders, and time-consuming or intuitively quick. Those who make decisions are private individuals and a group of people, such as a representative of an organization or the organization itself. The goal of the multi-criteria decision aiding is to assist decision makers in taking actions in the light of their knowledge [4].

Various methods exist to evaluate alternatives based on multiple decision criteria or conditions. At the same time, they make the decision process organized and transparent so they help stakeholders to understand the final decision. In the case of private decision making, for a decision maker who faces some difficulties, understanding her thinking on the issue is practical. The usefulness of the multi-criteria decision analysis is to reveal the decision making process. It is essential to consider a decision maker's thinking on the decision problem and to derive the result reflecting as it is. A system to help decision makers to choose the

V.-N. Huynh et al. (Eds.): IUKM 2023, LNAI 14375, pp. 85–96, 2023.
https://doi.org/10.1007/978-3-031-46775-2_8

method relevant for each case has been proposed in [3], where characteristics of decision problems are concerned with four features such as preference elicitation and preference recommendation. This study proposes a multi-criteria decision analysis method for a decision problem consisting of various types of decision criteria.

Our method considers both quantitative and qualitative criteria and the ideal and negative alternatives if there are in a decision maker's mind. In all types of decision criteria, the local evaluation of an alternative is denoted as the sum or average of its superiorities or preferences over all the other alternatives. The concept is based on pairwise comparison, which is given by comparing an alternative to the other alternative one by one.

On a quantitative criterion, we have objective numerical data of the alternatives and the ideal and negative ideal, if they are. The local evaluations are the sum of preference degrees of an alternative to all alternatives based on PROMETHEEiii [2]. On a qualitative criterion, we have subjective linguistic data given by the decision maker instead of numerical data. If the decision maker has her ideal and/or negative ideal, she compares each alternative to them. Otherwise, she compares all the pairs of alternatives. Based on AHP, the local evaluations are the geometric mean of the comparisons of an alternative to all alternatives [6].

The obtained local evaluations are normalized between 0 and 1 and aggregated into overall evaluations considering the priority weights of criteria. The PROMETHEE was applied to many fields, sometimes with the AHP [1]. In such hybrid models, the priority weights of criteria are obtained by the AHP, and the local evaluations of alternatives are obtained by the other methods such as PROMETHEE [7,8]. In this study, we use both the AHP and PROMETHEE to obtain the local evaluations reflecting the criteria types.

This paper is organized as follows. In the next section, a decision problem in the multi-criteria decision analysis is defined from the view of criteria variation. In Sect. 3, the problems to derive local evaluations under quantitative and qualitative decision criteria and with and without ideal and negative ideal are proposed. From the local evaluations under all criteria, the overall evaluations are obtained. Section 4 illustrates the proposed method with an example decision problem consisting of various decision criteria. We draw a conclusion in Sect. 5.

2   Decision Problem

Multi-criteria decision analysis is a method that compares the superiority or inferiority of alternatives under multiple decision criteria. In this paper, the goal is to assign an overall evaluation value to an alternative.

The hierarchical structure of a decision problem is illustrated in Fig. 1. There are n alternatives, $A_j, j = 1, \ldots, n$, and m criteria, $C^i, i = 1, \ldots, m$. A criteria is divided into sub-criteria $C^{ii'}, i' = 1, \ldots, m^i$. An overall evaluation value of an alternative is the aggregation of its local evaluation values under all criteria. We obtain the local evaluations from the given objective and subjective information.

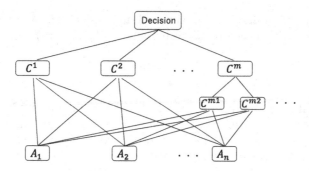

Fig. 1. Decision problem

For instance, we consider price and design as decision criteria when purchasing a car. The price is a quantitative criterion, so we have the price data of the alternatives numerically. The design is a qualitative criterion since the judgments on design are subjective and suitable for being given from their photographs or word of mouth linguistically.

A quantitative decision criterion, C^i, is measurable, so we easily have numerical data of alternatives: $\boldsymbol{x}^i = (x_1^i, \ldots, x_n^i), x_j^i \in \mathbb{R}, \forall j$. On the other hand, as for qualitative decision criteria, we access the decision maker's intuitive judgments, which are often linguistic data: $\boldsymbol{P}^i$. The judgment is about an alternative $\boldsymbol{P}^i = (p_1^i, \ldots, p_n^i)$, or a pair of alternatives, namely a comparison, $\boldsymbol{P}^i = (\boldsymbol{p}_1^i, \ldots, \boldsymbol{p}_n^i)$, where $\boldsymbol{p}_j^i = (p_{j1}^i, \ldots, p_{jn}^i)$. It is noted that the element, p_j^i and $p_{jj_1}^i, \forall j, j_1$ are linguistic values such as very good, bad, more preferable, and equal to. In this way, under each criterion, we have numerical values of alternative or linguistic values of alternatives or pairs of alternatives.

In addition to the set of alternatives, $A_j, \forall j$, a decision maker sometimes has her ideal and negative ideal in her mind under criterion C^{i_1}. We denote them as $A_+^{i_1}$ and $A_-^{i_1}$, which are not always in the list of alternatives. In the same way as the list of alternatives, we have numerical values or linguistic values about them. From these given data, first, we obtain local evaluations and then aggregate them into overall evaluations. In this study, we concern the comparability of alternatives under each criterion. Therefore, under a criterion, based on the comparisons of alternatives, we derive how much an alternative is relatively preferred to others as its local evaluation $\boldsymbol{y}^i = (y_1^i, \ldots, y_n^i), \forall i$ and $y_+^{i_1}, y_-^{i_1}, \forall i_1$.

Because of the scale differences among criteria, to aggregate the alternative's local evaluations under all criteria, we normalize the local evaluations between 0 and 1: $\bar{\boldsymbol{y}}^i = (\bar{y}_1^i, \ldots, \bar{y}_n^i), \forall i$, where $0 \leq \bar{y}_j^i \leq 1, \forall j$. If there are ideal and negative ideal alternatives, their local evaluations are 1 and 0: $\bar{y}_+^{i_1} = 1, \bar{y}_-^{i_1} = 0$. Otherwise, the local evaluations of the best and worst alternatives among n alternatives are 1 and 0.

The normalized evaluations are summed up considering the priority weights of decision criteria into the overall evaluation $\boldsymbol{z} = (z_1, \ldots, z_n)$.

$$z_j = \sum_i w^i \bar{y}_j^i, \forall j, \tag{1}$$

where w^i is criterion C^i's priority weight and the sum is one: $\sum_i w^i = 1$. The weighted sum of ordinal values assumes that one criterion compensates the other. The non-compensatory composite indicators have been proposed based on the PROMETHEE and Borda scoring procedure [5].

3   Local Evaluation Under a Decision Criterion

3.1   Quantitative Decision Criterion

Assume quantitative decision criterion C^i. We have the numerical values of alternatives, $\boldsymbol{x}^i$, and those of ideal and negative ideal alternatives, x_+^i, x_-^i, if a decision maker has them. There are n to $n + 2$ real values under the criterion C^i.

In the following, we consider the case without ideal and negative ideal alternatives for simplicity. We directly consider the given numerical data as local evaluations. Then, we normalize them so that the minimum and maximum are 0 and 1.

$$\bar{y}_j^i = \frac{x_j^i - \min_{j_1} x_{j_1}^i}{\max_{j_1} x_{j_1}^i - \min_{j_1} x_{j_1}^i}, \forall j, \tag{2}$$

where the more, the better is assumed. There are other types of criteria, such as the less, the better, and the non-linearity relation. In general, the relation between the given numerical data and the local evaluations under the criterion is denoted as a function: $\bar{y}_j^i = f(x_j^i)$, where $0 \leq \bar{y}_j^i \leq 1, \bar{y}_j^i \in \mathbb{R}$.

From the viewpoint of alternative comparability under a criterion, another indirect way is to aggregate the preference degrees of an alternative to all alternatives. Based on comparisons of pairs of alternatives, the local evaluation of an alternative is obtained by the Preference Ranking Organization Method for Enrichment Evaluation iii (PROMETHEEiii) [2].

Since we have numerical values of alternatives, the difference between one alternative from the other is obtained numerically, and the more positive difference, the more preferable. The preference degree of alternative A_j over the other alternative j_1 corresponds to the difference in their numerical numbers.

$$p_{jj_1}^i = p^i(x_j^i - x_{j_1}^i), \tag{3}$$

where function p^i transforms a difference into one's preference degree over the other.

Figure 2 illustrates an example function p^i: the larger the value is, the more preferable the alternative is. The parameters, $\alpha_*, \alpha^*, \beta_*$ and β^*, are determined by a decision maker. She takes tiny differences as equal and finds no difference between the considerable differences. She does care for the difference between

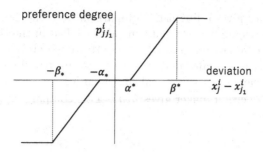

Fig. 2. One's preference degree over the other

$-\beta_*$ and $-\alpha_*$, and α_* and β^*. The direct way (2) is the case of $\alpha_* = \alpha^* = 0$ and $\beta_* = \beta^* = \max_j x_j^i - \min_j x_j^i$. The function p^i can be nonlinear such as step function and sigmoid curve.

The local evaluation of an alternative is the sum of its preference degrees over the other alternatives.

$$y_j^i = \sum_{j_1} p_{jj_1}^i, \forall. \tag{4}$$

The local evaluations $y_j^i, \forall j$ based on pairwise comparisons are normalized by (2) replacing x_j^i into y_j^i. In the case of the ideal and/or negative ideal alternatives, the two alternatives are added to the list of n alternatives (Table 1).

Table 1. Numerical data, preference degree, and local evaluation of quantitative criterion

	x^i		A_1	$\cdots$	A_{j_1}	$\cdots$	A_n	y^i
A_1	x_1^i	A_1	$p^i(x_1^i - x_1^i)$	$\cdots$	$p^i(x_1^i - x_{j_1}^i)$	$\cdots$	$p^i(x_1^i - x_n^i)$	y_1^i
$\vdots$	$\vdots$	$\vdots$						$\vdots$
A_j	x_j^i	A_j	$p^i(x_j^i - x_1^i)$	$\cdots$	$p^i(x_j^i - x_{j_1}^i)$	$\cdots$	$p^i(x_j^i - x_n^i)$	y_j^i
$\vdots$	$\vdots$	$\vdots$						$\vdots$
A_n	x_n^i	A_n	$p^i(x_n^i - x_1^i)$	$\cdots$	$p^i(x_n^i - x_{j_1}^i)$	$\cdots$	$p^i(x_n^i - x_n^i)$	y_n^i

3.2   Qualitative Decision Criterion

Some criteria are not suitable for numerical data. Here, we denote a qualitative criterion as C^i. Instead of numerical values, $\boldsymbol{x}^i$, we have linguistic values which a decision maker gives as her subjective judgments on alternatives. We first consider the case without ideal and negative ideal alternatives, and the next is the case with them.

A decision maker gives a linguistic comparison from the list in the left table of Table 2. The corresponding comparison value is filled in the right table and it is called pair wise comparison matrix in the AHP. Each element $a^i_{jj_1}$ represents how much more preferable alternative A_j is when compared to alternative A_{j_1}. Therefore, the elements are reciprocal and identical: $a^i_{jj} = 1, \forall j$ and $a^i_{jj_1} = 1/a^i_{j_1j}, \forall j, j_1$, so a decision maker gives $n(n-1)/2$ comparisons.

Table 2. Linguistic data, comparison and local evaluation of qualitative criterion

linguistic comparison	comparison value
extremely more (e.m.)	9
very strongly more (v.s.m.)	7
strongly more (s.m.)	5
moderately more (m.m.)	3
equal to (e.t.)	1
moderately less (m.l.)	1/3
strongly less (s.l.)	1/5
very strongly less (v.s.l.)	1/7
extremely less (e.l.)	1/9

$\boldsymbol{P}^i$	A_1	$\cdots$	A_{j_1}	$\cdots$	A_n	$\boldsymbol{y}^i$
A_1	a^i_{11}	$\cdots$	$a^i_{1j_1}$	$\cdots$	a^i_{1n}	y^i_1
$\vdots$						$\vdots$
A_j	a^i_{j1}	$\cdots$	$a^i_{jj_1}$	$\cdots$	a^i_{jn}	y^i_j
$\vdots$						$\vdots$
A_n	a^i_{n1}	$\cdots$	$a^i_{nj_1}$	$\cdots$	a^i_{nn}	y^i_n

We obtain the numerical value of alternatives y^i_j from the given pairwise comparison matrix $\boldsymbol{P}^i$ based on the Analytic Hierarchy Process (AHP) [6].

$$y^i_j = \sqrt[n]{\prod_{j_1} a^i_{jj_1}}, \forall i, j, \tag{5}$$

where the local evaluation of an alternative is a geometric mean of its superiorities over the others.

In the AHP, the eigenvector method is also used to derive the evaluation of alternatives from a pairwise comparison matrix. By the geometric mean and eigenvector methods, the evaluations are obtained as real values from the viewpoint of the most probable. On the other hand, from the possibilistic viewpoint, interval evaluations are obtained by the Interval AHP [10]. It is based on the idea that the intuitive comparisons are inconsistent because of uncertain evaluations. This study assumes the evaluations are real values, and we will extend to the interval evaluations in our future work.

As for the quantitative decision criterion in the previous section, a pair of alternatives is compared concerning their differences in (3) and the preference degrees of an alternative over all the others are summed up to the local evaluation in (4). Similarly, the local evaluation under a qualitative decision criterion is obtained as the average of the superiorities, which are preference degrees, in (5).

Next is the case with ideal and negative ideal alternatives. In the case without them, a decision maker has to compare all pairs of alternatives, so she gives $n(n-$

1)/2 comparisons. In the case with them, when she compares each alternative additionally to the ideal and negative ideal alternative, she needs to give $n(n + 1)/2$ comparisons [9]. In our proposal, a decision maker compares an alternative to the ideal and negative ideal alternatives shown in Table 3. The number of comparisons is decreased to $2n$. When she compares an alternative to the ideal, the given linguistic data is one in the lower half of the left table of Table 2: $a_{1+}^i \leq 1$. Similarly, the comparisons to the negative ideal are more than 1 in the upper half: $a_{1+}^i \geq 1$. By (5) with two comparisons for each alternative, the local evaluation is obtained at the right column of Table 3.

Table 3. Comparisons to ideal and negative ideal and local evaluations

P^i	ideal A_+^i	negative ideal A_-^i	local evaluation y^i
A_1	a_{1+}^i	a_{1-}^i	$y_1^i = \sqrt{a_{1+}^i a_{1-}^i}$
$\vdots$			$\vdots$
A_j	a_{j+}^i	a_{j-}^i	$y_j^i = \sqrt{a_{j+}^i a_{j-}^i}$
$\vdots$			$\vdots$
A_n	a_{n+}^i	a_{n-}^i	$y_n^i = \sqrt{a_{n+}^i a_{n-}^i}$

In addition to the list of n alternatives, we have two more alternatives, ideal A_+^i and negative ideal A_-^i. To normalize local evaluations of $n + 2$ local evaluations, we estimate their evaluations, y_+^i, y_-^i. However, the decision maker does not easily compare the ideal and negative ideal since both are in her mind and not the list of n alternatives. Therefore, we estimate the comparison of the ideal to the negative ideal, a_{+-}^i, which holds the reciprocity to another comparison of them: $a_{+-}^i = 1/a_{-+}^i \geq 1$.

The given comparisons are written using true local evaluation, $x_j^i, \forall j$, as follows.

$$a_{j+}^i = \frac{x_j^i}{x_+^i} \leq 1, \ a_{j-}^i = \frac{x_j^i}{x_-^i} \geq 1, \forall j. \tag{6}$$

The comparison between the ideal and negative ideal alternatives, a_{+-}^i, is estimated by two comparisons of each alternative A_j.

$$a_{+-}^i = \left\{ \frac{x_1^i/a_{1+}^i}{x_1^i/a_{1-}^i}, \ldots, \frac{x_n^i/a_{n+}^i}{x_n^i/a_{n-}^i} \right\} = \left\{ \frac{a_{j-}^i}{a_{j+}^i}, \forall j \right\} \in \left[\min_j \frac{a_{j-}^i}{a_{j+}^i}, \max_j \frac{a_{j-}^i}{a_{j+}^i} \right] \geq 1, \tag{7}$$

where the interval values indicate the possible differences between the ideal and negative ideal alternatives. The ideal and negative ideal are the best and worst ends in a decision maker's mind, whether realistic or not. Moreover, their evaluations are uncertain; in other words, they may more or less differ in which alternative the ideal is compared to. In this study, we assume the comparison between the ideal and negative ideal as the geometric mean of the estimations through all alternatives:

$$\hat{a}^i_{+-} = \sqrt[n]{\prod_j \frac{a^i_{j-}}{a^i_{j+}}}, \tag{8}$$

which is in the interval in (7). Correspondingly, the estimated comparison of the negative ideal to the ideal is the inverse because of the reciprocity: $\hat{a}_{-+} = 1/\hat{a}_{+-}$.

From the estimation, the local evaluations of the ideal and negative ideal alternatives are obtained.

$$y^i_+ = \sqrt{a^i_{++}\hat{a}^i_{+-}}, \ y^i_- = \sqrt{\hat{a}^i_{-+}a^i_{--}}. \tag{9}$$

where the comparisons between the identical alternatives, a^i_{++} and a^i_{--}, are 1. The obtained local evaluations of the ideal and negative ideal alternatives hold $y^i_- \le y^i_j \le y^i_+, \forall j$.

The numerically obtained local evaluations of n alternatives and ideal and negative ideal, $y^i_j, \forall j, y^i_+, y^i_-$, are normalized by (2). The normalized local evaluations of the ideal and negative ideal are 1 and 0, $\bar{y}^i_+ = 1, \bar{y}^i_- = 0$.

4   Numerical Example

Assume a decision maker is thinking of purchasing a new car. She narrowed down to five candidate cars as alternatives, considering five factors as decision criteria, which are price(yen), engine(cc), width(cm), accessory, and space. Table 4 shows the given information about her decision making.

Table 4. Alternatives and criteria for purchasing car

	C_1price (yen)	C_2engine (cc)	C_3width (cm)	C_4accessory A_1	A_2	A_3	A_4	A_5	C_5space ideal	negative
A_1	400	1500	160	1	v.s.m	m.m	s.m	e.m	e.t	e.m
A_2	385	1300	180	–	1	–	–	m.m	s.l.	v.s.m
A_3	417	2000	166	–	s.m.	1	m.m	v.s.m	m.l.	s.m
A_4	421	1800	170	–	m.m.	–	1	s.m	m.l.	v.s.m
A_5	399	1500	179	–	–	–	–	1	s.l.	s.m.
ideal	–	1800	190	–					car A^5_+	
negative	–	–	155	–					car A^5_-	

The former three criteria are objectively measurable, though the latter two criteria need subjective judgments. Moreover, she gave the ideal numerical values for C_2 engine and C_3 width and the ideal alternative for C_5 space. As for the negative ideal, she gave a numerical value for C_3 and an alternative for C_5. Table 4 is the information we had for this decision problem. In this way, in usual decision problems, the data types depend on decision criteria and decision makers' minds. Some decision criteria are quantitative, others are qualitative, and some have the ideal and/or negative ideal often outside her list of alternatives.

First, we obtained the local evaluations of five alternatives under three quantitative criteria. The given objective data are shown in Table 4. The preference degree of an alternative over the other is denoted as a function of their numerical differences. The functions of the three criteria are illustrated in Fig. 3.

The left figure is price C_1, and naturally the lower price is better for purchasing. The preference degrees of an alternative over the others by the function are shown in Table 5. For instance, since A_1 is more costly than A_2, its preference degree over A_2 is negative: $p_{12}^1 < 0$. A_1 is less costly than A_3 and A_4, so $p_{13}^1, p_{14}^1 > 0$. The price difference between A_1 and A_5 is tiny, so both preference degrees are zero: $p_{15}^1 = p_{51}^1 = 0$. A_1's local evaluation is obtained by (4) at the second from the right column of Table 5, and next to it, the normalized local evaluations are shown. The decision maker did not give her ideal or negative ideal price, so the best and worst prices among the five alternatives have the maximum and minimum normalized evaluations, 1 and 0.

Similarly, the normalized local evaluations under engine C_2 and width C_3 are obtained from the center and left functions in Fig. 3. As for C_3, the five alternatives range between 160 cm and 180 cm, though the decision maker accepts the range from 155 cm to 190 cm and prefers the wide car. As for C_2, the decision maker's ideal is 1800cc, and the five alternatives include the smaller and the larger ones. We use the deviations from the ideal to replace the given numerical values. When one alternative's deviation is more than the other's deviation, and the difference is negative, it mentions that the former alternative is preferable to the latter. The obtained normalized local evaluations are shown in the left columns of Table 6. If the decision maker gives her ideal and/or negative ideal, their normalized local evaluations are 1 and/or 0. If not, the best and worst of the five alternatives are assigned 0 and 1.

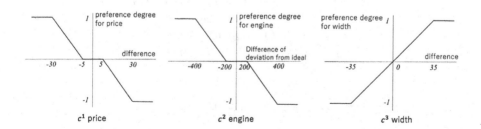

Fig. 3. Preference degrees of quantitative criteria

Next, we obtained the local evaluations under two qualitative decision criteria, accessory C_4 and space C_5. As for C_5, the decision maker gave her ideal,

Table 5. Local evaluations of price C_1

$p^1_{j_1 j_2}$	A_1	A_2	A_3	A_4	A_5	local evaluation	normalization
A_1 400	0	−0.40	0.48	0.64	0	0.72	0.60
A_2 385	0.40	0	1	1	0.36	2.76	1
A_3 417	−0.48	−1	0	0	−0.52	−2	0.06
A_4 421	−0.64	−1	0	0	−0.68	−2.32	0
A_5 399	0	−0.36	0.52	0.68	0	0.84	0.62

car A^+, and negative ideal, car A^-. Although describing her ideal is sometimes complicated, she may easily find some examples. She compares the alternatives to the two examples, and it is often easier for her than comparing all alternatives to each other.

Since these two criteria are unsuitable for numerical data, we asked her to give subjective judgments. The given linguistic data are shown in Table 4. She gave linguistic values in Table 2 by comparing a pair of alternatives. The number of comparisons for C_5 with the ideal and negative ideal is smaller than that for C_4 without them. The more alternatives are, the more different the numbers of comparisons, so it is worth considering the ideal and negative ideal if she has them in her mind.

The local evaluations under C_4 and C_5 are obtained from the corresponding numerical values by (5) and (9) and normalized. They are shown in Table 6. Since the ideal and negative ideal for C_5 are not included in the five alternatives, the maximum of the normalized evaluations is less than 1.

Table 6. Normalized local evaluations and overall evaluation

	local evaluation					overall evaluation		
	C_1	C_2	C_3	C_4	C_5	all	quantitative	qualitative
A_1	0.60*	0.33	0.14	1	0.67*	0.55*	0.36	0.84*
A_2	1*	0	0.71	0.06	0.23*	0.40	0.57*	0.15
A_3	0.06	0.67	0.31	0.48	0.26	0.36	0.35	0.37
A_4	0	1	0.43	0.20	0.31	0.39	0.48	0.26
A_5	0.62	0.33	0.69	0	0.19	0.37	0.55	0.09
A^i_+	–	1	1	–	1			
A^i_-	–	–	0	–	0			
correlation by excluding	0.55	0.84	0.91	0.36	0.96			

Finally, the overall evaluations are obtained as the weighted sum of the local evaluations by (1). For simplicity, assume that priority weights of all criteria are equal: $w^i = 0.2. \forall i$. They are shown in the right column of Table 6. For comparison, we assumed that we could handle quantitative or qualitative criteria, though our proposed model can handle both kinds of criteria simultaneously. The right two columns show the overall evaluations only with quantitative C_1, C_2 and C_3, and qualitative C_4 and C_5. Without considering qualitative or quantitative criteria, the obtained alternatives' ranks, as well as their overall scores, are different. In this viewpoint, it is necessary to handle both criteria in a decision aiding system.

We examined the effect of the ideal and/or negative ideal on an overall evaluation. In this example, A_1 was the best because of its highest local evaluations of C_4, which is qualitative and did not have an ideal and negative ideal. Alternative A_1 is the best among the five when we remove one decision criterion other than C_4. We excluded one criterion one by one and obtained the overall evaluations with four of the five criteria. At the bottom row of Table 6, we obtained correlations between the overall evaluations with five and four criteria. The minimum correlation is excluding C_4, and excluding C_3 or C_5 is highly correlated to all criteria. On C_3 and C_5, the decision maker gave her ideal and negative ideal outside the five alternatives, so the local evaluations under them never be 0 and 1. The local evaluations under the criterion with ideal and negative ideal tend to be in the smaller range than those without them. Namely, giving them reduced the effects of the criterion's local evaluations on the overall evaluations.

From a decision maker's viewpoint, she quickly gives her ideal and negative ideal when she is aware of the decision criteria. We picked up two criteria, quantitative C_1 without them and qualitative C_5 with them, and compared overall evaluations between A_1 and A_2. The corresponding local evaluations are denoted as $*$ in Table 6. Although both overall evaluations are almost equal, 0.63, A_2's local evaluation under C_5 is higher than that of A_1's. Considering her awareness, we might conclude that the decision maker could prefer A_5 to A_2 is better for the decision maker. In this way, the roles and effects of ideal and negative ideal on the overall evaluations need more discussion in future work.

5   Conclusion

We have proposed the multi-criteria decision aiding method to handle various types of criteria. Our method considers quantitative and qualitative criteria, and the ideal and negative ideal alternatives if there are in a decision maker's mind. We first derived local evaluations of alternatives under each criterion and then aggregated them into overall evaluations. In this study, we focused on the first step of obtaining normalized local evaluations regardless of quantitative or qualitative criteria and with or without the given ideal and negative ideal.

There are objective numerical data on a quantitative criterion, and the local evaluations are the sum of the preference degree of an alternative to all alternatives based on the PROMETHEEiii. On the other hand, we have linguistic data subjectively given by the decision maker on a qualitative criterion instead of numerical data. She compares a pair of alternatives, and if she has her ideal and/or negative ideal, she compares each alternative just to them. Based on AHP, the local evaluations are the geometric mean of the comparisons of an alternative to all alternatives. Whether a criterion is quantitative or qualitative, the local evaluation is based on comparing an alternative to the others. Then, the local evaluations are normalized in the range between the worst evaluation, 0, and the best one, 1. In the case of the ideal and/or negative ideal, their local evaluations are 1 and 0.

In this study, we denoted local evaluations under quantitative and qualitative criteria as real values, so their aggregation was simple. However, it is well-known that the subjectively given comparisons under a qualitative criterion are often inconsistent, and the uncertain local evaluation is, for instance, denoted as an interval value. In our future work, we will consider such a local interval evaluation. Furthermore, in aggregating local evaluations, the priority weights given by a decision maker are usually considered. The given ideal and negative ideal in the proposed method could play similar roles to those of priority weights. The criterion is worth considering because the high ideal results in low local evaluations of all alternatives. On the other hand, the high ideal can indicate the decision maker's awareness of the criterion, i.e., the criterion is essential. In our future work, we will consider using the given ideal and negative ideal to replace priority weights.

Acknowledgment. This work was partially supported by JSPS KAKENHI Grant Number JP19K04885.

References

1. Behzadian, M., Kazemzadeh, R., Albadvi, A., Aghdasi, M.: PROMETHEE: a comprehensive literature review on methodologies and applications. Eur. J. Oper. Res. **200**(1), 198–215 (2010)
2. Brans, J., Mareschal, B., Vinke, P.: Promethee: a new family of outranking methods. In: MCDM, IFORS' 84, Washington, pp. 408–421 (1984)
3. Cinelli, M., Kadziński, M., Miebs, G., Gonzalez, M., Słowiński, R.: Recommending multiple criteria decision analysis methods with a new taxonomy-based decision support system. Eur. J. Oper. Res. **302**(2), 633–651 (2022)
4. French, S., Maule, J., Papamichail, N.: Decision Behaviour, Analysis and Support. Cambridge University Press, Cambridge (2009)
5. Greco, S., Ishizaka, A., Tasiou, M., Torrisi, G.: The ordinal input for cardinal output approach of non-compensatory composite indicators: the PROMETHEE scoring method. Eur. J. Oper. Res. **288**(1), 225–246 (2021)
6. Saaty, T.L.: The Analytic Hierarchy Process. McGraw-Hill, New York (1980)
7. Savkovic, S., Jovancic, P., Djenadic, S., Tanasijevic, M., Miletic, F.: Development of the hybrid MCDM model for evaluating and selecting bucket wheel excavators for the modernization process. Expert Syst. Appl. **201**, 117199 (2022)
8. Siva Bhaskar, A., Khan, A.: Comparative analysis of hybrid MCDM methods in material selection for dental applications. Expert Syst. Appl. **209**, 118268 (2022)
9. Takahagi, E.: On pairwise comparisons including ideal and satisfaction poitsy. J. Soc. Bus. Math. **29**(1), 43–59 (2008). (in Japanese)
10. Tanaka, H., Sugihara, K., Maeda, Y.: Non-additive measures by interval probability functions. Inf. Sci. **164**, 209–227 (2004)

Exploring the Impact of Randomness in Roguelike Deck-Building Games: A Case Study of Slay the Spire

SangGyu Nam[1]([✉])[iD] and Pavinee Rerkjirattikal[2][iD]

[1] School of Information, Computer and Communication Technology, Sirindhorn International Institute of Technology, Thammasat University, Pathum Thani 12120, Thailand
sanggyu@siit.tu.ac.th
[2] School of Manufacturing Systems and Mechanical Engineering, Sirindhorn International Institute of Technology, Thammasat University, Pathum Thani 12120, Thailand
pavinee.rerk@dome.tu.ac.th

Abstract. Randomness in games can induce both player enjoyment and dissatisfaction, particularly in roguelike deck-building games, where the order of card draws significantly influences gameplay outcomes and player satisfaction. Players may experience frustration due to unfavorable draws leading to defeat, even when they have constructed a powerful deck. This paper presents a comprehensive framework for developing a believable and satisfactory pseudorandom deck generation algorithm in roguelike deck-building games. As a preliminary step, we analyze the impact of card draw randomness on battle outcomes using a case study of the popular roguelike deck-building game, "Slay the Spire." Through preprocessing and analyzing the battle log data via clustering analysis, we identify the pivotal of card draw order on different battle clusters. Our findings underscore the significant role of card order in determining battle results, thus providing a foundational basis for future advances in pseudorandom deck generation algorithms. Moreover, we discuss potential applications of our framework in enhancing player experiences and game design in roguelike deck-building games.

Keywords: Roguelike deck-building game · Clustering · Game entertainment · Randomness in games

1 Introduction

Randomness plays a pivotal role in shaping the gaming experience, offering diverse gameplay encounters, fostering player excitement, and extending the longevity of games. Players make decisions based on incomplete game information and encounter unforeseen consequences, making the gameplay experience distinct. However, this randomness can occasionally hinder gameplay enjoyment, leading to player frustration when unfavorable odds come into play [1]. Even the most skilled players may be unable to control the game's outcome, as it often depends on luck and probabilistic events. To address this challenge, pseudorandom techniques, a strategy aimed at subtly manipulating event probabilities without detection by players, have been adopted to create a more balanced and engaging experience, as exemplified by games like Dota 2 [2]. While some games, like League of Legends, have eliminated specific randomness aspects to ensure fairness and competitive integrity.

Nevertheless, removing randomness can make gameplay predictable and repetitive. Thus far, many research challenges and gaps exist in analyzing the influence of randomness and developing pseudorandom algorithms to address various elements of randomness across different game genres. This research proposes a framework for developing a pseudorandom deck generation algorithm emphasizing the famous game genre called roguelike deck-building. As a preliminary step, the impact of randomness is investigated using the case of the renowned game "Slay the Spire" (STS). The findings lay a foundation for developing believable and satisfactory pseudorandom deck generation algorithms in the subsequent research steps.

1.1   Randomness in Card Games

Card games have gained global popularity for their combination of skill and luck, providing enjoyment for players of various types and skill levels. Traditional games like Poker, Blackjack, Go-Stop, and newer trends like collectible card games and deck-building games, have all contributed to the diversity of card games. While each card game possesses unique features, they all share common attributes such as inherent randomness, turn-based gameplay, and the concept of hand cards.

Randomness is inherent in most card games, affecting decision-making and gameplay outcomes. One form of randomness is the hand cards, where players receive several random cards that significantly impact the gameplay. In games like Yu-Gi-Oh, Magic: The Gathering, and Hearthstone, players draw an additional card from the top of the deck during each turn, introducing randomness through the card order. Certain cards also have random effects, such as variable attack points or randomly discarding cards from players or opponents.

1.2   Roguelike Deck-Building Games: Gameplay Mechanics

Deck-building games are a genre where randomness significantly impacts gameplay variety, entertainment value, and player satisfaction. These games require players to construct powerful card decks while progressing through intermediate events and achieving the game objective. Roguelike deck-building games, a specific subset of this genre, add an extra layer of challenge by requiring players to restart the game upon defeat, with the primary goal being to defeat the final boss. Its general game flow (Fig. 1) is structured as follows:

1. In each round, players choose events from available paths, including battles, shops, and reinforcements.
2. Battles consist of multiple turns, where players draw hand cards from a shuffled deck to combat monsters (Fig. 2). After each turn, cards are discarded and drawn again. Cards are usually categorized into attack, defense, and buff types, with some cards having synergistic effects when combined. Each card has an energy cost, limiting the number of cards players can use per turn based on players' energy points.
3. Players can acquire new cards as battle rewards, purchase them, or obtain them through various events. They can optimize their deck by removing less valuable cards and reinforcing essential ones.
4. In the final round, players attempt to defeat the boss using a well-constructed deck.
5. If players are defeated at any point during the gameplay, they must restart the entire game process.

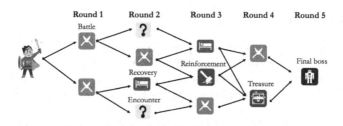

Fig. 1. An example of game composition in roguelike deck-building games.

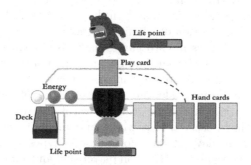

Fig. 2. The battle event of the roguelike deck-building games.

In deck-building games, players utilize a limited number of cards drawn from the top of a shuffled deck, where the sequence of card draws significantly influences gameplay dynamics. Even with a powerful deck, unfavorable card draw orders can lead to defeat, significantly impacting players' satisfaction and enjoyment. Thus, a card ordering algorithm is crucial to balance randomness, maintain an appropriate challenge level, and enhance the game's entertainment value.

The card randomization should ensure unpredictability while avoiding excessive favoritism that could compromise the game's inherent difficulty.

Building upon this premise, this research introduces a framework for developing a pseudorandom algorithm to generate believable and satisfactory initial decks in roguelike deck-building games, which remains an unexplored area in the existing literature. Preliminary research employs a case study of STS, the game that has popularized the deck-building genre since its 2017 launch. Despite genre popularity, its related research remains limited in the literature in scope for enhancing gameplay experience and entertainment value. By analyzing STS game logs, the impact of randomness on battle outcomes is investigated, laying the foundation for developing a pseudorandom deck generation algorithm.

This paper is organized as follows: Sect. 2 presents backgrounds, related works, and research gaps. Section 3 presents the proposed research framework for developing pseudorandom deck generation algorithms in roguelike deck-building games. Section 4 discusses preliminary analyses and their findings. Finally, Sect. 5 concludes and suggests future research directions.

2   Related Works

Randomness is integral to various aspects of gaming, such as item loot, game events, card draws, levels, and maps, contributing to diverse and engaging gameplay experiences. Nonetheless, players sometimes find excessive randomness a nuisance, highlighting the importance to strike a balance between randomness and difficulty [3].

Several studies have explored the impact of randomness in different game genres. Hammadi and Abdelazim [4] investigated the positive effects of randomness on cognitive development in an educational game for autistic children. In PlayerUnknown's Battlegrounds (PUBG), Galka and Strzelecki [5] examined the influence of random elements (e.g., item drop locations, play zones, weapon accuracy) on players' perception of winning chances.

The effects of randomness in card games have also been explored. Zhang et al. [6] examined the impact of input and output randomness on player satisfaction in collectible card games, revealing an adverse effect of drawing hard cards on player satisfaction. Trojanowski and Andersson [7] explored the correlation between player skill, randomness in acquired cards and items, and gameplay outcomes in STS. They observed that while randomness in items does not significantly influence gameplay outcomes for all player types, randomness in cards strongly affects outcomes, mainly based on player skill and adaptability to uncertain events. In contrast, our work specifically investigates the influence of the card draw order on battle outcomes, expanding beyond examining final gameplay results. Furthermore, we analyze encounters with diverse monster types to understand the impact of card draw order on battle results.

Previous research underscores the significant influence of randomness on gameplay and satisfaction, indicating the adverse effect of true randomness. To address these concerns, many games employ pseudorandom techniques to manipulate randomnesses, such as controlling item drop rates in Diablo and Pokemon egg acquisition rates and hatching durations. Temsiririrkkul et al. [8] highlight the importance of developing believable pseudorandomness algorithms that players cannot detect. They designed a pseudorandom dice-rolling mechanism based on features that led to biased outcomes, achieving enhanced believability when guided by properly-selected features. Other studies focus on utilizing pseudorandomness in generating dungeons [9,10], stages [11], and levels [12].

Despite the growing popularity of roguelike deck-building games, research on the role of randomness within this genre remains limited. Investigations into randomness impacts on battle results and player satisfaction is still an open research area. Moreover, the development of pseudorandom deck generation algorithms tailored for this genre is still lacking. To address this gap, we propose a comprehensive framework for devising a pseudorandom deck generation algorithm designed explicitly for roguelike deck-building games. This algorithm aims to generate initial card decks that strike a balance between believability and satisfaction, thereby enhancing the overall gameplay experience and entertainment value. This paper serves as an initial step within our framework, analyzing the impact of randomness in the order of card draws on battle outcomes, encompassing encounters with various monster types. The findings lay the foundation for developing a believable pseudorandom deck generation algorithm.

3   Research Framework

The illustrated research framework in Fig. 3 outlines the process of developing a pseudorandom deck generation algorithm for roguelike deck-building games. Comprising seven key steps, this framework serves as a comprehensive guide for our research. This paper presents the initial step within the framework, focusing on analyzing the influence of randomness on battle outcomes in roguelike deck-building games. By utilizing the STS game log data as a representative case study, these analyses offer insights into the impact of randomness on battle results—a pivotal factor affecting player dissatisfaction and gameplay enjoyment.

The outcomes of this research underscore the necessity for a believable and satisfactory algorithm for generating initial decks. Subsequently, a robust research platform is established by examining commercial deck-building games. By analyzing randomly generated decks, the randomness features that potentially influence player satisfaction are identified. Guided by these features, an algorithm is formulated and applied to generate decks, which are then evaluated through participant verification. Finally, the algorithm's effectiveness is benchmarked against randomly generated decks.

The subsequent section elaborates on the analyses carried out in this paper and presents their preliminary findings.

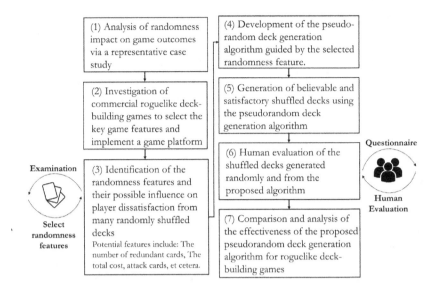

Fig. 3. Research framework for developing a pseudorandom deck generation algorithm in roguelike deck-building games.

4   Analysis of STS Battle Logs

This section analyzes game log data and outcomes for one week (23–30 September 2020), consisting of 267,767 battle entries from the game developer [13].

4.1   Data Preprocessing

The log data contains comprehensive player decisions, battle results, items, and deck details. Each feature is reviewed for inclusion, conditional exclusion, or modification. Specific feature details are outlined below:

1. **Battle log**: This feature is crucial for investigating the impact of initial decks on battle outcomes. The analysis takes into consideration the following variables:
   1.1. The number of battles (N): This variable captures the progression of the character, deck, and monsters throughout the game. It accounts for the varying effect of randomness in the card deck on battle outcomes based on the number of battles.
   1.2. Enemy types encountered (**Enemy**): Each monster exhibits distinct characteristics, and the drawn hand cards can yield favorable or unfavorable outcomes depending on the enemy type. Including this variable addresses such variability.
   1.3. Number of turns until battle completion: The sequence of card draws can impact the number of turns required to defeat a monster. Incorporating the variable representing the number of turns offers insights into the connection between card order and battle duration.

1.4. Damage sustained by the character during the battle (**Damage Taken**): This variable indicates how effectively players manage the battle and handle damage inflicted on their character.

2. **Character**: STS has four characters, each with a unique mechanism and set of cards. In this analysis, we concentrate solely on the Ironclad, which features relatively straightforward card mechanics compared to other options.

3. **Seeded run**: The roguelike nature of STS introduces randomness through seeded runs, where game levels and card rewards differ based on unique seeds. Players controlling the seed value can anticipate outcomes, potentially replaying the level for optimal results. To maintain a consistent analysis, data associated with seeded runs is excluded.

4. **Neow bonus**: At the start of each game run, an NPC named Neow offers players one of three random bonuses. One of these bonuses guarantees that the next three combats will involve enemies with only one life point, enabling players to defeat them without sustaining damage. Consequently, this bonus introduces noise to the battle results, rendering data linked to this bonus unfit for analysis and excluded.

5. **Ascension level**: The ascension level in STS signifies varying difficulty levels. As the ascension level escalates, the game becomes progressively more challenging. Lower ascension levels may allow players to effortlessly defeat enemies despite starting with a poorly shuffled deck. To ensure robust analysis, data from the maximum ascension level 20 is considered, given its complexity and relevance for experienced players.

6. **Enemy category**: In STS, normal monsters are linearly arranged throughout the game, with a boss monster positioned at the end of each of the four chapters. Additionally, there are elite monsters, a unique category of monsters that possess strength comparable to bosses, can appear at any point in the game, and offer better rewards. The battle outcomes differ depending on the monster category, which makes it necessary to classify data into four categories: normal monsters, elite monsters, boss monsters, and a combined category that includes all three types.

4.2 Clustering Battle Results

As previously highlighted, the impact of randomness on battle outcomes is substantial. We analyze the potential influence of randomness in an initial deck on battle results, utilizing the data from Sect. 4.1. Given the many battle outcomes, we employ the clustering technique to discern patterns and similarities. Clustering is an unsupervised machine learning method that effectively identifies structures within unclassified and unlabeled data. Principal Component Analysis (PCA) is then employed in our analysis. PCA helps with dimensional reduction, enhancing visualization and analytical capacity by identifying principal components, which are linear combinations of original features.

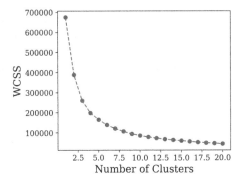

Fig. 4. WCSS value for each number of clusters from the combined monster categories data.

The initial step involves determining the optimal number of cluster groups. A range of potential clusters, from $k = 1$ to $k = 20$, is explored using the Within-Cluster Sum of Squares (WCSS) metric. This metric measures the sum of squared distances between data points and cluster centroids, identifying the point of significant change in WCSS, known as the "elbow point," which signifies the optimal cluster size. Results from the WCSS analysis for the combined monster categories data are depicted in Fig. 4. Similar trends emerge across analyses for other monster categories, consistently converging on an optimal cluster size of 5.

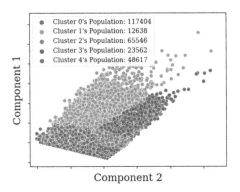

Fig. 5. PCA clusters using K-Means (N = 5).

This clustering process provides insights into the underlying structure of battle outcomes, revealing distinct patterns and relationships that might be influenced by the inherent randomness within the initial deck. Figure 5 illustrates the clustering result for the combined data of all monster categories with cluster populations. Further analysis examined the damage taken in each battle within each cluster. The damage taken in the Nth battle from each monster category is depicted in Fig. 6, displaying unique patterns associated with the clusters. Figure 6(a) shows the damage taken by each cluster in the combined monster categories data.

Clusters 0, 2, and 3 predominantly encompass battle outcomes from early game stages, where battles tend to be straightforward. Hence, minimal damage is expected, except for beginners or players with poorly shuffled decks. Clusters 0 and 2 demonstrate adequate battle results for players with reasonably shuffled decks. In contrast, Cluster 3 battles reflect significant damage, potentially arising from encounters with elite monsters in the early game or unfavorable card shuffling. Figure 7 highlights top 35 damage-inflicting monsters. An interesting case emerges in Fig. 7(c) where the "2 Louses," despite being a normal type, inflict significant damage in Cluster 3 battles. While a well-shuffled deck can easily defeat this monster, as shown in Fig. 7(a), it can deal substantial damage to players with unfavorable hand cards. The interplay between well-shuffled decks and favorable card draws highlights the role of card order randomness on battle outcomes.

Cluster 4 exhibits a distinctive pattern of sustaining minimal damage throughout battles. Although this aligns with early-game expectations, persistent adherence to this pattern in later stages might lead to monotonous gameplay lacking challenge. Factors such as having high-quality decks, valuable items, or overly favorable card draws may be one of the contributors. In the mid-to-late game, monsters with distinct traits may require specific card combinations to be defeated. Even with a customized deck for the particular monster, players may fail if they cannot draw suitable cards for combating that specific monster type. The order of card draws significantly influences battle outcomes, as evident in Cluster 1's notable standard error. This suggests that some players struggle to construct strong and flexible decks to conquer the battle. Alternatively, they may struggle to draw the appropriate cards even with proficient decks. For instance, consider the case presented in Fig. 7(b),(d). The monster "3 Byrds" has relatively weak power at the start but progressively grows stronger with each passing turn. Effectively defeating this monster requires a strategy focusing on attacks to finish the battle swiftly before their strength reaches critical levels. If players draw defensive cards in the initial turns, the "3 Byrds" can inflict significant damage to their characters, potentially resulting in their defeat.

Across all clusters, battle results exhibit similar patterns for all monster categories, with some insignificant outliers as shown in Fig. 6. Battle results from clusters 1, 3, and 4 are potentially influenced by the hand cards drawn from the shuffled deck. Although these clusters have small populations compared to others, battles with potential deck shuffling issues contribute to 31% of the total battles, highlighting instances where deck shuffling could impact outcomes. Considering that players engage in an average of 30–40 battles in each run, as shown in Fig. 6, players have a high chance of encountering battles with poorly shuffled decks that may lead them to fail the battle.

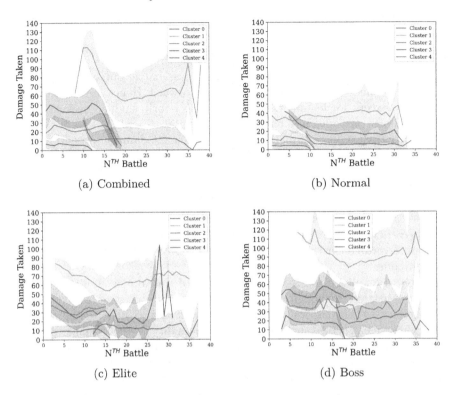

Fig. 6. The average and standard error of Damage Taken for N^{th} battle in each cluster by monster types: (a) Combined all types, (b) Normal type, (c) Elite type, (d) Boss type.

While STS is renowned for its challenging nature, with a win rate as low as 9% [14], one contributing factor to its difficulty is the element of luck involved in deck shuffling. Our research results highlight the impact of draw cards on battle outcomes across different clusters and monster types. The findings justify the necessity of a pseudorandom deck generation algorithm to ensure an engaging and enjoyable gameplay experience.

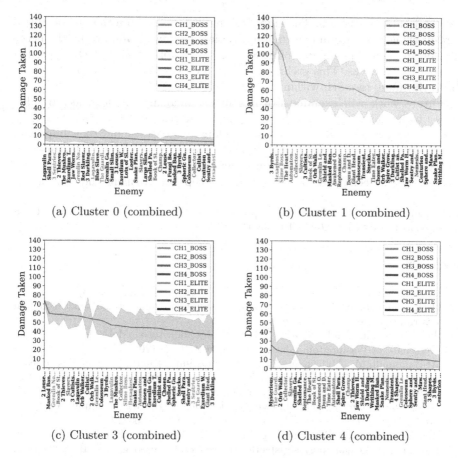

(a) Cluster 0 (combined)

(b) Cluster 1 (combined)

(c) Cluster 3 (combined)

(d) Cluster 4 (combined)

Fig. 7. The average and standard error of Damage Taken for each monster. The monsters are ordered by the average value of Damage Taken.

5   Conclusion

Randomness in games can evoke both enjoyment and frustration for players. In roguelike deck-building games, the order in which cards are drawn carries substantial influence over players' decisions, gameplay outcomes, and overall satisfaction. As players can only use the cards drawn from the top of the deck to engage in battles, unfavorable draws can lead to defeat even with a powerful deck. This paper presents a framework for developing a pseudorandom deck generation algorithm in roguelike deck-building games, aiming to enhance gameplay experiences by providing believable and satisfactory card decks.

A preliminary step of the framework is conducted in this paper. We investigate the influence of the order of drawing cards on battle outcomes using STS as a representative game in the genre. An STS battle log is analyzed using the PCA and Clustering techniques, with selective key features related to randomness and battle outcomes. The empirical findings underscore the pivotal role of card-drawing sequence in determining battle outcomes, underscoring the necessity for an algorithm capable of generating believable and satisfactory pseudo-random initial decks. This preliminary step provides insights that pave the way for forthcoming research within the framework.

Future work will examine features in commercial roguelike deck-building games to implement a representative game platform. By investigating numerous randomly shuffled decks, relevant randomness features and their impact on player satisfaction will be identified. These identified randomness features will be used to refine the algorithm in its development process to generate optimal pseudorandom decks, thereby maximizing gameplay enjoyment and satisfaction.

Acknowledgement. This work is supported by the SIIT Young Researcher Grant under contract No. SIIT 2022-YRG-SN02.

References

1. Costikyan, G.: Uncertainty in Games. MIT Press, Cambridge (2013)
2. Pseudo Random Distribution. https://liquipedia.net/dota2/Pseudo_Random_Distribution. Accessed 15 Jun 2023
3. Jaffe, A.: Understanding Game Balance with Quantitative Methods. Ph.D. thesis, University of Washington, Seattle, Washington, D.C. (2013)
4. Al-Hammadi, M., Abdelazim, A.: Randomness impact in digital game-based learning. In: 2015 IEEE Global Engineering Education Conference, pp. 806–811 (2015)
5. GaÅka, P., Strzelecki, A.: How randomness affects player ability to predict the chance to win at PlayerUnknown Battlegrounds (PUBG). Comput. Games J. **10**(1), 1–18 (2021). https://doi.org/10.1007/s40869-020-00117-1
6. Zhang, Y., Monteiro, D., Liang, H.N., Ma, J., Baghaei, N.: Effect of input-output randomness on gameplay satisfaction in collectable card games. In: 2021 IEEE Conference on Games, pp. 01–05 (2021)
7. Trojanowski, M., Andersson, J.: Are you lucky or skilled in slay the spire?: an analysis of randomness.? (Bachelor's thesis), Malmö University (2021)
8. Temsiririrkkul, S., Nomura, H., Ikeda, K.: Biased random sequence generation for making common player believe it unbiased. In: 2014 IEEE Games Media Entertainment, pp. 1–8 (2014)
9. Koesnaedi, A., Istiono, W.: Implementation drunkard's walk algorithm to generate random level in roguelike games. Int. J. Multi. Res. Publ. **5**(2), 97–103 (2022)
10. Smith, A.J., Bryson, J.J.: A logical approach to building dungeons: answer set programming for hierarchical procedural content generation in roguelike games. In: Proceedings of the 50th Anniversary Convention of the AISB (2014)
11. Nam, S., Ikeda, K.: Generation of diverse stages in turn-based role-playing game using reinforcement learning. In: 2019 IEEE Conference on Games, pp. 1–8 (2019)
12. Nam, S., Hsueh, C.H., Ikeda, K.: Generation of game stages with quality and diversity by reinforcement learning in turn-based RPG. IEEE Trans. Games **14**, 488–501 (2021)
13. Slay The Spire Play Log Data. https://drive.google.com/drive/folders/1c7MwTdLxnPgvmPbBEfNWa45YAUU53H0l. Accessed 05 Jun 2023
14. Slay the Spire statistical analysis. https://foxrow.com/slay-the-spire-statistical-analysis. Accessed 15 Jun 2023

Optimization and Statistical Methods

The a Priori Procedure for Estimating the Cohen's Effect Size Under Independent Skew Normal Settings

Cong Wang[1], Tonghui Wang[2(✉)], Xiangfei Chen[2], David Trafimow[3],
Tingting Tong[2], Liqun Hu[2], and S. T. Boris Choy[4]

[1] Department of Mathematical and Statistical Sciences,
University of Nebraska Omaha, Omaha, USA
[2] Department of Mathematical Sciences, New Mexico State University, Las Cruces,
USA
twang@nmsu.edu
[3] Department of Psychology, New Mexico State University, Las Cruces, USA
[4] Discipline of Business Analytics, The University of Sydney, Sydney, Australia

Abstract. Cohen's d is the most popular effect size index for traditional experimental designs, and it is desirable to know the minimum sample size necessary to obtain a sample value that is a good estimate of the population value. The present work addresses that lack with an application to independent samples under the umbrella of skew normal distributions. In addition to derivations of relevant equations, there is a link to a free and user-friendly computer program. Finally, we present computer simulations and a worked example to support the equations and program.

Keywords: Cohen's d · Effect size · Independent samples · Skew normal distributions

1 Introduction

In a typical experiment, the researcher randomly assigns participants to an experimental condition and a control condition, and desires to establish that there is a difference, hopefully a sizable one, between the two conditions. Not surprisingly, statisticians have developed many effect size indices, of which the most popular is Cohen's d, which is a standardized effect size for measuring the difference between two group means, and can be calculated by taking the difference between two means and dividing by the data's standard deviation.

Thus, for example, if Cohen's d equals 0.5, which Cohen (1988) [4] considered a medium-sized effect, the proper interpretation is that one of the means is half of a standard deviation greater than the other mean. An important issue for researchers who wish to determine suitable sample sizes for obtaining a sample Cohen's d that is a good estimate of the population Cohen's d, is that traditional power analysis is insufficient. Power analysis is useful for determining appropriate sample sizes to perform significance tests, but not for determining

V.-N. Huynh et al. (Eds.): IUKM 2023, LNAI 14375, pp. 111–122, 2023.
https://doi.org/10.1007/978-3-031-46775-2_10

appropriate sample sizes to obtain sample summary statistics that are good estimates of corresponding population parameters. It was to address this latter issue that the a priori procedure (APP) was developed by Trafimow (2017) [10] and Trafimow and MacDonald (2017) [12], see Trafimow (2019) [11] for a review). The idea of the APP is to determine the sample size necessary to meet specifications that the researcher sets for precision (the distance within which the sample statistic is of the corresponding population parameter) and confidence (the probability of meeting the precision specification). For example, Trafimow (2017) [10] showed that if one wishes to have a 0.95 probability of obtaining a sample mean within one-tenth of a standard deviation of the population mean, under normality, a minimum sample size of 385 participants is required. Since then, there has been many APP works being published, see [5,13,15–17]. Chen et al. (2021) [3] provided the first APP advance that specifically addresses the issue of necessary sample sizes to ensure that the researcher meets specifications for precision and confidence with respect to Cohen's d. Furthermore, Chen et al. (2021) [3] provided derivations of equations, and links to computer programs, that allow researchers to make the calculations regardless of whether the experimental design is between-participants (independent samples) or within-participants (dependent samples). A limitation is that Chen et al. (2021) [3] assumed normal distributions, but there have been many impressive demonstrations that most distributions are skewed, see Blanca et al. (2013) [2], Ho and Yu (2015) [7], and Micceri (1989) [8]. Therefore, it would be useful to expand the APP to handle the family of skew normal distributions, rather than remaining limited to normal distributions. The present goal is to address this lack by deriving the requisite equations, providing a link to a free and user-friendly program, and supporting these with computer simulations and a worked example.

2   Preliminaries

First, we need to list the properties of multivariate skew normal distribution briefly.

Definition 1 *(Azzalini and Dalla Valle (1996) [1]). A random vector $\mathbf{X}$ is said to have an n-dimensional multivariate skew normal distribution with vector of location parameters $\boldsymbol{\mu} = (\mu_1, \ldots, \mu_n)' \in \Re^n$, scale parameter of $n \times n$ nonnegative definite matrix Σ, and the vector of skewness (shape) parameters $\boldsymbol{\alpha} = (\alpha_1, \ldots, \alpha_n)' \in \Re^n$, denoted as $\mathbf{X} \sim SN_n(\mu, \Sigma, \alpha)$, if its pdf is*

$$f_{\mathbf{X}}(\mathbf{x}) = 2\phi_n(\mathbf{x}; \boldsymbol{\mu}, \Sigma)\Phi\left(\boldsymbol{\alpha}'\Sigma^{-1/2}(\mathbf{x} - \boldsymbol{\mu})\right), \tag{1}$$

where $\phi_n(\mathbf{x}; \boldsymbol{\mu}, \Sigma)$ is the pdf of the n-dimensional multivariate normal distribution with mean $\boldsymbol{\mu}$ and covariance matrix Σ, and $\Phi(z)$ is the cdf of the standard normal random variable.

The following lemma will be used in the proof of our main result.

Let $A = (1, \ldots, 1)'/n = \mathbf{1}_n/n \in \Re^n$ and $A = \mathbf{e}_i$ where $\mathbf{e}_i \in \Re^n$ denotes the vector with a 1 in the ith coordinate and 0's elsewhere, respectively, then it is easy to obtain the following result.

Lemma 1 *(Wang et al. (2016) [18]). Suppose that*

$$\mathbf{X} = (X_1, \ldots, X_n)' \sim SN_n(\boldsymbol{\mu}, \Sigma, \boldsymbol{\alpha})$$

with $\boldsymbol{\mu} = \xi \mathbf{1}_n$, $\Sigma = \omega^2 I_n$ and $\boldsymbol{\alpha} = \lambda \mathbf{1}_n$, where $\xi, \lambda \in \Re$ and $\omega > 0$. Then the following results hold.

(i) The sample mean $\bar{X} = \frac{1}{n}\sum_{i=1}^n X_i \sim SN\left(\xi, \frac{\omega^2}{n}, \sqrt{n}\lambda\right)$.
(ii) Random variables $X_1, \cdots, X_n$ are identically distributed:

$$X_i \sim SN\left(\xi, \omega^2, \lambda_*\right) \qquad i = 1, ..., n, \qquad \lambda_* = \frac{\lambda}{\sqrt{1 + (n-1)\lambda^2}}.$$

(iii) $\bar{X}$ and S^2 are independent, and $\frac{(n-1)S^2}{\omega^2} \sim \chi_{n-1}^2$, where

$$S^2 = \frac{1}{n-1}\sum_{i=1}^n (X_i - \bar{X})^2.$$

3   Sampling Distribution of the Sample Mean Under Two Independent Multivariate Skew Normal Settings

Consider two independent samples of sizes n and m, respectively, such that

$$\mathbf{X} = (X_1, \cdots, X_n)' \sim SN_n(\boldsymbol{\mu}_1, \Sigma_1, \boldsymbol{\alpha}_1), \mathbf{Y} = (Y_1, \cdots, Y_m)' \sim SN_m(\boldsymbol{\mu}_2, \Sigma_2, \boldsymbol{\alpha}_2),$$
$$\tag{2}$$

where $\boldsymbol{\mu}_1 = \xi_1 \mathbf{1}_n$, $\boldsymbol{\mu}_2 = \xi_2 \mathbf{1}_m$, $\boldsymbol{\alpha}_1 = \lambda_1 \mathbf{1}_n$, $\boldsymbol{\alpha}_2 = \lambda_2 \mathbf{1}_m$, $\Sigma_1 = \omega_1^2 I_n$ and $\Sigma_2 = \omega_2^2 I_m$. Let $\bar{X} = \sum_{i=1}^n X_i/n$ and $\bar{Y} = \sum_{j=1}^m Y_j/m$. Then with Lemma 1, the sampling distribution of mean difference, $\bar{X} - \bar{Y}$, is given below.

Proposition 1. *For distributions of $\mathbf{X}$ and $\mathbf{Y}$ given in Eq. (2), the pdf of $Z = \bar{X} - \bar{Y}$ is*

$$f_Z(z) = 4\phi(z; \xi_d, \tau^2)\Phi_2[B(z - \xi_d); \mathbf{0_2}, \boldsymbol{\Delta}], \tag{3}$$

where

$$\xi_d = \xi_1 - \xi_2, \qquad \tau^2 = \frac{\omega_1^2}{n} + \frac{\omega_2^2}{m}, \qquad \mathbf{b}' = (\lambda_1 \omega_1, -\lambda_2 \omega_2), \qquad B = \mathbf{b}'/\tau^2,$$

and $\boldsymbol{\Delta} = I_2 + diag\left(n\lambda_1^2, m\lambda_2^2\right) - \mathbf{b}\mathbf{b}'/\tau^2$. Note that here $\Phi_2(\mathbf{u}; \mathbf{0_2}, \boldsymbol{\Delta})$ is the cdf of the bivariate normal distribution with mean vector $\mathbf{0_2} = (0,0)'$ and the covariance $\boldsymbol{\Delta}$.

Remark 3.1. The distribution of Z with pdf given in Eq. (3) is called the closed skew normal distribution given by González-Farías et al. [6]. Also from Proposition 1, it is easy to obtain that $Z_{ij} = X_i - Y_j$ for $i = 1, \ldots, n$ and $j = 1, \ldots, m$, are identically distributed and the pdf of Z_{11} is given by

$$f_{Z_{11}}(z) = 4\phi(z; \xi_d, \tau_*^2)\Phi_2[B_{12}(z - \xi_d); \mathbf{0_2}, \boldsymbol{\Delta}_{12}],$$

where

$$\tau_*^2 = \omega_1^2 + \omega_2^2, \quad \lambda_{1*} = \frac{\lambda_1}{\sqrt{1 + (n-1)\lambda_1^2}}, \quad \lambda_{2*} = \frac{\lambda_2}{\sqrt{1 + (m-1)\lambda_2^2}},$$
$$\mathbf{b}' = (\lambda_{1*}\omega_1, -\lambda_{2*}\omega_2), \quad B_{12} = \mathbf{b}'/\tau_*^2, \quad \boldsymbol{\Delta}_{12} = I_2 + diag\left(n\lambda_{1*}^2, m\lambda_{2*}^2\right) - \mathbf{b}\mathbf{b}'/\tau_*^2.$$

Remark 3.2. For estimating the Cohen's effect size $\theta = (\xi_1 - \xi_2)/\omega$ using our APP procedure, we assume that $\omega_1 = \omega_2 = \omega$ and $m \geq n$, without loss of generality. Thus with $m = n$, the pdf of Z in Eq. (3), is reduced to

$$h_Z(z) = 4\phi(z; \xi_d, \tau^2)\Phi_2[B(z - \xi_d); \mathbf{0_2}, \boldsymbol{\Delta}], \tag{4}$$

where $\xi_d = \xi_1 - \xi_2$, $\tau^2 = \frac{2\omega^2}{n}$, $\mathbf{b_*} = (\lambda_1, -\lambda_2)'$, $B = \frac{n}{2\omega}\mathbf{b_*}$, and $\Delta = I_2 + \frac{n}{2}\mathbf{b_*b_*'}$.

4   Distribution of the Cohen's d

Suppose that there are two samples $\mathbf{X} = (X_1, \ldots, X_n)'$ and $\mathbf{Y} = (Y_1, \ldots, Y_m)'$ from independent multivariate skew normal populations given in Eq. (2). The population Cohen's effect size is the standardized location difference given by $\theta = \frac{\xi_1 - \xi_2}{\omega}$, where scale parameters are assume to be equal, $\omega_1 = \omega_2 \equiv \omega$. We would like to use the Cohen's d, defined by $T = \frac{\bar{X} - \bar{Y}}{S_p}$ as a point estimator of θ, where S_p^2 is the pooled sample variance given by

$$S_p^2 = \frac{(n-1)S_X^2 + (m-1)S_Y^2}{n + m - 2},$$

where S_X^2 and S_Y^2 are sample variances of $\mathbf{X}$ and $\mathbf{Y}$, respectively. Note that the distribution of $\bar{X} - \bar{Y}$ is obtained in Proposition 1. In the following, we will derive the distribution of the Cohen's d and its proof is given in Appendix.

Theorem 1. *With the assumptions given above, we have the following results.*

(a) The pdf of $T = \frac{\bar{X} - \bar{Y}}{S_p}$ is given by

$$g_T(t) = \frac{4}{2^{\nu/2-1}\sqrt{\nu}\Gamma(\nu/2)} \int_0^\infty v^\nu e^{-v^2/2} \phi\left(\frac{v}{\sqrt{\nu}}t; \theta, \tau^2\right)$$
$$\Phi_2\left[B\left(\frac{v}{\sqrt{\nu}}t - \theta\right); \mathbf{0_2}, \Delta\right] dv, \tag{5}$$

where $\nu = m + n - 2$, $\tau^2 = \frac{m+n}{mn}$, $\Delta = I_2 + diag\left(n\lambda_1^2, \ m\lambda_2^2\right) - \mathbf{b_*b_*'}/\tau^2$, $\mathbf{b_*} = (\lambda_1, -\lambda_2)'$, and $B = \frac{1}{\tau^2}\mathbf{b_*}$

(b) The mean and variance of T are, respectively,

$$\mu_T = a\left[\theta + \sqrt{\frac{2}{\pi}}(\delta_1 - \delta_2)\right]$$

and

$$\sigma_T^2 = \frac{\nu}{\nu - 2}\left[\left(\frac{m+n}{mn} - \frac{2}{\pi}(\delta_1^2 + \delta_2^2)\right) + \left(\theta + \sqrt{\frac{2}{\pi}}(\delta_1 - \delta_2)\right)^2\right]$$
$$- a^2\left(\theta + \sqrt{\frac{2}{\pi}}(\delta_1 - \delta_2)\right)^2,$$

where $a = \sqrt{\frac{\nu}{2}}\frac{\Gamma[(\nu-1)/2]}{\Gamma(\nu/2)}$, $\delta_1 = \frac{\lambda_1}{\sqrt{1+n\lambda_1^2}}$, and $\delta_2 = \frac{\lambda_2}{\sqrt{1+m\lambda_2^2}}$.

The density curves of T with different values of parameters are given in Fig. 1. From the left figure in Fig. 1, we can see that the shapes of density curves are effected by changes of shape parameters λ_1 and λ_2 for fixed $n = m = 80$ and $\theta = 0$. From the right figure in Fig. 1, we know that the shapes of density curves are not effected much by the changes of θ values, fixing other parameters.

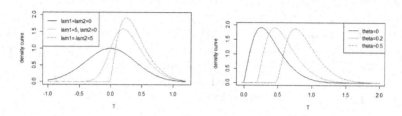

Fig. 1. The density curves of T (left) with $\theta = 0$ and $m = n = 80$ for $\lambda_1 = \lambda_2 = 0$, $\lambda_1 = 5, \lambda_2 = 0$, and $\lambda_1 = 5$, $\lambda_2 = -5$, and that (right) when $m = n = 80$ and $\lambda_1 = 5$, $\lambda_2 = -5$ for $\theta = 0.2$, 0.5, 0.8.

Remark 4.1. As a consequence of Theorem 1 and Remark 3.1, the variance of $T_{11} = \frac{X_1 - Y_1}{S_P}$ is

$$\sigma^2_{T_{11}} = \frac{\nu}{\nu - 2}\left[\left(2 - \frac{2}{\pi}(\delta_1^2 + \delta_2^2)\right) + \left(\theta + \sqrt{\frac{2}{\pi}}(\delta_1 - \delta_2)\right)^2\right] - a^2\left(\theta + \sqrt{\frac{2}{\pi}}(\delta_1 - \delta_2)\right)^2,$$

where $a = \sqrt{\frac{\nu}{2}}\frac{\Gamma[(\nu-1)/2]}{\Gamma[\nu/2]}$ (here $\nu = m + n - 2$), $\delta_1 = \frac{\lambda_1}{\sqrt{1+n\lambda_1^2}}$ and $\delta_2 = \frac{\lambda_2}{\sqrt{1+m\lambda_2^2}}$.

Let $Z_* = T - \mu_T$ and density curves of Z_* for different parameters when $m = n$ are given in Fig. 4. From Fig. 2, it is clear that values of θ are not effect the density curves of Z_* significantly.

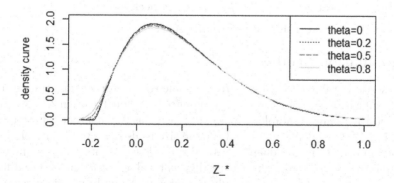

Fig. 2. The density curves of Z_* with $m = n = 80$, $\lambda_1 = 5$, $\lambda_2 = -5$ for $\theta = 0$, 0.2, 0.5, 0.8.

5   Setting up the APP Related to the Cohen's d

In this section, we will setup the APP for estimating $\theta = \frac{\xi_d}{\omega}$ on two independent samples. Since there are two unknown sample sizes m and n, we will assume, without loss of generality, that $m \geq n$ so that the degrees of freedom $m+n-2$ is equal or larger than $2(n-1)$. It has been shown that the length of the confidence interval constructed with $n+m-2$ degrees of freedom is smaller than the one with $2(n-1)$ degrees of freedom, See Wang et al. (2019) [14] and Chen et al. (2021) [3] for more details. Thus our APP for estimating Cohen's effect size θ will provide the minimum required sample size n for known shape parameters $\lambda_i\,(i = 1,\ 2)$ and, $\hat{\theta}$, an estimate of θ from previous data if it is available. The following is the main result on required sample size for estimating Cohen's effect size and its proof is given in Appendix.

Theorem 2. *Let c be the confidence level and f be the precision which are specified such that the error associated the unbiased estimator $\hat{T}$ of θ based on Remark 4.1 is $E = f\sigma_{T_{11}}$. More specifically,*

$$P\left[f_1\sigma_{T_{11}} \leq \hat{T} - \theta \leq f_2\sigma_{T_{11}}\right] = c. \tag{6}$$

Here, $\sigma_{T_{11}}$ is the standard deviation of T_{11} given in Remark 4.1, f_1 and f_2 are left and right precision satisfying $\max\{|f_1|, f_2\} \leq f$. Then the required sample size $n\,(n > 3)$ is obtained by

$$\int_L^U g_{Z_*}(z)dz = c \tag{7}$$

such that $U - L$ is the shortest. where $g_{Z_*}(z)$ is the pdf of $Z_* = \frac{T-\mu_T}{\sigma_T}$, $L = f_1\frac{\sigma_{T_{11}}}{\sigma_T}$, and $U = f_2\frac{\sigma_{T_{11}}}{\sigma_T}$.

Remark 1. The computer program that can be used to find the required sample size for different research goals is listed below. The input data are θ_0 specified Cohen's effect size from the previous data, shape values λ_1 and λ_2, the confidence level c, and the precision f. The output value is the required sample size n and respected left and right precision f_1 and f_2, respectively.

https://appcohensd.shinyapps.io/indcohensn/

6   Simulation Studies

In this section, Table 1 to Table 3 provide some results concerning sample size to meet specifications for precision and confidence based on Theorem 2. These tables show that the required sample sizes are decreasing as the differences in shapes are increasing. Meanwhile, the required sample size is also affected positively by the previous information of Cohen's effect size θ. Using Monte Carlo simulations, we count the relative frequencies for different values of θ and λ for different values of precision. All results are illustrated with a number of simulation runs $M = 10,000$. In each table, the corresponding coverage rate for each required sample size shows our method working well (Tables 1, 2 and 3).

Table 1. The values of sample size n and the corresponding coverage rates (c.r.) under different values of $\theta = 0$, 0.2, 0.5, 0.8, 1, precision $f = 0.1$, 0.15, 0.2, 0.25, and $\lambda_1 = \lambda_2 = 0.5$ for $c = 0.9$, 0.95.

c	f	$n_{\theta=0}$ (c.r.)	$n_{\theta=0.2}$ (c.r.)	$n_{\theta=0.5}$ (c.r.)	$n_{\theta=0.8}$ (c.r.)	$n_{\theta=1}$ (c.r.)
0.9	0.1	106 (0.9016)	107 (0.8983)	111 (0.8914)	124 (0.8980)	155 (0.9023)
	0.15	49 (0.9012)	49 (0.8981)	49 (0.8858)	50 (0.8861)	55 (0.8914)
	0.2	29 (0.8986)	29 (0.8921)	30 (0.8917)	33 (0.8932)	35 (0.8944)
	0.25	21 (0.9035)	21 (0.9080)	22 (0.9097)	24 (0.9128)	24 (0.9032)
0.95	0.1	153 (0.9478)	154 (0.9464)	157 (0.9401)	174 (0.9415)	221 (0.9478)
	0.15	70 (0.9504)	72 (0.9531)	75 (0.9497)	79 (0.9469)	82 (0.9447)
	0.2	41 (0.9489)	42 (0.9475)	44 (0.9496)	47 (0.9483)	49 (0.9499)
	0.25	28 (0.9482)	29 (0.9529)	30 (0.9511)	32 (0.9485)	34 (0.9515)

Table 2. The values of sample size n and the corresponding coverage rates (c.r.) under different values of $\theta = 0$, 0.2, 0.5, 0.8, precision $f = 0.1$, 0.15, 0.2, 0.25, and $\lambda_1 = 0.2$, $\lambda_2 = 0$ for $c = 0.9$, 0.95.

c	f	$n_{\theta=0}$ (c.r.)	$n_{\theta=0.2}$ (c.r.)	$n_{\theta=0.5}$ (c.r.)	$n_{\theta=0.8}$ (c.r.)
0.9	0.1	270 (0.9035)	270 (0.9027)	273 (0.8978)	284 (0.9035)
	0.15	120 (0.9057)	120 (0.9011)	121 (0.9004)	121 (0.8990)
	0.2	67 (0.9075)	67 (0.9033)	67 (0.9008)	69 (0.8983)
	0.25	43 (0.9034)	43 (0.9052)	43 (0.9039)	44 (0.8991)
0.95	0.1	384 (0.9521)	384 (0.9522)	385 (0.9490)	400 (0.9453)
	0.15	170 (0.9511)	171 (0.9535)	173 (0.9516)	176 (0.9434)
	0.2	95 (0.9534)	96 (0.9543)	97 (0.9477)	99 (0.9475)
	0.25	61 (0.9572)	61 (0.9514)	62 (0.9510)	67 (0.9482)

Table 3. The values of sample size n and the corresponding coverage rates (c.r.) under different values of $\theta = 0$, 0.2, 0.5, 0.8, precision $f = 0.1$, 0.15, 0.2, 0.25, and $\lambda_1 = -\lambda_2 = 0.5$ for $c = 0.9$, 0.95.

c	f	$n_{\theta=0}$ (c.r.)	$n_{\theta=0.2}$ (c.r.)	$n_{\theta=0.5}$ (c.r.)	$n_{\theta=0.8}$ (c.r.)
0.9	0.1	366 (0.9003)	368 (0.9009)	372 (0.8946)	384 (0.8887)
	0.15	157 (0.9019)	159 (0.9001)	161 (0.8993)	170 (0.9005)
	0.2	87 (0.9012)	89 (0.9010)	90 (0.8954)	97 (0.8983)
	0.25	54 (0.8991)	56 (0.8998)	57 (0.8996)	63 (0.9037)
0.95	0.1	483 (0.9511)	488 (0.9503)	494 (0.9513)	512 (0.9527)
	0.15	208 (0.9502)	212 (0.9516)	217 (0.9481)	229 (0.9485)
	0.2	114 (0.9505)	117 (0.9492)	122 (0.9508)	132 (0.9487)
	0.25	71 (0.9499)	73 (0.9499)	77 (0.9498)	83 (0.9497)

The following Fig. 3 shows the decreasing required sample size as the increase of shape parameters from $\lambda_1 = \lambda_2 = 0$ to 2.

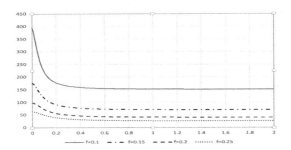

Fig. 3. The required sample size n for different precision $f = 0.1$, 0.15, 0.2, 0.25 and confidence level $c = 0.95$ when shape parameters $\lambda_1 = \lambda_2$ increase from 0 to 2.

7   An Example

In this section, we provide an example for illustration the use of the APP in Sect. 4 to real life data. The data set contains the annual salaries from Departments of Sciences (DS) and the remaining departments in the College of Arts and Sciences (RD), New Mexico State University, which are obtained from the Budget Estimate (2018/19) [9]. The total number of faculty salaries used in this data set is 394 (199 for DS and 195 for RD), and the ratio of the RD and DS is $r = 1.02$. The estimated distributions based on the data set are $SN(3.4131, 3.5768^2, 3.8487)$ and $SN(4.0995, 3.8794^2, 2.1394)$ (with unit $10,000$) for RD and DS via the Method of Moment Estimation, respectively. Histograms and their corresponding skew normal density fitting curves are, respectively, given in Fig. 4 and Fig. 5. We can see from both figures that the data sets are skewed to right, so that skew normal fittings are appropriate.

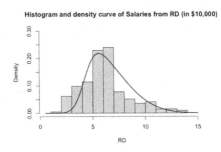

Fig. 4. The histogram and its corresponding density curved by the skew normal distribution with location $\xi_1 = 3.4131$, scale $\omega_1^2 = 3.5768^2$, and shape $\lambda_1 = 3.8487$.

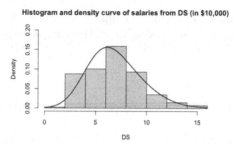

Fig. 5. The histogram and its corresponding density curved by the skew normal distribution with location $\xi_2 = 4.0995$, scale $\omega_2^2 = 3.8794^2$, and shape $\lambda_2 = 2.1394$.

The scale parameters are $\omega_1 = 3.5768$ and $\omega_2 = 3.8794$, which are sufficiently close to be considered the same. Suppose that the population shapes are $\lambda_1 = 3.8487$ and $\lambda_2 = 2.1394$ known, respectively, as well as the moment estimate of the Cohen's effect size, $\theta = 0.2412$ as a prior information. Then the required sample size $n = 68$ for $f = 0.15$ and $c = 0.95$, and the density curve for $\hat{T}$ is shown in Fig. 6.

Randomly choose a sample of size 68 from each group, and calculate the unbiased estimate of θ, $\hat{T} = 0.2423$ and $\sigma_{T_{11}} = 1.8417$. Then the 95% confidence interval given by Theorem 2 based on this sample can be obtained by [0.1160, 0.5186].

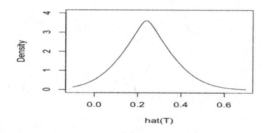

Fig. 6. The density curve of $\hat{T}$ for $n = 68$, $\theta = 0.2412$, $\lambda_1 = 3.8487$, and $\lambda_2 = 2.1394$.

8 Discussion

Our goal was to derive the requisite mathematics to facilitate extending the APP to apply to Cohen's d, under the large umbrella of skew normal distributions. The present article provides that extension and shows how it works for independent samples, with support from simulations and an worked example. Finally, we provided a free and user-friendly computer program. Thus, the present work should enhance the ability of researchers to use Cohen's d to index effect sizes, with increased assurance that sample Cohen's d values will be good estimates of population Cohen's d values, even when the samples are drawn from skew normal distributions.

Appendix

A. Proof of Theorem 1

Proof.(a) By Proposition 1 with $\omega_1 = \omega_2 = \omega$, we know that $\bar{X} - \bar{Y}$ is closed skew normal distributed and its pdf is given in Eq. (3) with $\tau = \frac{m+n}{mn}$ and $b' = \omega(\lambda_1, -\lambda_2)$. From Lemma 1, we know that

$$\frac{(n-1)S_X^2}{\omega^2} \sim \chi_{n-1}^2 \qquad \text{and} \qquad \frac{(m-1)S_Y^2}{\omega^2} \sim \chi_{m-1}^2$$

are independent so that

$$U \equiv \frac{(m+n-2)S_p}{\omega^2} \sim \chi_\nu^2$$

and the pdf of U

$$f_U(u) = \frac{1}{2^{\nu/2-1}\Gamma(\nu/2)} u^{\nu/2-1} e^{-u/2},$$

where $\nu = m + n - 2$. Since $Z = \bar{X} - \bar{Y}$ and S_p^2 are independent, we can define $T = \frac{Z}{S_p} = \frac{Z/\omega}{\sqrt{U/\nu}}$, and the density function is

$$g_T(t) = \int_0^\infty f_Z(\omega t u) f_U(\sqrt{\nu}u)\omega\sqrt{\nu}u du,$$

from which, let $v = \sqrt{\nu}u$,

$$g_T(t) = \frac{4}{2^{\nu/2-1}\sqrt{\nu}\Gamma(\nu/2)} \int_0^\infty v^\nu e^{-v^2/2} \phi\left(\frac{v}{\sqrt{\nu}}t; \ \theta, \ \tau^2\right)$$

$$\Phi_2\left[B\left(\frac{v}{\sqrt{\nu}}t - \theta\right); \ \mathbf{0}_2, \ \Delta\right] dv,$$

where $\tau^2 = \frac{m+n}{mn}$, $\mathbf{b}_* = (\lambda_1, -\lambda_2)'$, $B = \frac{1}{\tau^2}\mathbf{b}_*$ and $\Delta = I_2 + diag\left(n\lambda_1^2, \ m\lambda_2^2\right) - \mathbf{b}_*\mathbf{b}_*'/\tau^2$.
(b) Note that both $\bar{X}$ and $\bar{Y}$ are skew normal distributed and they are independent, the mean and variance of Z are

$$E(Z) = \xi_1 - \xi_2 + \sqrt{\frac{2}{\pi}}(\delta_1 - \delta_2)\omega \quad \text{and} \quad Var(Z) = \omega^2\left[\frac{n+m}{mn} - \frac{2}{\pi}(\delta_1^2 + \delta_2^2)\right],$$

respectively. Also it is easy to obtain

$$E\left(\frac{1}{S_p}\right) = \frac{\sqrt{\nu}}{\omega}E(\frac{1}{\sqrt{U}}) = \sqrt{\frac{\nu}{2\omega}}\frac{\Gamma[(\nu-1)/2]}{\Gamma(\nu/2)} \quad \text{and} \quad E\left(\frac{1}{S_p^2}\right) = \frac{\nu}{(\nu-2)\omega^2}.$$

(8)

Let $a = \sqrt{\frac{\nu}{2}} \frac{\Gamma[(\nu-1)/2]}{\Gamma(\nu/2)}$. Now the mean of T is

$$\mu_T = E(\bar{X} - \bar{Y})E\left(\frac{1}{S_p}\right) = a\left[\theta + \sqrt{\frac{2}{\pi}}(\delta_1 - \delta_2)\right].$$

For the variance of T, we have

$$E(T^2) = E(\bar{X} - \bar{Y})^2 E\left(\frac{1}{S_p^2}\right) = \{Var(\bar{X}) + Var(\bar{Y}) + [E(\bar{X} - \bar{Y})]^2\}E\left(\frac{1}{S_p^2}\right)$$

so that $\sigma_T^2 = E(T^2) - \mu_T^2$, the desired result follows after simplification.

B. Proof of Theorem 2

Proof. Let $\hat{T} = \frac{T}{a} - \sqrt{\frac{2}{\pi}}(\delta_1 - \delta_2)$, the unbiased estimator of θ. Then Eq. (6) can be rewritten to be

$$P\left[f_1 \sigma_{T_{11}} \le (T - \mu_T)/a \le f_2 \sigma_{T_{11}}\right],$$

then

$$P\left[f_1 a \frac{\sigma_{T_{11}}}{\sigma_T} \le Z_* \le f_2 a \frac{\sigma_{T_{11}}}{\sigma_T}\right],$$

where a is a constant related to n given in (ii) of Theorem 1. If we denote $Z_* = \frac{T - E(T)}{\sigma_T}$, then Eq. (7) holds in which $g_{Z_*}(z)$ is the density function of Z_*, and $g_{Z_*}(z) = \sigma_T g_T(z + \mu_T)$, where $g_T(t)$ is the density function of T given in the Theorem 1. If we have had an estimate $\hat{\theta}$ of θ by the previous information, then the required sample size n can be obtained by solving Eq. (7).

References

1. Azzalini, A., Dalla Valle, A.: The multivariate skew-normal distribution. Biometrica **83**(4), 715–726 (1996)
2. Blanca, M.J., Arnau, J., López-Montiel, D., Bono, R., Bendayan, R.: Skewness and kurtosis in real data samples. Methodol. Eur. J. Res. Meth. Behav. Soc. Sci. **9**(2), 78–84 (2013). https://doi.org/10.1027/1614-2241/a000057
3. Chen, X., Trafimow, D., Wang, T., Tong, T., Wang, C.: The APP procedure for estimating Cohen's effect size. Asian J. Econ. Bank. **5**(3), 289–306 (2021). https://doi.org/10.1108/AJEB-08-2021-0095
4. Cohen, J.: Statistical Power Analysis for the Behavioral Sciences, 2nd edn. Erlbaum, Hillsdale (1988)
5. Cao, L., Wang, C., Chen, X., Trafimow, D., Wang, T.: The a priori procedure (APP) for estimating the Cohen's effect size for matched pairs under skew normal settings. Int. J. Uncert. Fuzziness Knowl. Based Syst. (2023, accepted)
6. González-Farías, G., Domínguez-Molina, A., Gupta, A.K.: Additive properties of skew normal random vectors. J. Stat. Plan. Infer. **126**(2004), 521–534 (2004)

7. Ho, A.D., Yu, C.C.: Descriptive statistics for modern test score distributions: skewness, kurtosis, discreteness, and ceiling effects. Educ. Psychol. Measur. **75**, 365–388 (2015). https://doi.org/10.1177/0013164414548576

8. Micceri, T.: The unicorn, the normal curve, and other improbable creatures. Psychol. Bull. **105**, 156–166 (1989). https://doi.org/10.1037/0033-2909.105.1.156

9. New Mexico State University (2018/19) Budget estimate. [Salaries] (2018/19), Las Cruces, N.M. The University 1994/95

10. Trafimow, D.: Using the coefficient of confidence to make the philosophical switch from a posteriori to a priori inferential statistics. Educ. Psychol. Measur. **77**, 831–854 (2017). https://doi.org/10.1177/0013164416667977

11. Trafimow, D.: A frequentist alternative to significance testing, p-values, and confidence intervals. Econometrics **7**(2), 1–14 (2019). https://www.mdpi.com/2225-1146/7/2/26

12. Trafimow, D., MacDonald, J.A.: Performing inferential statistics prior to data collection. Educ. Psychol. Measur. **77**, 204–219 (2017). https://doi.org/10.1177/0013164416659745

13. Trafimow, D., Wang, T., Wang, C.: From a sampling precision perspective, skewness is a friend and not an enemy! Educ. Psychol. Measure. **79**, 129–150 (2018). https://doi.org/10.1177/0013164418764801

14. Wang, C., Wang, T., Trafimow, D., Chen, J.: Extending a priori procedure to two independent samples under skew normal setting. Asian J. Econ. Bank. **03**(02), 29–40 (2019)

15. Wang, C., Wang, T., Trafimow, D., Myüz, H.A.: Desired sample size for estimating the skewness under skew normal settings. In: Kreinovich, V., Sriboonchitta, S. (eds.) TES 2019. SCI, vol. 808, pp. 152–162. Springer, Cham (2019). https://doi.org/10.1007/978-3-030-04263-9_11

16. Wang, C., Wang, T., Trafimow, T., Zhang, X.: Necessary sample size for estimating the scale parameter with specified closeness and confidence. Int. J. Intell. Technol. Appl. Stat. **12**(1), 17–29 (2019). https://doi.org/10.6148/IJITAS.201903_12(1).0002

17. Wang, C., Wang, T., Trafimow, D., Li, H., Hu, L., Rodriguez, A.: Extending the a priori procedure (APP) to address correlation coefficients. In: Ngoc Thach, N., Kreinovich, V., Trung, N.D. (eds.) Data Science for Financial Econometrics. SCI, vol. 898, pp. 141–149. Springer, Cham (2021). https://doi.org/10.1007/978-3-030-48853-6_10

18. Wang, Z., Wang, C., Wang, T.: Estimation of location parameter on the skew normal setting with known coefficient of variation and skewness. Int. J. Intell. Technol. Appl. Stat. **9**(3), 45–63 (2016)

Fast – Asymptotically Optimal – Methods for Determining the Optimal Number of Features

Saeid Tizpaz-Niari, Luc Longpré, Olga Kosheleva[ID],
and Vladik Kreinovich[✉][ID]

University of Texas at El Paso, El Paso, TX 79968, USA
{saeid,longpre,olgak,vladik}@utep.edu

Abstract. In machine learning – and in data processing in general – it is very important to select the proper number of features. If we select too few, we miss important information and do not get good results, but if we select too many, this will include many irrelevant ones that only bring noise and thus again worsen the results. The usual method of selecting the proper number of features is to add features one by one until the quality stops improving and starts deteriorating again. This method works, but it often takes too much time. In this paper, we propose faster – even asymptotically optimal – methods for solving the problem.

Keywords: Machine learning · Data processing · Optimal number of features · Asymptotically optimal method

1 Formulation of the Problem

Selecting Optimal Number of Features: An Important Problem. In machine learning – and in data processing in general – an important problem is selecting the number of features; see, e.g., [2].

- When we only use very few of the available features, the results are not very good – since we do not use a significant portion of the available information.
- As we increase the number of features, the results get better and better.
- However, at some point, we exhaust useful features, and we start adding features that practically do not contribute to the desired decision. In such situations, new features mostly adds noise, so the performance deteriorates again.

This work was supported in part by the National Science Foundation grants 1623190 (A Model of Change for Preparing a New Generation for Professional Practice in Computer Science), HRD-1834620 and HRD-2034030 (CAHSI Includes), EAR-2225395, and by the AT&T Fellowship in Information Technology.
It was also supported by the program of the development of the Scientific-Educational Mathematical Center of Volga Federal District No. 075-02-2020-1478, and by a grant from the Hungarian National Research, Development and Innovation Office (NRDI).

V.-N. Huynh et al. (Eds.): IUKM 2023, LNAI 14375, pp. 123–128, 2023.
https://doi.org/10.1007/978-3-031-46775-2_11

In other words, the effectiveness E depends on the number of features n as follows:

- the value $E(n)$ first increases with n,
- but at some point, it starts decreasing with n.

We need to find the value n_0 at which the effectiveness $E(n)$ is the largest.

How This Problem is Solved Now. Of course, we can always do it if:

- first, we order the features in terms of their prospective usefulness – e.g., by the absolute value of the correlation between this feature and the value that we are trying to predict, and then
- we add features one by one in this order, and select the number of features that leads to the best prediction.

This is usually how people solve this problem now.

A Better Method is Needed. In many cases – e.g., for machine learning – the adding-features-one-by-one algorithm is very time-consuming, since for each number of features, we need to re-train the neural network, and this takes time.

It is thus desirable to have faster methods for finding the optimal value n_0.

What We Do in This Paper. In this paper, we first describe a straightforward asymptotically optimal method for finding the optimal number of features. Then, we show how to further speed up the corresponding computations. Specifically:

- in Sect. 2, we formulate the problem in precise terms;
- in Sect. 3, we describe a straightforward asymptotically optimal method for solving this problem;
- in Sect. 4, we describe the new method, and we show that this method is indeed faster than the straightforward method.

2   Let Us Formulate the Problem in Precise Terms

What We Are Given. We are given a number N – the overall number of available features. We have an algorithm that:

- given a natural number $n \leq N$,
- returns a real number $E(n)$ – the effectiveness that we get if we consider n most promising features.

What We Know. We know that the function $E(n)$ first strictly increases, then strictly decreases. In other words, there exits some threshold value n_0 – that is not given to us – for which:

- if $n < n' \leq n_0$, then $E(n) < E(n')$, and
- if $n_0 \leq n < n'$, then $E(n) > E(n')$.

What We Want to Compute. We want to compute the threshold value n_0, i.e., the value at which the effectiveness $E(n)$ attains the largest possible value.

Usual Method for Computing n_0. The usual method for computing n_0 is trying $n = 1$, $n = 2$, etc., until we reach the first value n for which $E(n) < E(n - 1)$. Then, we return $n_0 = n - 1$.

This method requires, in the worst case, N calls for the algorithm $E(n)$.

What We Want. We want to come up with a faster method for computing n_0.

3 Straightforward Asymptotically Optimal Method for Solving the Problem

Main Idea.

– For values $n < n_0$, we have $E(n) < E(n + 1)$.
– For values $n \geq n_0$, we have $E(n) > E(n + 1)$.

It is therefor reasonable to use bisection (see, e.g., [1]) to find the threshold value n_0.

Resulting Method. At each iteration, we have values $n_- < n_+$ for which $E(n_-) < E(n_- + 1)$ and $E(n_+) > E(n_+ + 1)$. Based on the properties described in the previous paragraph, this implies that $n_- < n_0 \leq n_+$.

We start with the values $n_- = 0$ and $n_+ = N - 1$. At each iteration, we take the midpoint

$$m = \left\lfloor \frac{n_- + n_+}{2} \right\rfloor$$

and check whether $E(m) < E(m + 1)$. Then:

– If $E(m) < E(m + 1)$, we replace n_- with the new value m.
– If $E(m) > E(m + 1)$, we replace n_+ with the new value m.

At each iteration, the width of the interval $[n_-, n_+]$ is decreased by half. We stop when this width becomes equal to 1, i.e., when $n_+ - n_- = 1$. Once we reach this stage, we return n_+ as the desired value n_0.

How Many Calls to the Algorithm $E(n)$ This Method Requires. We start with an interval $[0, N - 1]$ of width $\approx N$. At each stage, the width of the interval decreases by a factor of 2. Thus, after k iterations, we get an interval of width $2^{-k} \cdot N$. The number k of iterations needed to reach the desired interval of width 1 can be therefore determined from the formulas $2^{-k} \cdot N = 1$, so $k = \log_2(N)$.

On each iteration, we call the algorithm $E(n)$ twice: to find $E(m)$ and to find $E(m + 1)$. Thus, overall, this method requires $2 \cdot \log_2(N)$ calls to the algorithm $E(n)$.

This Method is Asymptotically Optimal. We need to find a natural number n_0 from the interval $[0, N]$. In general, by using b bits, we can describe 2^b different situations. Thus, the amount of information b that we need to determine n_0 must satisfy the inequality $2^b \geq N$, i.e., equivalently, $b \geq \log_2(N)$. To get each bit of information, we need to call the algorithm $E(n)$. Thus, to find n_0, we need to make at least $\log_2(N)$ calls to this algorithm.

The above algorithm requires $2 \cdot \log_2(N) = O(\log_2(N))$ calls. Thus, this method is indeed asymptotically optimal.

Natural Question. The straightforward method described in this section is asymptotically optimal, this is good. However, still, this method requires twice more calls to the algorithm $E(n)$ that the lower bound. Thus, a natural question is: can we make it faster?

Our answer to this question is "yes". Let us describe the new faster method.

4   New Faster Method: Description and Analysis

Preliminary Step. First, we compute the value $E(m)$ for the midpoint m of the interval $[0, N]$, so we form an interval $[n_-, n_+] \overset{\text{def}}{=} [0, N]$.

Iterations. At the beginning of each iteration, we have the values $n_- < n_+$ for which:

– we know the values $E(n_-)$, $E(n_+)$ and $E(m)$, where m is the midpoint of the interval $[n_-, n_+]$, and
– we know that $E(n_-) < E(m) > E(n_+)$.

We stop when $n_+ - n_- = 2$, in which case we return $n_0 \overset{\text{def}}{=} n_- + 1$.

At each iteration, we first select, with equal probabilities 0.5, whether we start with the left subinterval or with the right subinterval.

If we start with the left subinterval, then we compute the midpoint L of this subinterval, and compute the value $E(L)$. Then:

– If $E(L) > E(m)$, i.e., if $E(n_-) < E(L) > E(m)$, then we replace the interval $[n_-, n_+]$ with the new half-size interval $[n_-, m]$. In this case, the iteration is finished. So, if the stopping criterion is not yet satisfied, we start the new iteration.
– On the other hand, if $E(L) < E(m)$, then we compute the midpoint R of the right subinterval $[m, n_+]$, and compute the value $E(R)$. Then:
  • If $E(m) > E(R)$, i.e., if $E(L) < E(m) > E(R)$, then we replace the interval $[n_-, n_+]$ with the new half-size interval $[L, R]$.
  • On the other hand, if $E(m) < E(R)$, i.e., if $E(m) < E(R) > E(n_+)$, then we replace the interval $[n_-, n_+]$ with the new half-size interval $[m, n_+]$.

If we start with the right subinterval, then we compute the midpoint R of this subinterval, and compute the value $E(R)$. Then:

– If $E(m) < E(R)$, i.e., if $E(m) < E(R) > E(n_+)$, then we replace the interval $[n_-, n_+]$ with the new half-size interval $[m, n_+]$. In this case, the iteration is finished. So, if the stopping criterion is not yet satisfied, we start the new iteration.

- On the other hand, if $E(m) > E(R)$, then we compute the midpoint L of the left subinterval $[n_-, m]$, and compute the value $E(L)$. Then:
  - If $E(L) > E(m)$, i.e., if $E(n_-) < E(L) > E(m)$, then we replace the interval $[n_-, n_+]$ with the new half-size interval $[n_-, m]$.
  - On the other hand, if $E(L) < E(m)$, i.e., if $E(L) < E(m) > E(R)$, then we replace the interval $[n_-, n_+]$ with the new half-size interval $[L, R]$.

Why This Algorithm Works. If for some values $n_- < m < n_+$, we have $E(n_-) < E(m) > E(n_+)$, then:

- we cannot have $n_0 \leq n_-$, since then $n_- < m$ would imply $E(n_-) < E(m)$; thus, we must have $n_- \leq n_0$; and
- we cannot have $n_+ \leq n_0$, since then $m < n_+$ would imply $E(m) < E(n_+)$; thus, we mist have $n_0 \leq n_+$.

Thus, in this case, we must have $n_0 \in [n_-, n_+]$.

How Many Calls to the Algorithm $E(n)$ This Method Requires. Each iteration reduced the width of the original interval $[n_-, n_+] = [0, N]$ by half. So, similarly to the straightforward algorithm, we need $\log_2(N)$ iterations

On each iteration, we first make the first call to $E(n)$ and then the first comparison. We have two possible results of this comparison, so it is reasonable to assume that each comparison result occurs with probability 0.5. So, on each iteration:

- with probability 0.5, we require only one call to the algorithm $E(n)$, and
- with probability 0.5, we require two calls.

Thus, the expected number of calls on each iteration is $0.5 \cdot 1 + 0.5 \cdot 2 = 1.5$.

Since we make an independent random selection on each iteration, the numbers of calls on different iterations are independent random variables. Thus, due to the large numbers theorem (see, e.g., [3]), the overall number of calls will be close to the expected number of calls, i.e., to $1.5 \cdot \log_2(N)$. This is clearly faster than the number of calls $2 \cdot \log_2(N)$ required for the straightforward algorithms.

5 Conclusions

In training a neural network, it is important to select the proper number of features. If we select too few features, we may lose important information, and thus, decrease the prediction accuracy and effectiveness. On the other hand, if we include too many features, including ones that barely affect the predicted quantity, these barely-affecting features act, in effect, as noise, and the effectiveness of prediction decreases again. At present, to find the optimal number of features, practitioners sort them in the order of their prospective usefulness – e.g., by the absolute value of the correlation between the feature and the desired value, and then add features one by one in this order until we reach the largest possible effectiveness (and the smallest possible prediction errors). The main limitation of this process is that for each added feature, we need to re-train the neural network, and each re-training takes a significant amount of time. It is therefore desirable to be able to decrease the number of re-trainings needed to reach the most effective number of features.

In this paper, we present a method that drastically decreases the number of such re-trainings in comparison to the usual practice. Moreover, we show that this method is (asymptotically) optimal, i.e., it leads to the (asymptotically) smallest possible number of re-trainings.

Acknowledgements. The authors are greatly thankful to the anonymous reviewers for valuable suggestions.

References

1. Cormen, T.H., Leiserson, C.E., Rivest, R.L., Stein, C.: Introduction to Algorithms. MIT Press, Cambridge (2022)
2. Goodfellow, I., Bengio, Y., Courville, A.: Deep Learning. MIT Press, Cambridge (2016)
3. Sheskin, D.J.: Handbook of Parametric and Nonparametric Statistical Procedures. Chapman and Hall/CRC, Boca Raton (2011)

Why Inverse Layers in Pavement? Why Zipper Fracking? Why Interleaving in Education? a General Explanation

Edgar Daniel Rodriguez Velasquez[1,2], Aaron Velasco[3], Olga Kosheleva[4],
and Vladik Kreinovich[5(✉)]

[1] Department of Civil Engineering, Universidad de Piura in Peru (UDEP), Piura,
Peru
edgar.rodriguez@udep.pe, edrodriguezvelasquez@miners.utep.edu
[2] Department of Civil Engineering, University of Texas at El Paso,
El Paso, TX 79968, USA
[3] Department of Earth, Environment, and Resource Sciences,
University of Texas at El Paso, El Paso, TX 79968, USA
aavelasco@utep.edu
[4] Department of Teacher Education,
University of Texas at El Paso, El Paso, TX 79968, USA
olgak@utep.edu
[5] Department of Computer Science,
University of Texas at El Paso, El Paso, TX 79968, USA
vladik@utep.edu

Abstract. In many practical situations, if we split our efforts into two disconnected chunks, we get better results: a pavement is stronger if instead of a single strengthening layer, we place two parts of this layer separated by no-so-strong layers; teaching is more effective if instead of concentrating a topic in a single time interval, we split it into two parts separated in time, etc. In this paper, we provide a general explanation for all these phenomena.

Keywords: Pavement engineering · Fracking · Interleaving in education

1 Formulation of the Problem

General Idea. This research was motivated by the fact that in several application areas, there appears a similar empirical phenomenon, a phenomenon that, in each of these areas, is difficult to explain. In this paper, we provide a general explanation for this phenomenon. Let us list the examples of this phenomenon.

This work was supported in part by the National Science Foundation grants 1623190 (A Model of Change for Preparing a New Generation for Professional Practice in Computer Science), HRD-1834620 and HRD-2034030 (CAHSI Includes), EAR-2225395, and by the AT&T Fellowship in Information Technology.
It was also supported by the program of the development of the Scientific-Educational Mathematical Center of Volga Federal District No. 075-02-2020-1478, and by a grant from the Hungarian National Research, Development and Innovation Office (NRDI).

V.-N. Huynh et al. (Eds.): IUKM 2023, LNAI 14375, pp. 129–138, 2023.
https://doi.org/10.1007/978-3-031-46775-2_12

Pavement Engineering. Road pavement must be strong enough to sustain the traffic loads. To strengthen the pavement, usually, the pavement is formed by the following layers (see, e.g., [1]):

- First, on top of the soil, we place compacted granular material; this is called the *sub-base*.
- On top of the sub-base, we place granular material strengthened with cement; this layer is called the *base*.
- Finally, the top layer is the granular material strengthened by adding the liquid asphalt; this layer is called the *asphalt concrete layer*.

In this arrangement, the strength of the pavement comes largely from the two top layers: the asphalt concrete layer and the base.

Empirical evidence shows that in many cases, the inverse layer structure, where the base and sub-base are switched – so that the two strong layers are separated by a weaker sub-base layer – leads to better pavement performance; see, e.g., [3,6–8,10,11,16–18,20,22,25].

Fracking. Traditional methods of extracting oil and gas leave a significant portion of them behind. They were also unable to extract oil and gas that were concentrated in small amounts around the area. To extract this oil and gas, practitioners use the process called *fracking*, when high-pressure liquid is injected into the underground location, cracking the rocks and thus, providing the path for low-density oil and gas to move to the surface. Usually, several pipes are used to pump the liquid. Empirically, it turned out that the best performance happens when not all the pipes are active at the same time, but when there is always a significant distance between the active pipes. One way to maintain this distance – known as *zipper fracking* – is to activate, e.g., every other pipe, interchanging activations of pipes 1, 3, 5, etc., with activating the intermediate pipes 2, 4, 6, etc. (This particular technique is known as *Texas two-step*.) For more information, see, e.g., [19,26] and references therein.

Education. In education, best learning results are achieved when there is a pause between two (or more) periods when some topic is studied; this pedagogical practice is known as *interleaving*. Several studies show that interleaving enhances different types of learning, from learning to play basketball [9,13] to learning art [12] to learning mathematics [14,23,24], to training and re-training medical doctors [2,21]; see also [4,5,15].

2   Towards an Explanation

General Idea. What is the ideal situation?

- The ideal pavement would mean that all layers are strong.
- The ideal fracking would mean that all the pipes are active all the time.
- The ideal study process would mean that we study all the time.

So, a natural way to compare the quality of different strategies is to see which ones are closer to this ideal case.

A General Mathematical Description of the Problem. Let us formulate this general setting in precise terms.

In general, we have a certain range; this can be the range that describes:

- strength as a function of depth,
- study intensity as a function of time, etc.

From the mathematical viewpoint, we can always change the starting point to be 0. For example, for studying, we can measure time starting with the moment when we started the whole study process. In this case, the range will take the form $[0, T]$ for some $T > 0$. So, for simplicity, let us assume that this range has the form $[0, T]$.

Ideally, we should have full intensity at all points from this range:

- we should have full strength at all depth,
- we should have full study intensity at all moments of time, etc.

Again, from the mathematical viewpoint, we can re-scale intensity by taking this level as a new unit for measuring intensity. After this re-scaling, the value of the high level of intensity will be 1. So, the ideal case (I) is described by a function that takes the value 1 on the whole interval $[0, T]$; see Fig. 1.

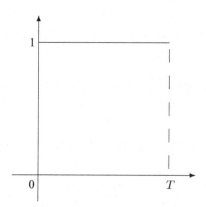

Fig. 1. Ideal case (I)

The problem is that in all the above applications, the ideal case is not realistic. In practice, we can have full strength only over a small portion of this range, a portion of overall size ε.

- We can have this strength portion concentrated on a connected (C) subrange (see Fig. 2) – as is the case, e.g., of the traditional pavement.
- Alternatively, we can divide this portion into two (or more) disconnected (D) subranges, as in Fig. 3.

In both cases, the value of intensity:

- is equal to 1 on a small part of the range, and
- is equal to 0 for all other values from the range.

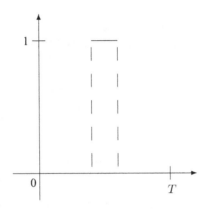

Fig. 2. Connected portion (C)

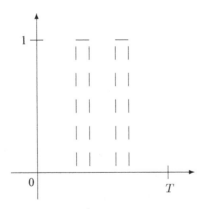

Fig. 3. Disconnected portion (D)

In all the above examples, the performance was better for the disconnected subranges. We will explain this by proving that, in some reasonable sense, the graph D corresponding to the disconnected portion is indeed closer to the graph I of the ideal case than the graph corresponding to the connected portion C, i.e., that $d(D, I) < d(C, I)$. In order to prove this, let us recall what is the natural way to describe distance $d(A, B)$ between two graphs A and B.

From the mathematical viewpoints, graphs are sets in a plane. So, to be able to describe distance between graphs, let us recall how to describe distance $d(A, B)$ between sets A and B.

How to Define the Distance $d(A, B)$ Between Two Sets A and B: Reminder. Let us start with the simplest case, when both sets are 1-element sets, i.e., when $A = \{a\}$ and $B = \{b\}$ for some points a and b. We assume that for two points a and b, distance $d(a, b)$ is already defined, In this case, it is reasonable to define $d(A, B) = d(\{a\}, \{b\}) \overset{\text{def}}{=} d(a, b)$.

A natural idea is to use Euclidean distance here:

$$d((x, y), (x', y')) = \sqrt{(x - x')^2 + (y - y')^2}.$$

Instead, we can use a more general ℓ^p-metric for some $p \geq 1$:

$$d((x, y), (x', y')) = (|x - x'|^p + |y - y'|^p)^{1/p}.$$

It is worth mentioning that our result remains valid whichever value $p \geq 1$ we select.

A slightly more complex case is when only one of the sets is a one-point set, e.g., $A = \{a\}$. In this case, it makes sense to define the distance $d(\{a\}, B)$ in such a way that this distance is 0 when $a \in B$. A reasonable idea is to take

$$d(A, B) = d(\{a\}, B) \overset{\text{def}}{=} \inf_{b \in B} d(a, b).$$

Finally, let us consider the general case, when both sets A and B may contain more than one point. In line with the general definition of a metric, we would like to have $d(A, B) = 0$ if and only if the sets A and B coincide, i.e., if and only if:

- every element the set A is also an element of the set B, and
- every element of the set B is also an element of the set A.

In other words, for us to declare that $d(A, B) = 0$:

- we must have $d(\{a\}, B) = 0$ for all $a \in A$, and
- we must have $d(\{b\}, A) = 0$ for all $b \in B$.

The usual way to achieve this purpose is – similarly to how we defined $d(\{a\}, B)$ – to define $d(A, B)$ as the largest of all these values; the resulting "worst-case" expression $d_w(A, B)$ is known as the *Hausdoff distance*:

$$d_w(A, B) \overset{\text{def}}{=} \max \left(\sup_{a \in A} d(\{a\}, B), \sup_{b \in B} d(\{b\}, A) \right). \tag{1}$$

In general, the worst case is not always the most adequate description. For example, if we have the set B almost equal to A, but with a very tiny additional part which is far away from the original set, the worst-case distance is huge, but in reality, the sets A and B are almost the same. To better capture the intuitive idea of distance between two sets, it is reasonable to consider not the *worst-case* values of $d(\{a\}, B)$ and $d(\{b\}, A)$, but their *average* values:

$$d_a(A, B) \stackrel{\text{def}}{=} \frac{1}{2} \cdot \frac{\int_A d(\{a\}, B) \, da}{\mu(A)} + \frac{1}{2} \cdot \frac{\int_B d(\{b\}, A) \, db}{\mu(B)}. \tag{2}$$

Let us see what these two definitions $d_w(A, B)$ and $d_a(A, B)$ say about the relation between our graphs I, C, and D.

What Are the Values $d_w(A, B)$ and $d_a(A, B)$ in Our Case. Both worst-case and average-case definitions are based on the values $d(\{a\}, B)$ and $d(\{b\}, A)$. So, to compute the distances between the corresponding graphs, let us first analyze what are the values $d(\{a\}, B)$ and $d(\{b\}, A)$ for our case.

Without losing generality, let us denote one of the graphs C or D by A, and the ideal graph I by B. Let us first consider the values $d(\{a\}, B) = d(\{a\}, I)$.

- Here, for points $a \in A$ corresponding to the portion of overall length ε, the intensity is equal to 1. So these points also belong to the graph I and thus, $d(\{a\}, I) = 0$.
- For all other points $a \in A$, the intensity is 0, i.e., this point has the form $(x, 0)$ for some $x \in [0, T]$. The set I is the straight line segment. So, the closest element to I is the projection of the point A on this straight line, i.e., the point $(x, 1)$. In this case, the shortest distance $d(\{a\}, I)$ from the point a and points $b \in I$ is equal to 1: $d(\{a\}, I) = 1$.

So, we have

$$\sup_{a \in A} d(\{a\}, I) = 1 \tag{3}$$

and

$$\frac{\int_A d(\{a\}, I) \, da}{\mu(A)} = \frac{0 \cdot \varepsilon + 1 \cdot (T - \varepsilon)}{T} = \frac{T - \varepsilon}{T}. \tag{4}$$

It should be mentioned that the values Eq. (3) and Eq. (4) are the same both:

- for the connected portion C and
- for the disconnected portion D;

these values only depend on the overall length of the portion.

Let us now consider the values $d(\{b\}, A)$, when $b \in I$, i.e., when $b = (x, 1)$ for some $x \in [0, 1]$, and A is C or D. By definition, $d(\{b\}, A)$ is the smallest of the values $d(a, b)$ when a is in the set A, i.e., when a is:

- either in the portion – in which case $a = (x', 1)$ for some $x' \in [0, T]$,
- or not in the portion – in which case $a = (x', 0)$ for some $x' \in [0, T]$.

In the second case, the distance is at least 1 – and can always be made smaller than or equal to 1 if we take the point $(x, \cdot) \in A$. In the first case, the distance is equal to $d(a, b) = d((x, 1), (x', 1)) = |x - x'|$. So:

- for points $b = (x, 1) \in I$ which are at most 1-close to the portion, the shortest distance $d(\{b\}, A)$ is equal to the distance z between x and the portion, while
- for all other points $b = (x, 1) \in I$, we have $d(\{b\}, A) = 1$.

And herein lies the difference between the connected case C and the disconnected case D. In the connected case, we have:

- one connected portion of length ε on which $d(\{b\}, A) = 0$, and
- two nearby intervals for which $d(\{b\}, A) < 1$,

see Fig. 4.

$$0 \qquad\qquad\qquad\qquad\qquad\qquad\qquad T$$

Fig. 4. Case of connected portion (C)

In this case, provided:

- that ε is sufficiently small, and
- that the portion is sufficiently separated from the endpoints 0 and T of the range,

we have

$$\sup_{b \in C} d(\{b\}, I) = 1 \tag{5}$$

and

$$\int_I d(\{b\}, C)\, db = 0 \cdot \varepsilon + 2 \int_0^1 z\, dz + (T - 2 - \varepsilon) \cdot 1 = 2 \cdot \frac{1}{2} + T - 2 - \varepsilon = T - 1 - \varepsilon,$$

thus

$$\frac{\int_I d(\{b\}, C)\, db}{\mu(I)} = \frac{T - 1 - \varepsilon}{T}. \tag{6}$$

In the disconnected case, we have:

- two connected subranges (of length $\varepsilon/2$ each) on which $d(\{b\}, A) = 0$, and
- two pairs of nearby intervals for which $d(\{b\}, A) < 1$,

see Fig. 5.

In this case, provided:

Fig. 5. Case of disconnected portion (D)

- that ε is sufficiently small, and
- that both subranges are sufficiently separated from each other and from the endpoints 0 and T of the range,

we have

$$\sup_{b \in D} d(\{b\}, I) = 1 \tag{7}$$

and

$$\int_I d(\{b\}, D)\, db = 0 \cdot \varepsilon + 4 \int_0^1 z\, dz + (T - 2 - \varepsilon) \cdot 1 = 4 \cdot \frac{1}{2} + T - 4 - \varepsilon = T - 2 - \varepsilon,$$

thus

$$\frac{\int_I d(\{b\}, C)\, db}{\mu(I)} = \frac{T - 2 - \varepsilon}{T}. \tag{8}$$

By combining the formulas (3), (5), and (7), we conclude that

$$d_w(C, I) = d_w(D, I) = 1.$$

Thus, if we only take into account the worst-case distance, then we cannot distinguish between the connected and disconnected cases.

However, if we use a more adequate average distance, then, by combining the formulas (4), (6), and (8), we get

$$d_a(C, I) = \frac{1}{2} \cdot \left(\frac{T - \varepsilon}{T} + \frac{T - 1 - \varepsilon}{T} \right) = \frac{T - 1/2 - \varepsilon}{T}, \tag{9}$$

while

$$d_a(D, I) = \frac{1}{2} \cdot \left(\frac{T - \varepsilon}{T} + \frac{T - 2 - \varepsilon}{T} \right) = \frac{T - 1 - \varepsilon}{T}. \tag{10}$$

Here clearly, $d_a(D, I) < d_a(C, I)$. In other words, the disconnected situation is closer to the ideal case than the connected one – which explains why in all above cases, the disconnected approach indeed leads to better results.

3 Conclusions and Future Work

This paper provides a mathematical explanation for a phenomenon that is observed in reality but runs against our intuition: that a "disconnected" control, when there is a pause between the two control stages or a gap between two control locations, is often more effective than the "connected" control, with no pauses or gaps. This is true in education, when learning the material during two time intervals, with a pause in between, is often more effective than when these two time intervals immediately follow each other. This is true in pavement engineering, where it is more effective to place two strong layers at some distance from each other rather than placing them next to each other – as it is usually done. This is true in fracking, where keeping a distance between two active pipes is more effective than activating two neighboring pipes.

In this paper, we provide a general mathematical explanation for this phenomenon – that for a naturally defined distance between settings, the discontinuous setting is closer to the ideal one than the continuous setting. The generality of this explanation makes us conjecture that a similar discontinuous arrangement is worth trying in many other cases, be it automatic control, medical therapy, or influencing people – and our first-approximation mathematical model will hopefully provide a way to compare different approaches.

Acknowledgements. The authors are greatly thankful to the anonymous reviewers for valuable suggestions.

References

1. American Association of State Highway and Transportation Officials (AASHTO): Mechanistic-Empirical Pavement Design Guide: A Manual of Practice, AASHTO, Washington, D.C. (2008)
2. Bachhel, R., Thaman, R.G.: Effective use of pause procedure to enhance student engagement and learning. J. Clin. Diagn. Res. **8**(8), XM01–XM03 (2014). https://doi.org/10.7860/JCDR/2014/8260.4691
3. Barksdale, R.D.: Performance of Crushed Stone Base Courses, Transportation Research Record 954. Transportation Research Board, Washington, D.C. (1984)
4. Birnbaum, M.S., Kornell, N., Bjork, E.L., Bjork, R.A.: Why interleaving enhances inductive learning: the roles of discrimination and retrieval. Mem. Cogn. **41**, 392–402 (2013)
5. Bokati, L., Urenda, J., Kosheleva, O., Kreinovich, V.: Why immediate repetition is good for short-term learning results but bad for long-term learning: explanation based on decision theory. In: Ceberio, M., Kreinovich, V. (eds.) How Uncertainty-Related Ideas Can Provide Theoretical Explanation for Empirical Dependencies, pp. 27–35. Springer, Cham, Switzerland (2021)
6. Buchanan, S.: Inverted pavements-what, why, and how?. In: Proceedings of the AFTRE Industry Education Webinar, Aggregates Foundation for Technology, Research, and Education (2010). https://www.vulcaninnovations.com/public/pdf/4-Inverted-Pavement-Systems.pdf
7. Georges, T.: Falling Weight Deflectometer (FWD) Test Results: Entrance Road of the Lafarge Quarry in Morgan County. Georgia, Final Report, Georgia Department of Transportation (2007)
8. Grau, R.W.: Evaluation of Structural Layers of Flexible Pavements, Miscellaneous Paper S-73-26. U.S. Army Engineer Waterways Experiment Station, Vicksburg, Mississippi (1973)

9. Hall, K.G., Dominguez, D.A., Cavazos, R.: Contextual interference effects with skilled basketball players. Percept. Mot. Skills **78**, 835–841 (1994)

10. Hoskins, B.E., McCullough, B.F., Fowler, D.W.: The development of a long-range rehabilitation plan for US-59, in District 11, Research Report No. 987-1, Austin, Texas (1991)

11. Johnson, C.W.: Comparative Studies of Combinations of Treated and Untreated Bases and Subbase for Flexible Pavements, Highway Research Board, Bulletin 289. National Research Council, Washington, D.C. (1960)

12. Kornell, N., Bjork, R.A.: Leaning concepts and categories: is spacing the 'enemy of induction'? Psychol. Sci. **19**, 585–592 (2008)

13. Landin, D.K., Hebert, E.P., Fairweather, M.: The effects of variable practice on the performance of a basketball skill. Res. Q. Exerc. Sports **64**, 232–236 (1993)

14. Le Blanc, K., Simon, D.: Mixed practice enhances retention and JOL accuracy for mathematical skills. In: Proceedings of the 49th Annual Meeting of the Psychonomic Society, Chicago, Illinois (2008)

15. Lerma, O., Kosheleva, O., Kreinovich, V.: Interleaving enhances learning: a possible geometric explanation. Geombinatorics **24**(3), 135–139 (2015)

16. Lewis, D.E.: Inverted Base Pavement at Lafarge Quarry Entrance Road: 445 Five-Year Field Evaluation. Final Report, Georgia Department of Transportation (GDOT), Georgia (2006)

17. Lewis, D.E., Ledford, K., Georges, T., Jared, D.M.: Construction and performance of inverted pavements in Georgia. In: Proceedings of the 91st Annual Meeting of the Transportation Research Board, Washington, D.C., January 22–26, 2012, Paper No. 12–1872, Poster Presentation in Session 639 (2012)

18. Moody, E.D.: Field Investigations of Selected Strategies to Reduce Reflective Cracking in Asphalt Concrete Overlays Constructed over Existing Jointed Concrete Pavements, Transportation Research Record 1449. Transportation Research Board, Washington, D.C. (1994)

19. Qian, B., Zhang, J., Zhu, J., Fang, Z., Kou, S., Chen, R.: Application of zipper-fracturing of horizontal cluster wells in the Changning shale gas pilot zone, Sichuan Basin. Natural Gas Industry B **2**(2–3), 181–184 (2015)

20. Rasoulian, M., Becnel, B., Keel, G.: Stone Interlayer Pavement Design, Transportation Research Record 1709. Transportation Research Board, Washington, D.C. (2000)

21. Richards, L.W., et al.: Use of the pause procedure in continuing medical education: a randomized controlled intervention study. Med. Teach. **39**(1), 74–78 (2017)

22. Rodriguez Velasquez, E.D.: Characterization and modeling of unbound and cementitoously stabilized materials for structural analysis of multilayer pavement systems, Ph. D. Dissertation, Department of Civil Engineering, University of Texas at El Paso, El Paso, Texas, USA (2023)

23. Rohrer, D., Taylor, K.: The shuffling of mathematics practice problems boosts learning. Instr. Sci. **35**, 481–498 (2007)

24. Taylor, K., Rohrer, D.: The effects of interleaved practice. Appl. Cogn. Psychol. **24**, 837–848 (2010)

25. Titi, H., Rasoulian, M., Martinez, M., Becnel, B., Keel, G.: Long-term performance of stone interlayer pavement. J. Transp. Eng. **129**(2), 118–126 (2003)

26. Zhen, W., et al.: Numerical analysis of zipper fracturing using a non-planar 3D fracture model. Front. Earth Sci. **10** (2022). https://doi.org/10.3389/feart.2022.808183

0-1 Combinatorial Optimization Problems with Qualitative and Uncertain Profits

Tuan-Anh Vu$^{(\boxtimes)}$, Sohaib Afifi, Éric Lefèvre, and Frédéric Pichon

University Artois, UR 3926, Laboratoire de Genie Informatique et d'Automatique de l'Artois (LGI2A), 62400 Bethune, France
{tanh.vu,sohaib.afifi,eric.lefevre,frederic.pichon}@univ-artois.fr

Abstract. Recent works have studied 0-1 combinatorial optimization problems where profits of items are measured on a qualitative scale such as "low", "medium" and "high". In this study, we extend this body of work by allowing these profits to be both qualitative and uncertain. In the first step, we use probability theory to handle uncertainty. In the second step, we use evidence theory to handle uncertainty. We combine their approaches with approaches in decision making under uncertainty that utilize the Maximum Expected Utility principle and generalized Hurwicz criterion, to compare solutions. We show that under probabilistic uncertainty and a special case of evidential uncertainty where the focal sets are rectangles, the task of identifying the non-dominated solutions can be framed as solving a multi-objective version of the considered problem. This result mirrors that of the case of qualitative profits with no uncertainty.

Keywords: Combinatorial optimization · Multiple objective optimization · Belief function · Decision making under uncertainty

1 Introduction

A 0-1 combinatorial optimization problem (01COP) can be seen as the selection of a subset of items from a given collection of subsets, with the objective of maximizing the total profits of the chosen items. Usually, these values are represented quantitatively using a vector in $\mathbb{R}^n_{\geq 0}$.

In many real-life situations, accurately assessing the exact numerical values of items can be challenging due to limited information availability. It is often much easier to make qualitative comparisons between these values. As an example, although most people will find it hard to determine the exact weights of a laptop and a smartphone, they can certainly say that the laptop is heavier.

Given an order between items, a mapping from the items to real values is called a representation of this order if it maintains the empirical relations among the items. The matroid optimization problem [6] is a special case of 01COPs in which the optimality of solutions is independent of the choice of representation. However, in measurement theory [7], it is known that in most cases, the optimality of solutions does depend crucially on the choice of representation, i.e., a solution is optimal for one representation but is not optimal for other representation.

Recently, in [9], the authors studied the Knapsack problem (KP) where profits of items are measured in a qualitative scale such as "low", "medium", "high". To deal with the above-mentioned issue, they provided a new way to compare solutions i.e., a solution x is preferred to a solution y if x has higher profit than y for any representation of the qualitative scale. They called a solution x non-dominated if there is no other solution which is strictly preferred to x and proceeded to enumerate all non-dominated solutions. In [9], they also observed a strong connection between KP with qualitative profits and multi-objective KP and this link is studied in greater details for 01COPs in a very recent paper [5].

In this paper, we further extend the works [5,9] by allowing profits of items to be both qualitative and uncertain. First, we utilize the traditional probabilistic framework to model uncertainty. Subsequently, following recent work encompassing a wide class of optimization problems [11], of which the 01COP is a subclass, we employ evidence theory [10], which is more general than probability theory, to represent uncertainty. It is worth noting that such evidential uncertainty, i.e., belief functions on ordinal variables, e.g. on the profit of some items, can be obtained from statistical data using, for instance, the approach described in [1].

In both cases, we adopt approaches in decision-making under uncertainty that utilize, respectively, the Maximum Expected Utility principle and the generalized Hurwicz criterion, to compare solutions, which still results in the concept of non-dominated solutions. Lastly, we show that under probabilistic uncertainty and a special case of evidential uncertainty where the so-called focal sets are rectangles, finding non-dominated solutions can be framed as solving a multi-objective version of the considered problem, which is similar to that of the case with no uncertainty.

The rest of this paper is organized as follows. Section 2 presents necessary background material. Section 3 quickly summarizes the works [5,9]. Section 4 presents the main results of the paper, where uncertainty is added and treated. The paper ends with a conclusion.

2   Preliminaries

In this section, we present necessary background for the rest of the paper. Throughout the paper, we denote by $[m]$ the set $\{1, \ldots, m\}$.

2.1   Evidence Theory

Let $\Omega = \{\omega_1, \ldots, \omega_q\}$ be the set, called frame of discernment, of all possible values of a variable ω. In evidence theory [10], partial knowledge about the true (unknown) value of ω is represented by a mapping $m : 2^\Omega \mapsto [0, 1]$ called mass function and such that $\sum_{A \subseteq \Omega} m(A) = 1$ and $m(\emptyset) = 0$, where mass $m(A)$ quantifies the amount of belief allocated to the fact of knowing only that $\omega \in A$.

A subset $A \subseteq \Omega$ is called a focal set of m if $m(A) > 0$. If all focal sets of m are singletons, then m is equivalent to a probability distribution. The mass function m gives rise to *belief* and *plausibility* measures defined as follows, respectively:

$$Bel(A) = \sum_{B \subseteq A} m(B) \text{ and } Pl(A) = \sum_{B \cap A \neq \emptyset} m(B), \ \forall A \subseteq \Omega. \tag{1}$$

We can also consider the set $\mathcal{P}(m)$ of all probability measures on Ω which are compatible with m, defined as $\mathcal{P}(m) = \{P : P(A) \geq Bel(A) \ \forall A \subseteq \Omega\}$.

2.2 Multi-objective Optimization Problem

A multi-objective optimization problem can be written as

$$\max \ \{f_1(x), \ldots, f_m(x)\} \tag{2}$$
$$x \in \mathcal{X}. \tag{3}$$

The notion of Pareto dominance is usually used for multi-objective optimization problems. The feasible solution x is said to Pareto dominate the feasible solution y, denoted by $x \succ_{Pareto} y$ if

$$f_i(x) \geq f_i(y) \ \forall i \in [m] \text{ and } \exists j \in [m] \text{ such that } f_j(x) > f_j(y). \tag{4}$$

As the objectives (2-3) are typically conflicting, there is usually no solution x that simultaneously maximizes all $f_i(x)$. Instead, we seek to find all so-called efficient feasible solutions of (2-3), defined as:

$$x \in \mathcal{X} \text{ such that } \nexists y \in \mathcal{X}, y \succ_{Pareto} x. \tag{5}$$

We refer to the book [4] for a comprehensive discussion on this subject.

2.3 0-1 Combinatorial Optimization Problem

A general 0-1 Combinatorial Optimization Problem (01COP) can be expressed as follows. Let $\mathcal{S}$ be a set of n items. Each item i has a profit r_i, represented as a vector $r \in \mathbb{R}^n_+$. The profit of a subset of $\mathcal{S}$ is obtained by summing the profits of the items within it. The goal of the decision-maker is to find a subset having maximum profit among a predefined collection $\mathcal{X} \subseteq 2^{\mathcal{S}}$ of subsets of $\mathcal{S}$. This problem can be modeled using a binary vector $x \in \{0,1\}^n$, where each element x_i indicates whether item i is included in the subset (1) or not (0). The 01COP can then be written as:

$$\max \ r^T x$$
$$x \in \mathcal{X} \subseteq \{0,1\}^n. \tag{01COP}$$

The Knapsack problem (KP) is one of the most important problems in the class 01COP, which will serve as a running example throughout the paper. It is defined as follows.

Example 1 (The 0-1 knapsack problem (01KP)). Suppose a company has a budget of W and needs to choose which items to manufacture from a set of n possible items, each with a production cost of w_i and fixed profit of r_i. The 01KP involves selecting a subset of items to manufacture that maximizes the total profit while keeping the total production costs below W. The 01KP can be formulated as

$$\max \sum_{i=1}^{n} r_i x_i$$

$$\sum_{i=1}^{n} w_i x_i \leq W \qquad\qquad \text{(01KP)}$$

$$x_i \in \{0,1\} \quad i \in [n].$$

3 01COP with Qualitative Levels

In this section, we quickly summarize the works in [5,9]. In many applications, we can only express profits of items on a finite scale of qualitative levels. More precisely, let $(\mathcal{L}, \prec) = \{l_1, \ldots, l_k\}$ be a fixed scale with k levels $l_1 \prec \ldots \prec l_k$. Profits of items are then represented by a fixed vector $r \in \mathcal{L}^n$.

Example 2. Consider 5 items whose profits are measured in the qualitative scale $\mathcal{L} = \{\text{"low", "medium", "high"}\}$. Their profits are recorded by a vector $r = \{\text{"low", "medium", "high", "high", "medium"}\}$ in $\mathcal{L}^5$.

In general, the set of items can include absolutely unprofitable items, resulting in the qualitative levels set $\mathcal{L}$ having a level that signifies "no profit at all". However, the decision-maker can always remove all such items from the outset. Due to this, we exclude the case involving the "no profit at all" level. A mapping $v : \mathcal{L} \to \mathbb{R}_{>0}$ is called a representation of $\mathcal{L}$ if

$$\forall i, j, \ l_i \prec l_j \Leftrightarrow v(l_i) < v(l_j). \qquad\qquad (6)$$

We denote by $\mathcal{V}$ the set of all representations of $\mathcal{L}$. Note that $\mathcal{V}$ is identified with a subset of $\mathbb{R}_{>0}^k$, that is

$$\mathcal{V} := \left\{ v \in \mathbb{R}_{>0}^k : v_{i+1} > v_i, \ \forall i \in [k-1] \right\}. \qquad\qquad (7)$$

In the following, to simplify the notation, we will use v_i instead of $v(l_i)$ for a representation v.

The rank cardinality vector of an $x \in \mathcal{X}$ is defined as:

$$g(x) = (g_1(x), \ldots, g_k(x)) \qquad\qquad (8)$$

where $g_j(x) = |\{i : x_i = 1 \text{ and } r_i = l_j\}|$. Hence, the j-th component of $g(x)$ is nothing but the total number of items in x with profit level l_j.

Let $v(x)$ be the profit of x with respect to a representation $v \in V$. By definition,

$$v(x) = \sum_{i=1}^{n} x_i v(r_i). \tag{9}$$

We can also compute $v(x)$ via its rank cardinality vector as

$$v(x) = \sum_{i=1}^{k} g_i(x) v_i. \tag{10}$$

The preferences between feasible solutions crucially depend on the choice of v as illustrated in the next example.

Example 3. Consider the following KP with 5 items and $W = 6$. The profits and weights of items are given in Table 1. Let $x = (1,1,1,0,0)$ (selecting items 1,2 and 3) and $y = (0,0,1,1,0)$ (selecting items 3 and 4) be two feasible solutions. If a representation v is chosen such that $v(l_1) = 2, v(l_2) = 3, v(l_3) = 4$, x is preferred to y as $v(x) = 8 > v(y) = 7$. However, if v is chosen such that $v(l_1) = 2, v(l_2) = 3, v(l_3) = 6$, y is preferred to x as $v(x) = 8 < v(y) = 9$.

Table 1. Profits and weights of items

items	1	2	3	4	5
w	2	2	2	3	4
r	l_1	l_2	l_2	l_3	l_3

To avoid the issue encountered in Example 3, the preference between feasible solutions is defined as follows in [9]:

Definition 1. *Let $x, y \in \mathcal{X}$ be two feasible solutions. Then,*

1. *x weakly dominates y, denoted by $x \succeq y$, if for every $v \in V$, it holds that $v(x) \geq v(y)$.*
2. *x dominates y, denoted by $x \succ y$, if x weakly dominates y and there exists $v^* \in V$ such that $v^*(x) > v^*(y)$.*
3. *$x^* \in \mathcal{X}$ is called efficient or non-dominated, if there does not exist any $x \in \mathcal{X}$ such that $x \succ x^*$.*

In [9], it is shown that the relation $\succeq$ in Definition 1 is a preorder, *i.e.*, it is reflexive and transitive. At first glance, Definition 1 appears to require checking every representation of $\mathcal{L}$ to determine the dominance relation between two feasible solutions. However, there exists a rapid and straightforward test based on the following key result.

Lemma 1 (see [9]). *Let x, y be two feasible solutions. We have $x \succeq y$ iff $\sum_{i=j}^{k} g_i(x) \geq \sum_{i=j}^{k} g_i(y)$ for all $j \in [k]$.*

Lemma 1 is of great importance as it establishes the link between 01COP with qualitative levels and multi-objective optimization. This link was first observed for the KP in [9] and has been systematically studied in [5] for 01COP. Indeed, from Lemma 1 and material in Sect. 2.2, it is easy to see that x^* is an efficient solution according to Definition 1 if and only if it is an efficient solution of the following problem:

$$\max \left\{ \sum_{i=1}^{k} g_i(x), \sum_{i=2}^{k} g_i(x), \ldots, \sum_{i=k}^{k} g_i(x) \right\} \tag{11}$$

$$x \in \mathcal{X} \tag{12}$$

Note that Problem (11-12) can be rewritten so that its objective functions are linear. Indeed, for each $i \in [k]$, define vector $c^i \in \{0,1\}^n$ as follow:

$$c_j^i = 0 \text{ if } r_j \neq l_i \text{ and } c_j^i = 1 \text{ otherwise.} \tag{13}$$

Hence, c^i is nothing but a vector that records positions of the qualitative level l_i in r, and thus we have $(c^i)^T x = g_i(x) \; \forall i \in [k]$. Problem (11-12) is then rewritten as

$$\max \left\{ (\sum_{i=1}^{k} c^i)^T x, (\sum_{i=2}^{k} c^i)^T x, \ldots, (c^k)^T x \right\} \tag{14}$$

$$x \in \mathcal{X} \tag{15}$$

Therefore, methods in multi-objective optimization can be readily applied to find efficient solutions of Problem 01COP with qualitative profits.

Example 4 (Example 3 continued). In the KP in Example 3, the position vectors are $c^1 = (1,0,0,0,0), c^2 = (0,1,1,0,0)$, and $c^3 = (0,0,0,1,1)$. To find non-dominated solutions according to Definition 1 of the KP, we need to solve the following multi-objective optimization problem:

$$\max \{x_1 + x_2 + x_3 + x_4 + x_5, \; x_2 + x_3 + x_4 + x_5, \; x_4 + x_5\} \tag{16}$$

$$x \in \{0,1\}^5 : 2x_1 + 2x_2 + 2x_3 + 3x_4 + 4x_5 \leq 6 \tag{17}$$

4   01COPs with Uncertain Qualitative Profits

In this section, we extend the approaches presented in [5,9] to address the case where profits are uncertain and qualitative. Note that Lemma 1 is originally proved for rank cardinality vectors in $\mathbb{Z}_{\geq 0}^k$ (as shown in the original proof in [9] or a simplified version in [5]). In our extended setting, we will require a generalized version of this lemma that can accommodate vectors in $\mathbb{R}_{\geq 0}^k$. Therefore, we present the generalized version here. Note that the proof in [9] can be easily modified to fit the generalized version. However, we present a new proof of

Lemma 1 based on the duality theory of linear programming, which conceptually differs from the proofs presented in [5,9].

Let A be a $m \times n$ matrix. Let us recall from linear programming that if the primal problem is

$$\min \left\{ v^T b : v^T A \geq c, v \in \mathbb{R}_{\geq 0}^m \right\}. \tag{18}$$

then its dual problem is

$$\max \left\{ c^T u : Au \leq b, u \in \mathbb{R}_{\geq 0}^n \right\}. \tag{19}$$

Lemma 2. *Let $g(x), g(y)$ be two vectors in $\mathbb{R}_{\geq 0}^k$. Then,*

$$\sum_{i=1}^{k} g_i(x)v_i \geq \sum_{i=1}^{k} g_i(y)v_i \ \forall v \in V \Leftrightarrow \sum_{i=j}^{k} g_i(x) \geq \sum_{i=j}^{k} g_i(y) \ \forall j \in [k]. \tag{20}$$

Proof. Let $f(v) = \sum_{i=1}^{k} (g_i(x) - g_i(y))v_i$. Then,

$$\sum_{i=1}^{k} g_i(x)v_i \geq \sum_{i=1}^{k} g_i(y)v_i \ \forall v \in V \Leftrightarrow f(v) \geq 0 \ \forall v \in V \tag{21}$$

$$\Leftrightarrow f(v) \geq 0 \ \forall v \in \overline{V} := \left\{ v \in \mathbb{R}_{\geq 0}^k : v_{i+1} \geq v_i, \ \forall i \in [k-1] \right\}, \tag{22}$$

since f is continuous and $\overline{V}$ is the closure of V. Let $z^* = \min \left\{ f(v) : v \in \overline{V} \right\}$. So z^* is the optimal value of the linear programming problem (P):

$$\min \ \sum_{i=1}^{k} (g_i(x) - g_i(y))v_i$$
$$v_1 \geq 0 \tag{P}$$
$$v_{i+1} - v_i \geq 0, \ \forall i \in [k-1]$$

Note that $f(v) \geq 0 \ \forall v \in \overline{V}$ iff $z^* \geq 0$. Furthermore, $z^* \geq 0$ iff Problem (P) is bounded, *i.e.*, $z^* \neq -\infty$.

Indeed, for the sake of contradiction, suppose that Problem (P) is bounded, and yet there exists a v^* such that $f(v^*) = z^* < 0$. For any positive scalar λ, we have $\lambda v^* \in \overline{V}$, and thus $f(\lambda v^*) = \lambda z^* < z^*$, which contradicts the optimality of z^*. By duality, we have $z^* \neq \infty$ iff the dual Problem (D) has the finite optimal value, or in this case Problem (D) is feasible:

$$\max \ 0^T u$$
$$u_i - u_{i+1} \leq g_i(x) - g_i(y), \ \forall i \in [k-1]$$
$$u_k \leq g_k(x) - g_k(y) \tag{D}$$
$$u \geq 0.$$

It is easy to see that Problem (D) is feasible iff $\sum_{i=j}^{k} g_i(x) \geq \sum_{i=j}^{k} g_i(y) \ \forall j \in [k]$. Hence, we get the desired result. □

4.1   Under Probabilistic Uncertainty

In this section, we assume that information about the qualitative levels of items is given by a probability distribution P on a subset $\mathcal{R}$ of $\mathcal{L}^n$. Each $r \in \mathcal{R}$ is called a scenario. Given $v \in \mathcal{V}$, let $v^r(x)$ be the profit of x under scenario r. The expected utility of a feasible solution $x \in \mathcal{X}$ with respect to v is defined as:

$$E_P^v(x) := \sum_{r \in \mathcal{R}} P(r)v^r(x) \tag{23}$$

According to the Maximum Expected Utility principle [8], it is reasonable to compare solutions based on their expectations. Furthermore, for similar reasons as those that lead to Definition 1, i.e., the preference between two solutions x and y should not depend on the choice of v, we define, for any $x, y \in \mathcal{X}$,

$$x \succeq_P y \text{ iff } E_P^v(x) \geq E_P^v(y) \; \forall v \in \mathcal{V}. \tag{24}$$

Let $g_i^r(x)$ be the number of items in x with qualitative level l_i under scenario r. The next result shows how to check whether $x \succeq_P y$.

Proposition 1. $x \succeq_P y \Leftrightarrow \sum_{i=j}^k \bar{g}_i(x) \geq \sum_{i=j}^k \bar{g}_i(y) \; \forall j \in [k]$, where $\bar{g}_i(x) := \sum_{r \in \mathcal{R}} P(r)g_i^r(x)$.

Proof. By definition in Equation (23), we have

$$E_P^v(x) = \sum_{r \in \mathcal{R}} P(r) \sum_{i=1}^k g_i^r(x)v_i = \sum_{i=1}^k \left(\sum_{r \in \mathcal{R}} P(r)g_i^r(x) \right) v_i. \tag{25}$$

Equivalently,

$$E_P^v(x) = \sum_{i=1}^k \bar{g}_i(x)v_i. \tag{26}$$

Therefore, $x \succeq_P y \Leftrightarrow \sum_{i=1}^k \bar{g}_i(x)v_i \geq \sum_{i=1}^k \bar{g}_i(y)v_i \; \forall v \in \mathcal{V}$. The desired result follows by applying Lemma 2. □

Note that $\bar{g}_i(x)$ can be interpreted as the expected number of items in x with profit level i.

From Proposition 1, non-dominated solutions according to $\succeq_P$ are efficient solutions of the following problem:

$$\max \left\{ \sum_{i=1}^k \bar{g}_i(x), \sum_{i=2}^k \bar{g}_i(x), \ldots, \sum_{i=k}^k \bar{g}_i(x) \right\} \tag{27}$$

$$x \in \mathcal{X}. \tag{28}$$

Note that each objective of Problem (27-28) is still linear. Indeed, let $c^{ri} \in \{0,1\}^n$ be a vector that records positions of qualitative level l_i in scenario r, defined as:

$$c_j^{ri} = 0 \text{ if } r_j \neq l_i \text{ and } c_j^{ri} = 1 \text{ otherwise.} \tag{29}$$

Therefore,

$$g_i^T(x) = (c^{ri})^T x \tag{30}$$

and $\bar{g}_i(x) = \left(\sum_{r \in \mathcal{R}} P(r) c^{ri} \right)^T x.$

Example 5 (Example 3 continued). Assume now that the information about the profits of items in Example 3 are given by two scenarios r^1 and r^2 in Table 2 with $P(r^1) = 0.8$ and $P(r^2) = 0.2$. We can see that $c^{r^1 1} = (1,0,0,0,0), c^{r^1 2} = (0,1,1,0,0), c^{r^1 3} = (0,0,0,1,1), c^{r^2 1} = (0,0,1,0,1), c^{r^2 2} = (0,1,0,1,0)$ and $c^{r^2 3} = (1,0,0,0,1)$. For any feasible solution x, we have

Table 2. Profits of items under two scenarios

items	1	2	3	4	5
w	2	2	2	3	4
r^1	l_1	l_2	l_2	l_3	l_3
r^2	l_3	l_2	l_1	l_2	l_1

$$\bar{g}_1(x) = (0.8c^{r^1 1} + 0.2c^{r^2 1})^T x = 0.8x_1 + 0.2x_3 + 0.2x_5. \tag{31}$$

Similarly, $\bar{g}_2(x) = x_2 + 0.8x_3 + 0.2x_4$ and $\bar{g}_3(x) = 0.2x_1 + 0.8x_4 + 0.8x_5$. Hence, finding non-dominated solutions boils down to solving the following multi-objective KP.

$$\max \left\{ \begin{array}{c} x_1 + x_2 + x_3 + x_4 + x_5, \\ 0.2x_1 + x_2 + 0.8x_3 + x_4 + 0.8x_5, \\ 0.2x_1 + 0.8x_4 + 0.8x_5, \end{array} \right\} \tag{32}$$

$$x \in \{0,1\}^5 : 2x_1 + 2x_2 + 2x_3 + 3x_4 + 4x_5 \leq 6 \tag{33}$$

4.2 Under Evidential Uncertainty

A more general approach than the one in Sect. 4.1 is to use evidence theory to represent uncertainty. Let m be a mass function on a subset $\mathcal{R}$ of $\mathcal{L}^n$. Let $\mathcal{F}$ be the set of focal sets of m. Following [11], the lower and upper expected values of a feasible solution $x \in \mathcal{X}$ with respect to a $v \in \mathcal{V}$ are defined as:

$$\underline{E}^v(x) := \sum_{F \in \mathcal{F}} m(F) \min_{r \in F} v^r(x), \tag{34}$$

$$\overline{E}^v(x) := \sum_{F \in \mathcal{F}} m(F) \max_{r \in F} v^r(x). \tag{35}$$

For a fixed v, we may remark that the interval $\left[\underline{E}^v(x), \overline{E}^v(x) \right]$ is the range of $E_P^v(x)$ for all compatible probability measures P in $\mathcal{P}(m)$ [2].

As in [11], solutions can be compared according to the generalized Hurwicz criterion [2], defined by $H_\alpha^v(x) = \alpha \overline{E}^v(x) + (1 - \alpha)\underline{E}^v(x)$ for some chosen optimism/pessimism degree $\alpha \in [0, 1]$. Furthermore, as in Sects. 3 and 4.1, we wish to compare solutions regardless of the choice of representation:

$$x \succeq_{hu}^\alpha y \text{ iff } H_\alpha^v(x) \geq H_\alpha^v(y) \ \forall v \in \mathcal{V}. \tag{36}$$

We first consider the case where the focal sets of m take a special form.

Rectangular Focal Sets. A subset $F \subseteq \mathcal{L}^n$ is called a rectangle iff it can be expressed as the Cartesian product of sets, that is, $F = \times_{i=1}^n F^{\downarrow i}$, where $F^{\downarrow i} \subseteq \mathcal{L}$. Since $\mathcal{L}$ is a linear order, we can associate two scenarios RF, rF for each focal set F defined as:

$$RF_i = \max F^{\downarrow i} \text{ and } rF_i = \min F^{\downarrow i}, \ \forall i \in [n] \tag{37}$$

In this case, it is easy to compute $\overline{E}^v(x)$, $\underline{E}^v(x)$ for a given v as shown in the Proposition 2.

Proposition 2. *When focal sets of m are rectangles, for any $v \in \mathcal{V}$ we have*

$$\underline{E}^v(x) = \sum_{F \in \mathcal{F}} m(F) v^{rF}(x) \tag{38}$$

$$\overline{E}^v(x) = \sum_{F \in \mathcal{F}} m(F) v^{RF}(x). \tag{39}$$

Proof. For any $r \in F$, by (37) we have

$$v^{rF}(x) = \sum_{i=1}^n x_i v(rF_i) \leq v^r(x) = \sum_{i=1}^n x_i v(r_i) \leq \sum_{i=1}^n x_i v(RF_i) = v^{RF}(x) \tag{40}$$

Hence, inequality (40) together with Eqs (34)-(35) lead to the desired result. $\square$

Similarly to the probabilistic case in Sect. 4.1, we are able to derive a characterization for $x \succeq_{hu}^\alpha y$:

Proposition 3. $x \succeq_{hu}^\alpha y \Leftrightarrow \sum_{i=j}^k \overline{g}_i^\alpha(x) \geq \sum_{i=j}^k \overline{g}_i^\alpha(y) \ \forall j \in [k]$ *where*

$$\overline{g}_i^\alpha(x) := \sum_{F \in \mathcal{F}} m(F) \left(\alpha g_i^{RF}(x) + (1 - \alpha) g_i^{rF}(x) \right). \tag{41}$$

Proof. By Proposition 2, we have

$$H_\alpha^v(x) = \sum_{F \in \mathcal{F}} m(F) \left((1 - \alpha) \sum_{i=1}^k g_i^{rF}(x) v_i + \alpha \sum_{i=1}^k g_i^{RF}(x) \right) v_i. \tag{42}$$

Exchanging the summation leads to

$$H_\alpha^v(x) = \sum_{i=1}^k \left[\sum_{F \in \mathcal{F}} m(F) \left(\alpha g_i^{RF}(x) + (1-\alpha) g_i^{rF}(x) \right) \right] v_i = \sum_{i=1}^k \overline{g}_i^\alpha(x) v_i. \quad (43)$$

Hence $H_\alpha^v(x) \geq H_\alpha^v(y) \; \forall v \Leftrightarrow \sum_{i=1}^k \overline{g}_i^\alpha(x) v_i \geq \sum_{i=1}^k \overline{g}_i^\alpha(y) v_i \; \forall v$. The result follows then from Lemma 2. $\qquad \square$

From Proposition 3, we obtain that non-dominated solutions according to $\succeq_h^\alpha$ are efficient solutions of the following problem:

$$\max \left\{ \sum_{i=1}^k \overline{g}_i^\alpha(x), \sum_{i=2}^k \overline{g}_i^\alpha(x), \ldots, \sum_{i=k}^k \overline{g}_i^\alpha(x) \right\} \quad (44)$$

$$x \in \mathcal{X}. \quad (45)$$

Similar to Problem (27-28), each objective of Problem (44,45) is also linear. At first glance, the assumption that focal sets are rectangles may seem restrictive. Still, it can appear in numerous practical situations. In the next example, we provide such a situation.

Example 6 (Example 3 continued). Assume that the profits of items are unknown, and an expert predicts that the profit vector is $r = \{l_2, l_3, l_1, l_2, l_2\}$. However, the expert is not entirely reliable, and from results of his past predictions, we know that the probability of him being correct is 0.8. If the prediction is accurate, the profit vector is indeed r. On the other hand, when the prediction is wrong, we are completely ignorant about the true profit, which could be any vector in $\{l_1, l_2, l_3\}^5$. This piece of information can be naturally modeled using a mass function m with two focal sets: $F_1 = \{(l_2, l_3, l_1, l_2, l_2)\}$ with a mass of 0.8, and $F_2 = \{l_1, l_2, l_3\}^5$ with a mass of 0.2 (Table 3). Let us choose $\alpha = 0.5$. For any feasible solution x, we can compute that

Table 3. Profits of items in two focal sets

items	1	2	3	4	5
w	2	2	2	3	4
F_1	l_2	l_3	l_1	l_2	l_2
F_2	$\{l_1, l_2, l_3\}$	$\{l_1, l_2, l_3\}$	$\{l_1, l_2, l_3\}$	$\{l_1, l_2, l_3\}$	$\{l_1, l_2, l_3\}$

$$\overline{g}_1^\alpha(x) = 0.1x_1 + 0.1x_2 + 0.9x_3 + 0.1x_4 + 0.1x_5$$
$$\overline{g}_2^\alpha(x) = 0.8x_1 + 0.8x_4 + 0.8x_5$$
$$\overline{g}_3^\alpha(x) = 0.1x_1 + 0.9x_2 + 0.1x_3 + 0.1x_4 + 0.1x_5.$$

So, finding non-dominated solutions according to the $\succeq_{hu}^{\alpha}$ can be formulated as solving the following multi-objective KP:

$$\max \left\{ \begin{array}{l} x_1 + x_2 + x_3 + x_4 + x_5, \\ 0.9x_1 + 0.9x_2 + 0.1x_3 + 0.9x_4 + 0.9x_5, \\ 0.1x_1 + 0.9x_2 + 0.1x_3 + 0.1x_4 + 0.1x_5, \end{array} \right\} \tag{46}$$

$$x \in \{0,1\}^5 : 2x_1 + 2x_2 + 2x_3 + 3x_4 + 4x_5 \le 6 \tag{47}$$

Arbitrary Focal Sets. In this case, it is hard to derive a similar result as in Lemma 2. As a first result in this direction, we give a sufficient condition for $x \succeq_{hu}^{\alpha} y$, with $x, y \in \mathcal{X}$. Let $\mathcal{R}^* := \{r \in \mathcal{R} : \exists F \in \mathcal{F} \text{ such that } r \in F\}$.

Proposition 4. *If for each $r \in \mathcal{R}^*$, we have $\sum_{i=j}^{k} g_i^r(x) \ge \sum_{i=j}^{k} g_i^r(y)$ for all $j \in [k]$ then $x \succeq_{hu}^{\alpha} y$.*

Proof. Immediate from (34-35).   □

Clearly, the condition stated in Proposition 4 is very stringent as it requires that for each scenario in $\mathcal{R}^*$, x weakly dominates y. Hence, in future research, it would be valuable to find more relaxed conditions or, ideally, establish a characterization similar to Lemma 2.

5   Conclusion

In this paper, we have investigated 0-1 Combinatorial Optimization Problems (01COPs), where the profits of items can be both qualitative and uncertain. We have combined approaches from [5,9] with decision-making under uncertainty methodologies [2] to compare solutions. Our main result is that under probabilistic uncertainty and a special case of evidential uncertainty where focal sets are rectangles, we still can find non-dominated solutions by solving a multi-objective version of the original 01COP. Going forward, we plan to study deeper the case of evidential uncertainty with arbitrary focal sets, aiming to provide more comprehensive insights and understanding. Another interesting direction is to adapt the approach in [3] where the authors compared acts by means of Sugeno integrals.

References

1. Denœux, T.: Constructing belief functions from sample data using multinomial confidence regions. Int. J. Approx. Reason. **42**(3), 228–252 (2006)
2. Denoeux, T.: Decision-making with belief functions: a review. Int. J. Approx. Reason. **109**, 87–110 (2019)
3. Dubois, D., Prade, H., Sabbadin, R.: Decision-theoretic foundations of qualitative possibility theory. Eur. J. Oper. Res. **128**(3), 459–478 (2001)
4. Ehrgott, M.: Multicriteria Optimization, vol. 491. Springer (2005). https://doi.org/10.1007/3-540-27659-9

5. Klamroth, K., Stiglmayr, M., Sudhoff, J.: Ordinal optimization through multi-objective reformulation. Eur. J. Oper. Res. **311**, 427–443 (2023)
6. Oxley, J.G.: Matroid Theory, vol. 3. Oxford University Press, USA (2006)
7. Roberts, F.S.: Meaningfulness of conclusions from combinatorial optimization. Discret. Appl. Math. **29**(2–3), 221–241 (1990)
8. Savage, L.J.: The Foundations of Statistics. Courier Corporation (1972)
9. Schäfer, L.E., Dietz, T., Barbati, M., Figueira, J.R., Greco, S., Ruzika, S.: The binary knapsack problem with qualitative levels. Eur. J. Oper. Res. **289**(2), 508–514 (2021)
10. Shafer, G.: A Mathematical Theory of Evidence. Princeton University Press (1976)
11. Vu, T.A., Afifi, S., Lefèvre, E., Pichon, F.: Optimization problems with evidential linear objective. Int. J. Approx. Reason. **161**, 108987 (2023)

Forecasting Precious Metals Prices Volatility with the Global Economic Policy Uncertainty Index: The GARCH-MIDAS Technique for Different Frequency Data Sets

Roengchai Tansuchat[1,2], Payap Tarkhamtham[2], Wiranya Puntoon[2,3], and Rungrapee Phadkantha[2,3](✉)

[1] Faculty of Economics, Chiang Mai University, Chiang Mai 50200, Thailand
[2] Center of Excellence in Econometrics, Faculty of Economics, Chiang Mai University, Chiang Mai, Thailand
rungrapee.ph@gmail.com
[3] Office of Research Administration, Chiang Mai University, Chiang Mai 50200, Thailand

Abstract. This study aims to assess how global economic policy uncertainty influences the volatility of precious metals prices especially in the case of "gold and silver" that occupy the two largest shares in the global precious metals market. It covers four periods of interest, namely the pre subprime crisis (2000–2007), pre COVID-19 pandemic (2008–20019), during COVID-19 pandemic (2020–2022), and full sample (2000–2022) to evaluate the impacts of economic policy uncertainty in different sample periods. The GARCH-MIDAS methodology is employed in this study to accommodate the incorporation of datasets with disparate frequencies, specifically monthly data on Global Economic Policy Uncertainty (GEPU) and daily data on precious metals prices, for examining their relationship and yielding more insightful results.

Keywords: Precious metals · economic policy uncertainty · GARCH · MIDAS · GARCH-MIDAS

1 Introduction

Investors commonly consider precious metals, like gold and silver, as safe-haven assets that they can rely on during times of economic instability or when there is a lack of clarity surrounding economic policy. Therefore, the prices of precious metals can be subject to significant fluctuations during times of economic policy uncertainty, as investors may seek to minimize potential risks and fluctuations in the economy by hedging with these assets [1]. Accurate modeling of volatility and correlation dynamics is crucial for improving investment decisions in these markets. Various factors, such as changes in economic activity, supply and demand dynamics, and geopolitical events, can influence the volatility and correlation of precious metals markets. In addition to the previously mentioned factors, economic policy uncertainty (EPU) can also have an impact on the prices of precious metals. EPU is characterized as a risk resulting from undefined government policies and regulatory frameworks in the near future [2]. This factor has been the focus of studies conducted by [17] and [15]. As a result, it is crucial to develop precise modeling of volatility and correlation dynamics in precious metals markets to make informed investment decisions. Investors and portfolio managers can create more accurate models and make better investment decisions by comprehending the elements that influence these dynamics.

Global Economic Policy Uncertainty (GEPU) refers to the amalgamation of fiscal, regulatory, and monetary policies established by a country's central bank and government. The uncertainty associated with GEPU arises from unpredictable changes in tax and regulatory policies, which can heighten the risk associated with certain commodities, particularly speculative assets, during periods of high uncertainty. Given the ongoing COVID-19 pandemic, financial markets and policy decisions are experiencing an elevated level of uncertainty. Numerous studies have been conducted to investigate the effects of GEPU on various financial indicators, including stock markets, exchange rates, oil and stocks, bonds, institutional conditions, unemployment, commodity markets, cryptocurrency, financial inclusion, environmental quality, and monetary policy. These studies collectively illustrate that GEPU exerts an influence on the overall economy and brings about modifications in these economic variables (see: [3, 6, 16]). Although previous studies have established that Global Economic Policy Uncertainty (GEPU) has an impact on various economic indicators, its association with the commodity market, particularly precious metals, was initially discussed by [4]. The presence of GEPU can influence the investment and consumption decisions made by economic agents, thereby influencing the prices of these precious metals in the commodity market. Furthermore, GEPU can also affect the risk premium, interest rates, and supply and demand, all of which have implications for the economy and investment. These fluctuations can have consequences on financing and production expenditures. In times of crises or economic downturns, commodities futures are frequently employed within the commodities market to mitigate downside risk. Consequently, through portfolio rebalancing, GEPU can alter the structure of the commodity market. The various factors of global economic policy uncertainty (GEPU) may cause investors to become more cautious in their investments, which can lead to increased volatility in financial markets.

The GARCH-MIDAS model is employed to shed light on the relationship between economic policy uncertainty and the volatility of precious metals prices. It is recognized that directly introducing low-frequency economic uncertainty to explain high-frequency price volatility of precious metals may not be appropriate. Therefore, the GARCH-MIDAS model is utilized as a powerful tool to bridge this gap and provide insights into how economic uncertainty influences the volatility of precious metals market. [8] proposed the GARCH-MIDAS model within the MIDAS framework as a means to assess time-varying market volatility. This model splits the conditional variance into long-term and short-term components. The long-term component examines the impact of low-frequency variables on the conditional variance. The GARCH-MIDAS model is a combination of the MIDAS framework developed by [12] and the component model proposed by [9]. One significant advantage of this model is that it allows for the direct examination of how macroeconomic variables affect market volatility by linking daily observations of stock returns with macroeconomic variables that are collected at lower frequencies. In this study, we use the recently developed GARCH-MIDAS methodology to investigate how macroeconomic factors that are prone to uncertainty affect the precious metals market volatility. Therefore, our research examines variance predictability and seeks to determine whether including policy-sensitive economic variables can enhance the capacity of the conventional volatility models for predicting. Moreover, we divided the study into four periods: the full sample period (2000–2022), the pre-crisis period (2000–2007), the pre-COVID-19 period (2008–2019), and the COVID-19 pandemic period (2020–2022) to assess the impacts of various sample ranges containing different degrees of uncertainty.

The remainder of the study is structured as follows. In Sect. 2, we go over the methodology. Data used in the study are presented in Sect. 3. We present the findings in Sect. 4 and the conclusion in Sect. 5.

2   Methodology

2.1   GARCH-MIDAS

The GARCH (1,1) model is a widely used model to estimate and forecast volatility in financial markets. It assumes that the conditional variance of the asset returns follows an autoregressive process that depends on the past squared errors. To assess the influence of global economic policy uncertainty (GEPU) on the volatility of the precious metals market, we employ the GARCH-MIDAS model, which was introduced by researchers [10] and [21]. This model effectively addresses the disparity in data frequency between the daily prices of precious metals and the monthly GEPU values. To aid in explaining the GARCH-MIDAS model, we initially outline the GARCH (1,1) model utilizing data of similar frequency

$$r_t - E_{t-1}(r_t) = \varepsilon_t \tag{1}$$

$$\varepsilon_t = \sqrt{\sigma_t^2 Z_t} \tag{2}$$

$$\sigma_t^2 = \alpha_0 + \alpha_1 \varepsilon_{t-1}^2 + \beta \sigma_{t-1}^2 \tag{3}$$

where r_t is the natural logarithmic rate of the precious metals return, $E_{t-1}(r_t)$ is its conditional mean, ε_t is the residual, Z_t is the innovation, and σ_t^2 is the conditional variance. ω, α, β are the model coefficients.

In contrast to the GARCH model utilizing data of identical frequency, the GARCH-MIDAS model incorporates a long-run composition equation. Within the framework of the GARCH-MIDAS model, the volatility of the precious metals market is deconstructed into two primary elements: long-term volatility and short-term volatility. $r_{i,t}$ is the return on day i of period t (month/quarter/year). Short-run volatility changes at the daily frequency i, and long-run volatility changes at the period frequency t. Assuming $\sqrt{\sigma_t^2} = \sqrt{\tau_t \times g_{i,t}}$, in Eq. (2), the new return equation can be specified as follows:

$$r_{i,t} = E_{t-1}(r_{i,t}) + \sqrt{\tau_t \times g_{i,t}} Z_{i,t} \quad \forall i = 1, ..., N \tag{4}$$

where $g_{i,t}$ is daily variation, related to short-run factors and τ_t denotes the long-run component. $Z_{i,t}|\Phi_{t-1}, t \sim N(0, 1)$ with $\Phi_{t-1,t}$ represents the set of information determined by the day $(i-1)$ of period t. $g_{i,t}$ is assumed to follow the GARCH (1,1) process.

$$g_{i,t} = (1 - \alpha - \beta) + \alpha \frac{(r_{i-1,t} - \mu)^2}{\tau_t} + \beta g_{i-1,t} \tag{5}$$

where μ is the unconditional mean of the precious metals return. As for τ_t, [10] expressed by employing techniques like smoothing realized volatility or macroeconomic variables, in alignment with the principles of MIDAS regression and MIDAS filtering:

$$\ln \tau_t = m_E + \theta_E \sum_{k=1}^{K} \varphi_k(\omega_1, \omega_2) X_{t-k} \tag{6}$$

Here X_{t-k} is the low-frequency data representing such the explanatory variable under consideration as global economic policy uncertainty (GEPU). K is the maximum lags. The element pertaining to long-run τ_t is predetermined by $E_{\tau-1}[(r_{i,t} - \mu)^2] = \tau_t E_{t-1}(g_{i,t}) = \tau_t E_{t-1}(g_{i,t})$ is regarded as an expectation without conditions of $g_{t,t}$ and thus $E_{-1}(g_{i,t}) = 1$. $\phi_k(\omega_t)$ is a weight equation based on the beta function and is described as:

$$\phi_k(\omega_i) = \frac{(1 - k/K)^{\omega_i - 1}}{\sum_{j=1}^{K} \left(\frac{j}{k}\right)^{\omega_i}} \tag{7}$$

Equations (4–7) jointly construct the GARCH-MIDAS model with the GEPU.

2.2   Model Performance and Evaluation

To assess the performance of the models, an 80% in-sample calibration is conducted. The models are estimated using a calibration window, and the estimated parameters are then utilized to predict the 20% out-of-sample variance. In order to evaluate the accuracy of volatility predictions generated by the GARCH-MIDAS model, different variants of the loss function are employed, following the approach adopted by [13]. One commonly used loss function is the root mean squared error (RMSE), which is defined as:

$$RMSE = \sqrt{\frac{1}{S} \sum_{s=1}^{S} (\sigma_{s+1}^2 - E(\sigma_{s+1}^2))^2} \tag{8}$$

where σ_{s+1}^2 is the actual daily total variance on day $s + 1$, $E_s(\sigma_{s+1}^2)$ is the predicated daily total variance for day $s + 1$, and S is the length of prediction interval.

3   Data

To study the influence of economic policy uncertainty on the price volatility of the two precious metals with the largest market share, gold and silver, we divided the study into four periods of the pre subprime crisis (2000–2007), pre COVID-19 pandemic (2008–20019), during COVID-19 pandemic (2020–2022), and full sample (2000–2022). Regarding the data sources, we used monthly data on policy uncertainty from the Global Economic Policy Uncertainty Index (https://www.policyuncertainty.com/) and daily data on precious metals for this study (https://www.lbma.org.uk/). The daily returns of precious metals prices are calculated as follows:

$$r_t = \ln\left(\frac{P_t}{P_{t-1}}\right) \tag{9}$$

where t is in units of trading days.

Table 1 provides a comprehensive summary of descriptive statistics for the Global Economic Policy Uncertainty Index (GEPU) and the daily returns for two precious metals, gold and silver. It presents the mean, maximum value, minimum value, standard deviation, skewness, kurtosis, J-B test for normality, and ADF test for stationarity for each variable, as well as the results of each sample period. The GEPU index and its changes are observed on a monthly basis, with sampling frequencies lower than the daily returns of precious metals prices. The mean values of all-time series, except for the GEPU index, hover around zero. The GEPU index, however, exhibits a mean within the range of approximately 140–280.The distributions of the GEPU index and its changes are positively skewed and leptokurtic. The two precious metals show both negative and positive skewness, along with leptokurtosis. Furthermore, we assess the stationarity of each time series using the augmented Dickey-Fuller (ADF) test [7]. Our findings indicate that the GEPU index and the two precious metals' price returns are stationary. Therefore, all data are suitable for analysis in the next step. In terms of volatility, silver exhibits higher volatility compared to gold across all ranges. The greatest volatility is typically observed during times of crisis. Conversely, during the COVID-19 epidemic, the volatility tends to be lower, as evidenced by the lower standard deviation. The standard deviation serves as a measure of the risk associated with each precious metal, implying that investors seeking less risky investments may prefer gold. On the other hand, those willing to take on more risks might opt for silver.

Overall, Table 1 provides a good starting point for analyzing the relationship between policy uncertainty and the market volatility of precious metals. By comparing the descriptive statistics for each variable, one can identify the precious metals that are most and least expensive, and most and least volatile, and have most and least normal distributions. Be aware that this study uses the Minimal Bayes factor (MBF) to determine whether the results are significant. MBF can be used as an alternative to the p-values for testing the significance of results ([14] and [18]). The understanding of MBF (Minimum Bayes Factor) values is contingent upon the value's magnitude, with distinct ranges of values signifying varying degrees of evidence, outlined as follows: 0.33 < MBF < 1: "weak evidence", 0.1 < MBF < 0.33: "moderate evidence", 0.033 < MBF < 0.1: "substantial evidence", 0.01 < MBF < 0.033: "strong evidence", 0.003 < MBF < 0.01: "very strong evidence", MBF < 0.003: "decisive evidence". These intervals imply that as the MBF value diminishes, the level of evidence supporting the alternative hypothesis (i.e., rejecting the null hypothesis) grows stronger. Therefore, smaller MBF values indicate stronger evidence in favor of the alternative hypothesis.

Table 1. Descriptive statistics.

Precious Metals	Mean	Maximum	Minimum	Std. Dev.	Skewness	Kurtosis	Jarque-Bera	ADF test	MBF ADF-test
Full sample									
GEPU	144.7409	437.2473	51.7902	77.2860	1.1666	3.7950	69.8731***	−4.286***	0.000
Gold	0.0003	0.0964	−0.0891	0.0109	−0.0986	10.2381	12659***	−78.343***	0.000
Silver	0.0003	0.1828	−0.1959	0.0199	−0.5934	13.9657	29374***	−80.784***	0.000
Pre subprime crisis									
GEPU	81.6415	170.2364	51.7903	24.1685	1.4907	5.5181	60.9202***	−3.779***	0.000
Gold	0.0005	0.0964	−0.0550	0.0103	0.3326	9.4309	3504.154***	−47.726***	0.000
Silver	0.0005	0.1032	−0.1608	0.0172	−0.9707	11.9782	7073.603***	−46.864***	0.000
Pre-COVID-19 pandemic									
GEPU	153.3322	335.3522	79.8540	53.9498	1.1891	3.8437	38.2038***	−4.469***	0.000
Gold	0.0002	0.0955	-0.0891	0.0114	-0.3383	10.7527	7638.418***	−56.407***	0.000
Silver	0.0001	0.1828	−0.1869	0.0211	−0.4081	13.7277	14598.870***	−60.635***	0.000
During the COVID-19 pandemic									
GEPU	278.6409	437.2473	186.9300	61.5518	0.4162	2.5046	1.4073**	−7.4302***	0.000
Gold	0.0002	0.0679	−0.0540	0.0100	0.0968	7.8054	728.581***	−26.178***	0.000
Silver	0.0004	0.1021	−0.1959	0.0214	−0.7117	14.3071	4091.132***	−25.821***	0.000

4   Results

4.1   Forecasting Capability of GEPU for Prices of Precious Metals (Gold and Silver)

Table 2 presents the results obtained from applying the GARCH-MIDAS (Mixed Data Sampling) approach to analyze the volatility of precious metals prices, specifically gold and silver. The findings are based on four different sample periods. Upon analyzing the results, it is evident that the coefficients of ARCH (α), and GARCH (β) are all positive and statistically significant while the constant long-term term (m) is negative. This suggests that these factors contribute to the volatility of precious metals prices. Additionally, the sum of α and β, which represents the level of volatility persistence, is consistently found to be less than one ($\alpha + \beta < 1$) across all four sample periods. This indicates that there is a high degree of volatility persistence in the prices of precious metals. Regarding the influence of the global economic policy uncertainty (GEPU), the results indicate that the impact on precious metals price volatility is both mean-reverting ($\alpha + \beta < 1$) and variable. This implies that the effect of the GEPU on precious metals prices is not fixed and permanent, but rather changes over time. The significance of the slope parameter (θ) within the MIDAS model lends additional support to these observations. During both the complete dataset and the subprime crisis timeframe, the slope coefficient (θ) reflects the ability to forecast the volatility of daily precious metals prices using monthly GEPU values. Upon analyzing the outcomes, we identify a notable and statistically meaningful positive slope coefficient for each precious metal. This suggests that GEPU indeed holds a positive and significant impact on the volatility of precious metals prices, aligning with findings akin to those in [11]. On the other hand, during the pre-subprime crisis period, we find the slope coefficient to be negatively significant for only silver which means the impact of the pre subprime crisis, economic uncertainty had a statistically significant negative effect on precious silver returns. In the same way, during the COVID-19 pandemic period, the GEPU has a statistically significant negative effect on gold and silver price returns. The results are consistent with the previous studies of [19] and [20], which consider gold as a safe haven during stressful times such as economic crises. However, it is noteworthy that during the COVID-19 pandemic, both gold and silver have not been secure assets. This is similar to the study conducted by [5] which states that gold is weak as a safe haven during the outbreak of COVID-19. In summary, the outcomes indicate elevated volatility in precious metals prices across all four analyzed timeframes. This volatility is shaped by the coefficients of both ARCH and GARCH, alongside the impact of global economic policy uncertainty.

Figures 1, 2, 3 and 4 show the estimated volatility of precious metals price corresponding to the global economic policy uncertainty (GEPU) for four sample periods. The orange line represents the realized volatility of precious metals – a proxy of volatility, and the blue line represents the estimated volatility by the MIDAS-GARCH model. Both volatilities display a parallel pattern, and the proportion of long-term volatility in relation to overall volatility changes over time. Therefore, it can be concluded that the price of precious metals, especially gold and silver, may be affected by the GEPU.

Table 2. Results of GEPU's predictability for precious metals prices (Gold and Silver).

	α	β	θ	w	m
Full Sample					
Gold	0.0570**	0.9337***	0.2241	17.3115***	-8.9186***
	[0.0262]	[0.0348]	[4.238]	[2.5578]	[0.4726]
Silver	0.0541***	0.9401***	0.7625***	1.0012**	-7.6674***
	[0.0114]	[0.0140]	[0.1844]	[0.4003]	[0.3518]
Pre subprime crisis					
Gold	0.0692***	0.9185***	0.2797	1.2134***	-9.0265***
	[0.0144]	[0.0186]	[0.7206]	[0.2853]	[0.1354]
Silver	0.0676***	0.9227***	-11.0177***	1.0170***	-7.9705***
	[0.0225]	[0.0252]	[0.2568]	[0.1804]	[0.2752]
Pre-COVID-19 Pandemic					
Gold	0.0570***	0.9353***	3.5026***	1.0010***	-8.6258***
	[0.0210]	[0.0249]	[0.7858]	[0.2081]	[0.3263]

(*continued*)

Table 2. (*continued*)

	α	β	θ	w	m
Silver	0.0594*	0.9233***	3.6274***	1.0010***	-7.6093***
	[0.0308]	[0.0421]	[0.1094]	[0.2661]	[0.3180]
During COVID-19 Pandemic					
Gold	0.0974***	0.8644***	-6.6142***	2.0056***	-9.2813***
	[0.0328]	[0.0417]	[0.2131]	[0.1400]	[0.2583]
Silver	0.1193	0.8602***	-3.6160***	6.1822***	-7.7849***
	[0.0799]	[0.0966]	[0.1963]	[0.2372]	[0.4295]

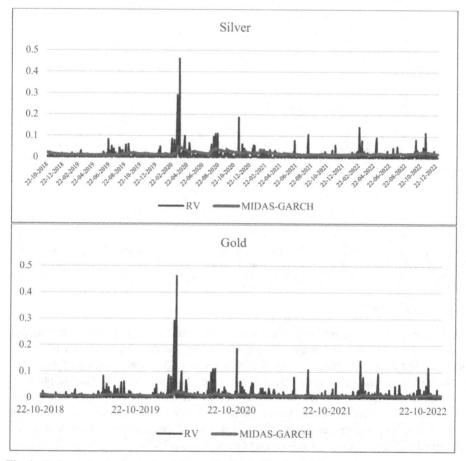

Fig. 1. Estimated out-of-sample of the realized volatility for "full sample" of precious metals price in relation to the global economic policy uncertainty.

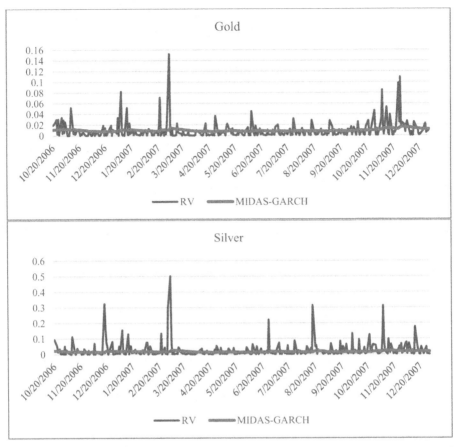

Fig. 2. Estimated out-of-sample of realize volatility for "pre subprime crisis" of precious metals price relation to the global economic policy uncertainty.

4.2 RMSE Results

Table 3 presents the evaluation results of the model's performance for the three specific time periods. We assessed both the performance within the observed data 80% (in-sample) and the performance outside the observed data 20% (out-of-sample) using the RMSE (root mean square error) statistic as a measure of adequacy. The adoption of the RMSE aligns with the framework utilized in the study by [10]. This test allows us to determine smaller RMSE values, which indicate better forecasts. The findings demonstrate that the model performed well across all three samples, both within the observed data and in terms of forecasting future time horizons.

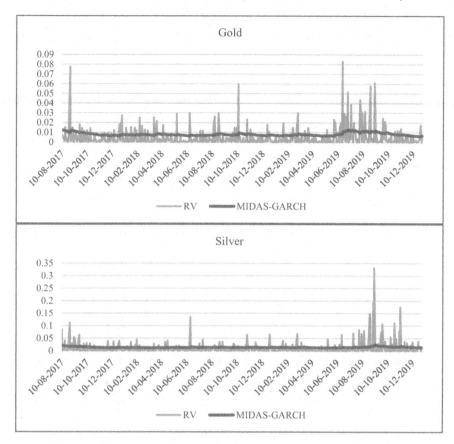

Fig. 3. Estimated out-of-sample of realize volatility for "pre-COVID-19" of precious metals price relation to the global economic policy uncertainty.

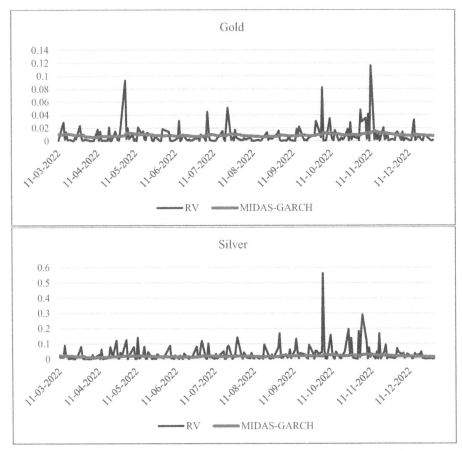

Fig. 4. Estimated out-sample of realize volatility for "during COVID-19 pandemic" of precious metals price relation to the global economic policy uncertainty.

Table 3. RMSE results.

Precious Metals	In Sample	Out-of-Sample
Full sample		
Gold	0.0036	0.0023
Silver	0.0062	0.0047
Pre crisis		
Gold	0.0024	0.0016
Silver	0.0116	0.0054

(*continued*)

Table 3. (*continued*)

Precious Metals	In Sample	Out-of-Sample
Crisis		
Gold	0.0043	0.0009
Silver	0.0171	0.0026
During COVID-19 Pandemic		
Gold	0.0025	0.0015
Silver	0.0160	0.0059

5 Conclusions

The empirical findings of this research demonstrate a statistically significant prediction of precious metals price movements based on the level of uncertainty pertaining to global economic policies. The findings are based on four sample periods. The coefficients of ARCH (α) and GARCH (β) are positive and statistically significant, indicating their contribution to the volatility of precious metals prices. The constant long-term term is negative, suggesting its influence on volatility. The level of volatility persistence, represented by the sum of ARCH and GARCH coefficients, is consistently found to be less than one across all sample periods, indicating high volatility persistence in precious metals prices. Furthermore, the results indicate that the Economic Policy Uncertainty (GEPU) index exerts a substantial influence on the volatility of precious metals prices in all periods, especially during the COVID-19 pandemic. These findings imply that the price volatility of precious metals tends to escalate during periods characterized by economic downturns or crises. In addition, during the pre-subprime crisis period, the impact of economic uncertainty had a statistically significant negative effect on silver returns. Similarly, during the COVID-19 pandemic time, economic policy uncertainty had a statistically significant negative effect on gold and silver price returns. These findings suggest that precious metals prices exhibit higher volatility in all sample periods, influenced by the ARCH and GARCH coefficients as well as the global economic policy uncertainty. The results also indicate the changing nature of precious metals as a safe haven asset during economic crises, including the COVID-19 pandemic.

Acknowledgments. The authors thank the "Center of Excellence in Econometrics, Chiang Mai University" for its financial support. This research work was partially supported by "Chiang Mai University".

References

1. Alqaralleh, H., Canepa, A.: The role of precious metals in portfolio diversification during the Covid19 pandemic: a wavelet-based quantile approach. Resour. Policy **75**, 102532 (2022)
2. Al-Thaqeb, S.A., Algharabali, B.G.: Economic policy uncertainty: a literature review. J. Econ. Asymmetries **20**, e00133 (2019)

3. Asgharian, H., Hou, A.J., Javed, F.: The importance of the macroeconomic variables in forecasting stock return variance: a GARCH-MIDAS approach. J. Forecast. **32**(7), 600–612 (2013)

4. Baker, S.R., Bloom, N., Davis, S.J.: Measuring economic policy uncertainty. Q. J. Econ. **131**(4), 1593–1636 (2016)

5. Chemkha, R., BenSaïda, A., Ghorbel, A., Tayachi, T.: Hedge and safe haven properties during COVID-19: evidence from Bitcoin and gold. Q. Rev. Econ. Finance **82**, 71–85 (2021)

6. Dai, P.F., Xiong, X., Zhang, J., Zhou, W.X.: The role of global economic policy uncertainty in predicting crude oil futures volatility: evidence from a two-factor GARCH-MIDAS model. Resour. Policy **78**, 102849 (2022)

7. Dickey, D.A., Fuller, W.A.: Distribution of the estimators for autoregressive time series with a unit root. J. Am. Stat. Assoc. **74**(366a), 427–431 (1979)

8. Engle, R.: Anticipating Correlations: A New Paradigm for Risk Management. Princeton University Press (2009)

9. Engle, R.F., Lee, G.: A long-run and short-run component model of stock return volatility. In: Cointegration, causality, and forecasting: A Festschrift in honour of Clive WJ Granger, pp. 475–497 (1999)

10. Engle, R.F., Ghysels, E., Sohn, B.: Stock market volatility and macroeconomic fundamentals. Rev. Econ. Stat. **95**(3), 776–797 (2013)

11. Gao, W., Aamir, M., Shabri, A.B., Dewan, R., Aslam, A.: Forecasting crude oil price using Kalman filter based on the reconstruction of modes of decomposition ensemble model. IEEE Access **7**, 149908–149925 (2019)

12. Ghysels, E., Santa-Clara, P., Valkanov, R.: Predicting volatility: getting the most out of return data sampled at different frequencies. J. Econometrics **131**(1–2), 59–95 (2006)

13. Hansen, P.R., Lunde, A.: A forecast comparison of volatility models: does anything beat a GARCH (1, 1)? J. Appl. Economet. **20**(7), 873–889 (2005)

14. Held, L., Ott, M.: How the maximal evidence of p-values against point null hypotheses depends on sample size. Am. Stat. **70**(4), 335–341 (2016)

15. Huynh, T.L.: The COVID-19 risk perception: a survey on socioeconomics and media attention. Econ. Bull. **40**(1), 758–764 (2020)

16. Khaskheli, A., Zhang, H., Raza, S.A., Khan, K.A.: Assessing the influence of news indicator on volatility of precious metals prices through GARCH-MIDAS model: a comparative study of pre and during COVID-19 period. Resour. Policy **79**, 102951 (2022)

17. Li, S., Lucey, B.M.: Reassessing the role of precious metals as safe havens–What colour is your haven and why? J. Commod. Mark. **7**, 1–14 (2017)

18. Maneejuk, P., Yamaka, W.: Significance test for linear regression: how to test without P-values? J. Appl. Stat. **48**(5), 827–845 (2021)

19. Mokni, K., Al-Shboul, M., Assaf, A.: Economic policy uncertainty and dynamic spillover among precious metals under market conditions: does COVID-19 have any effects? Resour. Policy **74**, 102238 (2021)

20. Raza, S.A., Masood, A., Benkraiem, R., Urom, C.: Forecasting the volatility of precious metals prices with global economic policy uncertainty in pre and during the COVID-19 period: Novel evidence from the GARCH-MIDAS approach. Energy Econ **120**, 106591 (2023). https://doi.org/10.1016/j.eneco.2023.106591

21. Zhou, Z., Fu, Z., Jiang, Y., Zeng, X., Lin, L.: Can economic policy uncertainty predict exchange rate volatility? New evidence from the GARCH-MIDAS model. Financ. Res. Lett. **34**, 101258 (2020)

Lasso and Ridge for GARCH-X Models

Woraphon Yamaka[1], Paravee Maneejuk[1(✉)], and Sukrit Thongkairat[1,2]

[1] Center of Excellence in Econometrics, Faculty of Economics, Chiang Mai University,
Chiang Mai 50200, Thailand
mparavee@gmail.com
[2] Department of Statistics, Faculty of Science, Chiang Mai University,
Chiang Mai 50200, Thailand

Abstract. This paper examines the efficacy of the least absolute shrinkage and selection operator (Lasso) and Ridge algorithms in improving the volatility forecasting of the Generalized Autoregressive Conditional Heteroskedasticity (GARCH) models with exogenous covariates. Our study proposes a novel parameter estimation approach by combining a quasi-maximum-likelihood estimator (QMLE) with the Lasso and Ridge regularization methods. The other objective is to identify the most critical predictors that can enhance the volatility forecasting performance of various GARCH models. To demonstrate the effectiveness of our proposed algorithms, we conduct an empirical analysis of the US stock market. Our results indicate that by imposing a specific constraint on the penalty parameters, the Lasso and Ridge estimators can significantly enhance the volatility forecasting performance of different GARCH-type models.

Keywords: GARCH-type models · Lasso · Ridge · Volatility forecasting

1 Introduction

Volatility has long been used as a measure of risk. Historically, it was represented through standard deviation or variance, that is separated into diversifiable and non-diversifiable components. However, these volatility measurements have limitations from not allowing for the identification of patterns in asset volatility and the specification of time-varying and clustering properties that may occur in the future. As a result, researchers have introduced various new perspectives for the consideration and forecast of volatility.

The early model for volatility investigation is the autoregressive conditional heteroskedasticity (ARCH) proposed by Engle in 1982, which was consequently used to describe the crucial view of stochastic volatility offered by Taylor in 1986 and Hull and White in 1987. Later, the Generalized Autoregressive Conditional Heteroskedasticity (GARCH) model was developed from ARCH. GARCH-type models are now among the most powerful and widely used tools for predicting financial volatility. Many studies, such as those by Hammoudeh and Yuan [10], Narayan and Narayan [16], Sadorsky

Supported by Chiang Mai University.

[19], Tully and Lucey [20], and Hammoudeh et al. [9], have applied GARCH models to different commodities, particularly oil. These models can handle the volatility persistence, asymmetry, and clustering displayed by many time-series data.

However, Poon and Granger [8], and Byun and Chon [5] revealed that volatility forecasting with GARCH-type models is not always optimal, as they tend to perform poorly in out-of-sample forecast estimation despite their ability to produce satisfactory in-sample forecasts. Additionally, Xu et al. [22] argued that volatility forecasting based on the GARCH-type models solely depends on the dynamics of financial time series themselves, rather than other information or exogenous variables. According to the Efficient Market Hypothesis (EMH), all available information is fully reflected in asset prices, and researchers and practitioners cannot obtain more accurate prediction results.

In recent times, the notion of market efficiency as a reliable indicator of risk has been subject to scrutiny. This is because empirical evidence suggests that market inefficiencies do exist, and they are influenced by various macroeconomic and microeconomic factors [1, 3, 22]. Consequently, recent studies have proposed incorporating conditional information from other economic variables to improve volatility prediction accuracy [11, 25]. To this end, researchers and practitioners often employ exogenous regressors in the specification of volatility dynamics to better model and forecast the volatility of economic and financial time series. Among the most popular models used in this regard is the GARCH-X model.

Despite the effectiveness of including additional regressors in explaining the volatilities of stock return series, exchange rate returns series, or interest rate series, which results in improved in-sample fit and out-of-sample forecasting performance [11], there are concerns regarding the choice of covariates in the GARCH-X model. Empirical studies have shown that the selection of covariates can span a wide range of various economic or financial indicators. However, the theoretical properties of the estimator associated with the GARCH-X model, namely the quasi maximum likelihood estimator (QMLE), are not yet fully understood. Of particular interest is how the persistence of the chosen covariate affects the QMLE, given the wide range of different choices of covariates. It is natural to be concerned that different degrees of persistence of the chosen covariates may lead to different behavior of the QMLE and associated inferential tools [11]. Furthermore, if an inappropriate set of predictors is considered, multicollinearity may result, further complicating the estimation and interpretation of the GARCH-X model.

This study proposes the use of the Lasso approach [21] for variable selection and the Ridge regression to handle multicollinearity in the GARCH modeling. In the literature, these methods have been extensively studied in the regression framework [15, 23, 24]. The recent popularity of these approaches is mainly due to their ability to handle situations where the number of parameters to be estimated is larger than the available sample size [14]. By using the Lasso and Ridge regression in the GARCH models, we aim to address concerns of variable selection, multicollinearity, and overfitting. These issues have been widely discussed in the literature, and our proposed approach seeks to improve the accuracy of volatility forecasting by selecting relevant covariates and accounting for the correlation among them.

The main objective of this study is to extend the application of Lasso and Ridge regularization methods to GARCH-X models. Specifically, we propose to penalize the QMLE of GARCH-X models using Lasso and Ridge penalties, and then estimate the models using quasi maximum likelihood estimation. To assess the effectiveness of this approach, we conduct a real data analysis using the S&P500, a benchmark index of the US stock market, and including various macroeconomic variables as predictors of S&P500 volatility. We compare the forecasting performance of the GARCH-X models estimated using the quasi maximum penalized likelihood approach with those done by the conventional maximum likelihood estimation to evaluate the effectiveness of the proposed approach.

The structure of this paper is organized as follows. In Sect. 2, we outline the methodology for extending Lasso and Ridge to the GARCH-X models, including a review of the concept and specifications of GARCH-X models, as well as estimation and model evaluation techniques. Section 3 presents the descriptive statistics of the data used in the study. Section 4 provides and discusses the main results of the empirical analysis, comparing the forecasting performance of the GARCH-X models based on the quasi maximum penalized likelihood and conventional maximum likelihood procedures. Finally, in Sect. 5, we provide concluding remarks and suggestions for future research.

2 Methodology

In this study, we consider several types of GARCH models to examine their performance in combination with the Lasso and Ridge penalties. Specifically, we utilize the standard GARCH model specification, as well as the EGARCH model introduced by Nelson [17] and the GJR-GARCH model proposed by Glosten, Jagannathan, and Runkle [7]. For ease of comparison, we focus on the GARCH(1,1) specification. By evaluating the performance of these different models under different penalty schemes, we can gain an insight into the effectiveness of the Lasso and Ridge approaches for variable selection and dealing with multicollinearity in GARCH-X models.

2.1 GARCH-X-Type Models

GARCH models are widely used in financial time series analysis to model asset volatility. Bollerslev [4] introduced the GARCH(p,q) model as a generalization of the ARCH(q) process to address the issue of long memory. The model is expressed by a set of conditional variance equations that depend on the past squared errors and the past conditional variances. The GARCH(1,1) specification is commonly used for comparison purpose in empirical studies. It can be expressed as follows:

$$\sigma_t^2 = \alpha_0 + \sum_{i=1}^{q} \alpha_i \varepsilon_{t-i}^2 + \sum_{j=1}^{p} \beta_j \sigma_{t-j}^2 + \sum_{k=1}^{K} \theta_k x_{k,t}^2 \tag{1}$$

where $\varepsilon_t = y_t - u$ is the error term and y_t is the observed time series variable. The estimation of parameters in the GARCH-X models is subject to certain restrictions to ensure that the resulting conditional variances are positive and stationary. These constraints are $\alpha_0 > 0, \alpha_i \geq 0, i = 1, \ldots, q, \ \beta_j \geq 0, j = 1, \ldots, q, \theta_i \geq 0, k = 1, \ldots, K$. The incorporation of additional predictor variable $x_{k,t}$ can help to capture more information that may explain the volatility of the asset under investigation.

This paper also focuses on the EGARCH and GJR-GARCH models, which have been extended to capture several characteristics of financial time series, such as thick-tailed returns and volatility clustering, by introducing additional parameters that control the behavior of the conditional variance. The EGARCH model introduces an asymmetric effect of shocks on volatility, while the GJR GARCH model allows for an asymmetric response of volatility to negative and positive shocks. The E-GARCH-X model is written as follows.

$$\log \sigma_t = \alpha_0 + \sum_{i=1}^{q} \alpha_i \left[\left| \frac{\varepsilon_{-i}}{\sigma_{t-i}} \right| - \gamma_i \left| \frac{\varepsilon_{-i}}{\sigma_{t-i}} \right| \right] + \sum_{j=1}^{p} \beta_j \log \sigma_{t-j} + \sum_{k=1}^{K} \theta_k x_{k,t}^2 \qquad (2)$$

where γ_i is the symmetric effect. The GJR-GARCH model is designed to capture asymmetry in the ARCH process by including an additional term. This term is a function of the lagged squared residuals and an indicator function that takes the value of 1 if the residual is negative and 0 otherwise. By incorporating this term, the GJR-GARCH model can more accurately capture the asymmetric effects of news and events on asset volatility, including the impact of bad news versus good news. The GJR-GARCH-X model extends this further by incorporating additional predictor variables. Its equation can be represented as follows:

$$\sigma_t = \alpha_0 + \sum_{i=1}^{p} \alpha_i \varepsilon_{t-i}^2 + \sum_{i=1}^{q} \gamma_i \varepsilon_{t-i}^2 I_{t-i} + \sum_{j=1}^{p} \beta_j \sigma_{t-j} + + \sum_{k=1}^{K} \theta_k x_{k,t}^2 \qquad (3)$$

where $I_{t-i} = 1$ if $\varepsilon_{t-1} < 0$ and 0 otherwise. In the GJR-GARCH model, negative or positive values of ε_{t-1} indicate bad or good news.

2.2 Estimation

To simplify the estimation of the QMLE, the exogenous variables are assumed to be stationary, which avoids the need for further restrictions on the dynamics of $x_{k,t}$ [6]. Furthermore, we require that the exogenous variables be independent of both the error term and the lagged conditional variance to avoid simultaneity biases in the estimation of the GARCH-X-type models $\left(E \left(\varepsilon_t \mid x_{k,t-1} \right) = 0 \text{ and } E \left(\varepsilon_t^2 \mid x_{k,t-1} \right) = 1 \right)$. We assume that the exogenous variable follows a specific distribution, typically the normal distribution $\left(\varepsilon_t \sim N \left(0, \sigma_t^2 \right) \right)$. This assumption simplifies the estimation of the QMLE, and the resulting log-likelihood function can be written as follows

$$L_n(\phi, \theta) = \sum_{t=1}^{T} \frac{1}{2} \left[-\log 2\pi - \log(\sigma_t) - \frac{(y_t - u)^2}{\sigma_t} \right] \qquad (4)$$

Note that σ_t is the conditional variance of the GARCH-X-type models in Eqs.(1-3) and $\phi = \{\alpha, \beta, u, \gamma\}$. Then, we can add penalties of Lasso and Ridge to the QMLE to account for potential overfitting and improve the model's predictive performance. Lasso and Ridge are regularization techniques that can be used to shrink the coefficients of the exogenous variables towards zero. The penalized QMLE can be defined as follows:

Log Penalized Ridge Likelihood

$$\tilde{L}_{\text{Ridge}}(\Theta) = L_n(\phi) + \sum_{i=1}^{d} \lambda \theta^2 \tag{5}$$

where $\Theta = (\phi, \theta)$ and λ is the regularization parameter. When the penalty parameter in ridge regression takes large values, the optimization function penalizes the coefficients by shrinking them towards zero. This helps to reduce problems caused by the complexity of the model and multicollinearity among the predictor variables.

Log Penalized Lasso Likelihood. Lasso regression is similar to ridge regression, but it has the additional benefit of aiding in feature selection as well as reducing overfitting. In Lasso regression, the penalty term is the absolute value of the magnitude of the coefficients. This penalty term encourages smaller coefficients and can set some coefficients to exactly zero, effectively eliminating those variables from the model. This makes Lasso regression a useful tool for identifying the most important predictors in a model.

$$\tilde{L}_{\text{Lasso}}(\Theta) = L_n(\phi) + \sum_{i=1}^{d} \lambda |\theta_i| \tag{6}$$

2.3   Model Evaluation

To evaluate the performance of the estimators, we compare the forecasting accuracy of the GARCH-X type models. A well-estimated model should yield more accurate forecasts. Therefore, it is reasonable to compare the performance of the penalized QMLE with the conventional QMLE using forecasting criteria. In this study, we employ three different loss functions as forecasting criteria:

1. Mean absolute percentage error (MAPE)

$$MAPE = \frac{1}{N} \sum_{i=1}^{N_i} \left| \frac{\sigma_i - \sigma_i}{\sigma_i} \right| \tag{7}$$

2. Root-mean-square error (RMSE)

$$RMSE = \sqrt{\frac{1}{N} \sum_{i=1}^{N} (\sigma_i - \sigma_i)^2} \tag{8}$$

3. Mean absolute error (MAE)

$$MAE = \frac{1}{N} \sum_{i=1}^{N} |\sigma_i - \sigma_i| \tag{9}$$

where $\hat{\sigma}_i$ is the estimated forecasting volatility and σ_i is the realized volatility. Note that the realized volatility is measured by $\sigma_i = |y_i - E(y_i)|$. N is the number of out-of-sample data. We divided the dataset for each variable into two subsets: in-sample and out-of-sample. The in-sample subset includes data from 3rd January 2000 to 31st December 2015 and is used for model training. The out-of-sample subset includes data from 30th June 2016 to 31st December 2019 and is used for testing the model's prediction ability.

3   Data Description

This study utilizes a time series dataset spanning from January 2000 to September 2019, consisting of monthly closing prices retrieved from the Thomson Reuters database. To make the data mathematically convenient, we transform the price into a log return. The dataset structure used in this research is presented in Table 1, which shows the descriptive statistics of the variables. To investigate the factors affecting the volatility of S&P500 index (y), we consider nine input variables, including Crude oil WIT (x_1), Crude oil Brent (x_2), Gold (x_3), Silver (x_4), Platinum (x_5), Consumer Price Index (x_6), Consumer Confidence Index (x_7), money supply M2 (x_8), and Unemployment Rate (x_9). Interestingly, we observe that the mean value of all variables is very close to zero, except for the Consumer Confidence Index and Unemployment Rate which are negative, indicating a slight difference from the other variables.

Table 1. Descriptive statistics

(2020/01/01-2022/01/01)	Baseline	Min-Variance	A2C	PPO	DDPG	SAC	TD3
Cumulative returns	3.60%	−6.16%	9.77%	9.66%	17.58%	10.50%	13.34%
Annual return	1.87%	−3.25%	4.97%	4.92%	8.80%	5.33%	6.74%
Annual volatility	22.79%	20.82%	25.75%	26.00%	25.75%	25.46%	26.08%
Sharpe ratio	0.20	−0.05	0.32	0.32	0.46	0.33	0.38
Max drawdown	−35.99%	−36.10%	−39.21%	−38.04%	−37.11%	−36.28%	−38.02%
Daily value at risk	−2.85%	−2.63%	−3.21%	−3.24%	−3.20%	−3.17%	−3.25%

Note: MBF is Minimum Bayes Factor.

In addition, the data series exhibit non-normality with negative skewness and leptokurtic kurtosis due to values being higher than three. To confirm the normal distribution property, we conducted the Jarque-Bera (J-B) test, which is a statistical inference test. According to Maneejuk and Yamaka (2020), we used the Minimum Bayes Factor (MBF) for null hypothesis tests. MBF values between 1 and 1/3 are considered weak evidence, between 1/3 and 1/10 are moderate evidence, between 1/10 and 1/30 are strong evidence, between 1/100 and 1/300 are robust evidence, and if the MBF is less than 1/300, it is considered decisive. Our J-B test resulted in the MBF values close to zero, indicating decisive evidence for non-normal distribution.

4   Empirical Results

In this section, we employ the GARCH-X models (GARCH-X, EGARCH-X, and GJR-GARCH-X) with QMLE to predict the volatility of the S&P500. We then explore the effectiveness of Lasso and Ridge penalty methods in forecasting volatility, and analyze the predictability of each feasible combination of macroeconomic variables to identify factors influencing stock market volatility. The estimated results from the in-sample data are reported in Table 2. The GARCH-X model results show that the impact of the exogenous regressors on the conditional variance of the model is weak, as the coefficient values are close to zero. However, the EGARCH-X model shows that all regressors have a significant influence on the volatility of S&P500, indicating their predictive role.

The use of penalized models, such as Ridge and Lasso, provides insightful information regarding the data. The coefficients obtained from the Ridge estimator differ significantly from those of the non-penalized models, and some regressors are removed from the model by the Lasso penalty. Notably, Gold and money supply are prominently removed from the volatility model. These findings may suggest that the penalty terms of the Ridge and Lasso estimators strongly impact the estimations, and their use may lead to different results compared to the non-penalized models.

Table 3 shows the out-of-sample forecasting performance of the three GARCH-X models (GARCH, EGARCH, and GJR-GARCH) without penalty and with Lasso and Ridge penalty. The evaluation criteria are Root Mean Square Error (RMSE), Mean Square Error (MSE), and Mean Absolute Error (MAE).

The traditional GARCH models without any penalty have the highest RMSE and MSE values across all three variants. The EGARCH model shows the best performance among the traditional models, with the lowest RMSE and MSE values. However, both Lasso and Ridge penalty models significantly improve the forecasting accuracy compared to the traditional models.

In terms of RMSE, the Lasso penalty models outperform all other models by a significant margin, with values as low as 0.0023 for EGARCH. This suggests that the Lasso penalty successfully identified the most relevant variables for predicting the volatility of S&P500, while discarding irrelevant ones. The Ridge penalty models also show improved performance over the traditional models, but with smaller improvements compared to the Lasso models.

Similar to RMSE, the MSE and MAE results also indicate that the Lasso models outperform the traditional and Ridge penalty models. In contrast, the Ridge penalty models provide only a moderate improvement in forecasting accuracy over the traditional models.

Overall, the results suggest that the Lasso penalty method is highly effective in improving the forecasting accuracy of GARCH-X models, while the Ridge penalty method provides only moderate improvements. Finally, we made comparisons with realized volatility (RV), as shown in Figs. 1, 2 and 3.

Figures 1, 2 and 3 provide a visual comparison between the classical GARCH-X-type models, the GARCH-X-type models with Lasso penalty, and the GARCH-X-type models with Ridge penalty, against the realized volatility. It is evident that the classical GARCH-X-type models exhibit a tendency to overestimate volatility across several periods. In contrast, the Lasso and Ridge models display an improved ability to capture the patterns of volatility, resulting in a closer proximity to the true volatility or RV (red line).

The observed out-of-sample forecasting performance suggests that the Lasso and Ridge regularization methods are effective in improving forecasting accuracy, which is likely attributed to their capacity to control model variance and enhance model stability. Furthermore, the close tracking of the black and yellow lines representing the Lasso and Ridge models, respectively, suggests that the two approaches provide similar performance in capturing actual volatility.

Table 2. Estimation results from in-sample data.

Parameter	GARCH-X	EGARCH-X	GJRGARCH-X	Ridge GARCH-X	Ridge EGARCH-X	Ridge GJR-GARCH-X	Lasso GARCH-X	Lasso EGARCH-X	Lasso GJR-GARCH-X
μ	0.0105	0.0103	0.0055	0.0015	0.0011	0.0277	0.0010	0.0011	0.0156
ω	0.0003	-2.0045	0.0006	0.0005	0.0001	0.0001	0.0001	0.0001	0.0013
α	0.6499	-0.1236	0.0062	0.1176	0.0772	0.0648	0.0324	0.0409	0.0001
β	0.4545	0.7081	0.3781	0.4798	0.0006	0.2721	0.5067	0.0009	0.4092
γ	-	0.5029	0.9995	-	0.0061	0.0033	-	0.0002	0.1204
θ_1	0.0001	-5.2458	0.0001	0.0336	0.1542	0.0076	0.0118	0.1167	0.0076
θ_2	0.0001	0.7160	0.0001	0.0124	0.1038	0.1416	0.0106	0.0507	0.0136
θ_3	0.0001	6.6544	0.0001	0.0755	0.0096	0.0196			
θ_4	0.0001	3.4554	0.0001	0.0726	0.6874	0.0336	0.0428	-0.5435	-0.0140
θ_5	0.0001	0.8550	0.0001	0.0472	0.1933	0.0245	0.0857	0.2044	
θ_6	0.0001	0.7848	0.0001	0.0494	0.1128	0.0068	0.0330	0.0989	-0.0463
θ_7	0.0001	0.9153	0.0001	0.1242	0.0918	0.1528	0.0530	-0.0368	0.0653
θ_8	0.0001	1.6863	0.0001	0.0396	0.0645	0.1536			
θ_9	0.0001	0.3112	0.0001	0.0620	0.1336	0.0094		0.1321	

Table 3. Out-of-sample forecasting comparison.

	RMSE		
	GARCH	EGARCH	GJRGARCH
Traditional	0.0577	0.0479	0.0586
Lasso	0.0033	0.0023	0.0034
Ridge	0.0498	0.0410	0.0486
	MSE		
	GARCH	EGARCH	GJRGARCH
Traditional	0.0080	0.0408	0.0088
Lasso	0.0001	0.0017	0.0001
Ridge	0.0055	0.0316	0.0056
	MAE		
	GARCH	EGARCH	GJRGARCH
Traditional	0.0202	0.0554	0.0205
Lasso	0.0004	0.0031	0.0004
Ridge	0.0146	0.0425	0.0150

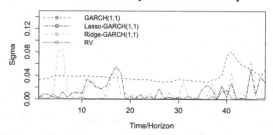

Fig. 1. The out-of-sample forecasts of GARCH models, including traditional, Lasso, and Ridge, compared to the realized volatility.

These findings suggest that incorporating penalized models, specifically Lasso and Ridge, into the GARCH framework can result in improved forecasting performance, particularly for the volatility of the S&P500. Future studies may consider the application of other regularization methods and alternative specifications of the GARCH model to further enhance forecasting accuracy. Moreover, extending the analysis to other stock markets may yield valuable insights into the predictive ability of macroeconomic variables across diverse contexts.

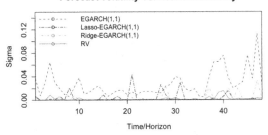

Fig. 2. The out-of-sample forecasts of E-GARCH models, including traditional, Lasso, and Ridge, compared to the realized volatility.

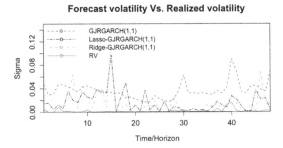

Fig. 3. The out-of-sample forecasts of GJR-GARCH models, including traditional, Lasso, and Ridge, compared to the realized volatility.

5   Conclusion

Volatility of the stock market is an issue of significant importance to investors and policymakers alike. It is a complex phenomenon that can be influenced by a variety of factors, including macroeconomic variables. GARCH models are commonly used to forecast market volatility, but incorporating macroeconomic variables in these models can be challenging due to the curse of dimensionality. However, the use of penalized regression techniques such as Ridge and Lasso can mitigate this issue.

This study aims to investigate the effectiveness of penalized GARCH models in predicting stock market volatility given various macroeconomic variables as the determinants. Consequently, it also seeks to identify the key macro-factors that affect stock market volatility. The results of the study demonstrate the use of Ridge and Lasso penalties further improves the accuracy of the model's forecasts. Lasso, in particular, is an effective tool for identifying the most relevant variables.

The study's findings indicate that the stock market is influenced by various macroeconomic variables, including oil prices, exchange rates, and interest rates. However, it is important to note that the analysis is based on a single market (the S&P 500), and therefore the results may not be generalizable to other markets. Additionally, the study only considers a limited set of macroeconomic variables, and there may be other variables that have an effect on stock market volatility. Finally, it is worth noting that the analysis is based on historical data, while the relationship between macroeconomic variables and stock market volatility may evolve over time. Future research could focus on expanding the analysis to include more markets and a wider range of macroeconomic variables. It would also be interesting to investigate the impact of incorporating more complex machine learning techniques, such as neural networks, in forecasting stock market volatility [12]. Finally, a real-time analysis of the relationship between macroeconomic variables and stock market volatility could provide valuable insights for investors and policymakers.

Acknowledgements. This research work was partially supported by Chiang Mai University.

References

1. Albu, L.L., Lupu, R., Călin, A.C.: Stock market asymmetric volatility and macroeconomic dynamics in Central and Eastern Europe. Procedia Econ. Finan. **22**, 560–567 (2015)
2. Arouri, M., Estay, C., Rault, C., Roubaud, D.: Economic policy uncertainty and stock markets: long-run evidence from the US. Financ. Res. Lett. **18**, 136–141 (2016)
3. Bahloul, W., Gupta, R.: Impact of macroeconomic news surprises and uncertainty for major economies on returns and volatility of oil futures. Int. Econ. **156**, 247–253 (2018)
4. Bollerslev, T.: Generalized autoregressive conditional heteroskedasticity. J. Econom. **31**(3), 307–327 (1986)
5. Byun, S.J., Cho, H.: Forecasting carbon futures volatility using GARCH models with energy volatilities. Energy Econ. **40**, 207–221 (2013)
6. Dittmann, I., Granger, C.W.: Properties of nonlinear transformations of fractionally integrated processes. J. Econom. **110**(2), 113–133 (2002)
7. Glosten, L.R., Jagannathan, R., Runkle, D.E.: On the relation between the expected value and the volatility of the nominal excess return on stocks. J. Financ. **48**(5), 1779–1801 (1993)
8. Poon, S.H., Granger, C.W.: Forecasting volatility in financial markets: a review. J. Econ. Lit. **41**(2), 478–539 (2003)
9. Hammoudeh, S., Dibooglu, S., Aleisa, E.: Relationships among US oil prices and oil industry equity indices. Int. Rev. Econ. Financ. **13**(4), 427–453 (2004)
10. Hammoudeh, S., Yuan, Y.: Metal volatility in presence of oil and interest rate shocks. Energy Econ. **30**(2), 606–620 (2008)
11. Han, H., Kristensen, D.: Asymptotic theory for the QMLE in GARCH-X models with stationary and nonstationary covariates. J. Bus. Econ. Stat. **32**(3), 416–429 (2014)
12. Liao, R., Yamaka, W., Sriboonchitta, S.: Exchange rate volatility forecasting by hybrid neural network Markov switching Beta-t-EGARCH. IEEE Access **8**, 207563–207574 (2020)
13. Maneejuk, P., Yamaka, W.: Significance test for linear regression: how to test without P-values? J. Appl. Stat. **48**(5), 827–845 (2021)
14. McDonald, G.C.: Ridge regression. Wiley Interdiscip. Rev. Comput. Stat. **1**(1), 93–100 (2009)
15. Medeiros, M.C., Mendes, E.F.: Adaptive LASSO estimation for ARDL models with GARCH innovations. Economet. Rev. **36**(6–9), 622–637 (2017)
16. Narayan, P.K., Narayan, S.: Modelling oil price volatility. Energy Policy **35**(12), 6549–6553 (2007)
17. Nelson, D.B.: Conditional heteroskedasticity in asset returns: a new approach. Econom. J. Econ. Soc. **59**, 347–370 (1991)
18. Roubaud, D., Arouri, M.: Oil prices, exchange rates and stock markets under uncertainty and regime-switching. Finan. Res. Lett. **27**, 28–33 (2018)

19. Sadorsky, P.: Modeling and forecasting petroleum futures volatility. Energy Econ. **28**(4), 467–488 (2006)
20. Tully, E., Lucey, B.M.: A power GARCH examination of the gold market. Res. Int. Bus. Financ. **21**(2), 316–325 (2007)
21. Tibshirani, R.: Regression shrinkage and selection via the lasso. J. R. Stat. Soc. Ser. B Stat Methodol. **58**(1), 267–288 (1996)
22. Xu, Q., Bo, Z., Jiang, C., Liu, Y.: Does Google search index really help predicting stock market volatility? Evidence from a modified mixed data sampling model on volatility. Knowl. Based Syst. **166**, 170–185 (2019)
23. Yamaka, W.: Sparse estimations in kink regression model. Soft. Comput. **25**, 7825–7838 (2021)
24. Yoon, Y.J., Lee, S., Lee, T.: Adaptive LASSO for linear regression models with ARMA-GARCH errors. Commun. Stat. Simul. Comput. **46**(5), 3479–3490 (2017)
25. Zhou, Z., Fu, Z., Jiang, Y., Zeng, X., Lin, L.: Can economic policy uncertainty predict exchange rate volatility? New evidence from the GARCH-MIDAS model. Financ. Res. Lett. **34**, 101258 (2020)

Mean-Variance Portfolio Allocation Using ARMA-GARCH-Stable and Artificial Neural Network Models

Nguyen T. Anh[1,2], Mai D. Lam[3], and Bao Q. Ta[1,2(✉)]

[1] Department of Mathematics, International University, Ho Chi Minh City, Vietnam
baotq@hcmiu.edu.vn
[2] Vietnam National University, Ho Chi Minh City, Vietnam
[3] Faculty of State Management of Economic Affairs and Public Finance, National Academy of Public Administration, Hanoi, Vietnam

Abstract. Optimal portfolio allocation is one of the most important practical problems in financial engineering. The portfolio optimization problem and traditional approach were first initiated by Markowitz in which investors construct a portfolio with a minimum risk for a specified return. The most critical assumption of the Markowitz model is that returns are assumed to be normally distributed. In practice, returns usually do not follow a normal distribution and have heavy tails. Furthermore, forecasting returns and volatilities of assets are essential tasks in the financial market and portfolio optimization. In this paper, we use stable distribution to capture non-normal and skewness properties of asset returns and utilize the traditional econometric ARMA-GARCH model incorporating the Artificial Neural Networks model to predict returns and volatilities of assets on the Vietnam stock market. We then construct an optimal portfolio selection problem. Our main results indicate that the proposed model outperforms the market index (VN30) and traditional econometric models in terms of both risk and return.

Keywords: ARMA-GARCH · Stable distribution · Artificial Neural Network · Portfolio optimization

1 Introduction

Portfolio allocation has become increasingly popular among academic researchers and investors due to its practicality and effectiveness, especially during periods of crisis. The foundation of portfolio optimization is traced back to the pioneering work of Markowitz in 1952 [10], where investors strive not only to maximize portfolio returns but also to minimize the associated risks. Later many studies have tried to improve the model by changing the risk-return measures. For instance, Sharpe proposed maximum of the Sharpe ratio, which is the ratio between mean and standard deviation of portfolio's return.

© The Author(s), under exclusive license to Springer Nature Switzerland AG 2023
V.-N. Huynh et al. (Eds.): IUKM 2023, LNAI 14375, pp. 177–186, 2023.
https://doi.org/10.1007/978-3-031-46775-2_16

A crucial work in the portfolio optimization is the prediction of future returns and their volatilities. There are several common statistical models used to forecast returns and volatilities such as traditional ARMA and GARCH models [2,6]. These econometrics models are utilized for risk management and portfolio optimization, see, e.g., [5,12]. Due to the complex non-linear dependence structure of financial time series these models may not capture this dependence structure. So these tasks are still challenging. The recent advances in machine learning algorithms provide us powerful tools and flexible techniques to address and overcome these limitations, see, e.g. [15,17]. However, traditional econometric models still play an important role in prediction. In fact, some empirical research show that hybrid models (a combination of econometric models and Machine learning models) give better performance than single models [3,7,16].

It is observed that asset returns do not follow normal distribution. The early research of Mandelbrot [9] showed that α-stable distribution is more suitable to model distributions of asset returns. The α-stable distributions is a family of distributions with four parameters $(\alpha, \beta, \gamma, \delta)$ that include Gaussian distributions as a special case when $\alpha = 2, \beta = 0$. These distributions characterize skewness and heavy tails. Consequently, they are more useful to model realistic behavior of returns. Recently, a number of authors (see, e.g., [8,11,14]) utilize this family of distributions to construct optimal portfolios. In this paper we propose hybrid ARMA-GARCH and α-stable distribution and Artificial Neural Network models to forecast asset returns and volatilities, and then construct optimal portfolio allocation. The findings show that the combination of traditional ARMA-GARCH and α-stable distribution and Artificial Neural Network models outperforms the Vietnam market index in terms of both risk and return.

The paper is organized as follows. In the next Sect. 2 we describe formal statement for optimal portfolio selection. In Sect. 3 we give a brief introduction to ARMA-GARCH model, and α-stable distribution and Artificial Neural Network. Section 4 we describe data, and present the empirical results from Vietnam stock market with the 30 largest capitalization stocks. Finally, Sect. 5 concludes the paper and gives future studies.

2   Optimal Portfolio Allocation

In the classical Markowitz model, there are two crucial factors: the expected return and the variance of the portfolio. The portfolio's expected return quantifies its potential profitability, while the variance measures the level of risk. A portfolio with a high variance means that there exists significant fluctuations, which can lead to cause a big loss. So prudent investors seek to minimize the risk measure of their portfolios. In this paper we focus on the risk minimization problem.

Let P_t is the price of an asset S at time t. The log-return of the asset S is defined by $r_t = \log(P_t/P_{t-1})$. Consider an N-dimensional portfolio consisting of asset returns r_{tj} with expected returns $\mu_{tj} = \mathbb{E}(r_{tj}), i = 1, \ldots, N$. Denote $r_t = (r_{t1}, \ldots, r_{tN})$ the vector of returns and $\boldsymbol{\mu}_t = (\mu_{t1}, \ldots, \mu_{tN})$ the vector of expected returns, and μ_0 a fixed level of expected return, and $\boldsymbol{w}_t = (w_{t1}, \cdots, w_{tN})$ the vector of asset weights. The mean-variance portfolio optimization problem is formulated as follows

$$
\begin{aligned}
\underset{\boldsymbol{w}_t}{\text{minimize}} \quad & \boldsymbol{w}_t^T \Sigma \boldsymbol{w}_t \\
\text{subject to} \quad & \boldsymbol{w}_t^T \boldsymbol{\mu}_t \geq \mu_0, \\
& \boldsymbol{w}_t^T \boldsymbol{1} = 1, \\
& w_{tj} \geq 0, j = 1, \ldots, N.
\end{aligned}
\tag{1}
$$

where Σ is the covariance matrix of returns and $\boldsymbol{1} := (1, 1, \ldots, 1) \in \mathbb{R}^N$.

To evaluate the efficiency of the models, the Sharpe ratio and Sortino ratio are commonly utilized as a measures to assess the performance of an investment while taking into account its associated risk. Denote r_p, σ_p the return and standard deviation of the portfolio. The Sharpe ratio is defined as

$$
SR_p := \frac{\mathbb{E}(r_p) - r_f}{\sigma_p}.
\tag{2}
$$

Besides Sharpe ratio one can also use the Sortino ratio which is defined as

$$
Sp := \frac{\mathbb{E}(r_p) - r_f}{\sigma_d},
\tag{3}
$$

where σ_d is standard deviation of the negative asset return, it is the so-called the downside risk.

3   The Models

3.1   ARMA-GARCH Model

Consider a log-return r_t of a stock price. Denote $\mathcal{F}_t$ the information set available up to time t generated by the process r_t. From the efficient markets hypothesis, the return r_t can be decomposed as

$$
r_t = \mu_t + a_t
\tag{4}
$$

where the first term μ_t is the conditional mean given $\mathcal{F}_{t-1}$, i.e., $\mu_t = \mathbb{E}(r_t|\mathcal{F}_{t-1})$, and the second term a_t is the innovation component, it is assumed to be a random variable with zero mean and has conditional variance

$$
Var(a_t|\mathcal{F}_{t-1}) = Var(r_t|\mathcal{F}_{t-1}) = \sigma_t.
\tag{5}
$$

The term σ_t is called volatility of the return and it is time-varying stochastic. The innovation component $a_t = \sigma_t Z_t$, where Z_t are i.i.d random variables with $\mathbb{E}(Z_t) = 0$ and $Var(Z_t) = 1$. The random variables Z_t are called standardized residuals. From Eqs. (4) and (5), we see that μ_t is usually modeled by ARMA(p, q) models. The ARMA models are commonly used in financial econometrics to capture the autocorrelation and moving average components of asset returns. The volatility σ_t is often modeled by GARCH(L_1, L_2) models which are particularly useful for modeling the volatility clustering and time-varying volatility observed in financial data. In ARMA and GARCH models, the numbers p, q, and L_1, L_2 plays as lagged forecast terms in the ARMA and GARCH models, see [13] for more detailed treatments of these models. In practical applications it is shown that models with smaller order is sufficient to model the data. So in this paper we use ARMA(1,1)-GARCH(1,1) to model our datasets. More precisely, we have mathematical expressions of ARMA(1,1)-GARCH(1,1) model.

$$r_t = \phi_0 + \phi_1 r_{t-1} + \theta_1 a_{t-1} + a_t, \tag{6}$$
$$a_t = \sigma_t Z_t, \tag{7}$$
$$Z_t \sim i.i.d, \tag{8}$$
$$\sigma_t^2 = \omega + \alpha_1 a_{t-1}^2 + \beta_1 \sigma_{t-1}^2, \tag{9}$$

where ϕ_0 is constant, the coefficients satisfy the conditions: $|\phi_1| < 1, |\theta_1| < 1$, $\phi + \theta \neq 0$, and $\omega > 0, \alpha_1 \geq 0, \beta_1 \geq 0, \alpha_1 + \beta_1 < 1$.

3.2 Stable Distributions

In practical application most data sets in finance and insurance are skewness and heavy tails. So non-normal distribution models are proposed for modeling these kinds of data sets. There are some proposed distributions such as mixture distributions, t-distributions, extreme value distributions, and stable distribution. However, stable distributions possess mathematically appealing properties, for instance the property of stability, that makes them a practical substitute for normal distributions in many financial engineering and risk management problems. Mandelbrot [9], and Fama[4] showed that stable distributions are suitable models for accurately capturing the heavy-tailed (leptokurtic) returns.

Let X be a non-degenerate random variable and $X_1, \ldots, X_n$ be a independent and identically distributed copies of X. The random variable X is said to be stable if and only if there exists $c_n > 0$ and d_n such that

$$X_1 + \cdots + X_n = c_n X + d_n, \quad n > 1$$

Now denote $r_t(k) = \log(P_t/P_{t-k})$ the gross log-return over the most recent k periods, where P_t is the stock price at time t. We have the decomposition

$$r_t + r_{t-1} + \cdots + r_{t-k+1} = r_t(k)$$

from which we see that the stable distributions are suitable to approximate for distributions of returns.

In general, stable distributions are determined by four parameters which are stable index or characteristic exponent α, skewness parameter β, scale parameter γ and location parameter δ. We say that a random variable X has stable law

$S(\alpha, \beta, \gamma, \delta)$ if and only if it has the characteristic function:

$$\mathbb{E}(\exp(itX)) = \begin{cases} \exp\left(-\gamma^\alpha |t|\left[1 - \beta \tan(\pi\alpha/2)sign(t)\right] + i\delta t\right) & \alpha \neq 1 \\ \exp\left(-\gamma |t|[1 - \beta\frac{2}{\pi}sign(t)\log|t|] + i\delta t\right) & \alpha = 1 \end{cases} \quad (10)$$

The stable index $\alpha \in (0,2]$ controls the tail behavior, hence it can be seen as the degree of leptokurtosis. It is seen that when $\alpha = 2$ then the stable distribution becomes to normal distribution with variance $2\gamma^2$. The skewness parameter $\beta \in [-1,1]$ determines the density's skewness.

3.3 Artificial Neural Networks

Artificial neural networks (ANN) is a computational mathematical model inspired from biological neural networks. One of the most powers of ANN is that it can perform highly complex and nonlinear processing. In financial applications, many datasets have fat tails and contain non-linear components. So some regression methods and ARMA models fail to predict accurately financial time series. The ANN can incorporate different models including linear regression model, binary probit model and various others. By utilizing neural network technique, we can potentially capture complex patterns and nonlinear relationships that may be challenging for traditional econometric models. Recently, ANN models become one of the most attractive models used in finance such as portfolio management, risk management, credit rating, forecasting exchange rates, and predicting stock values.

The ANN model consists of layers including input, output layers, and hidden layers, we refer to Aggarwal [1] for more detail treatments of Neural Networks and Deep Learning. The mathematical expression of ANN model with single hidden layer is given as follows

$$Y = \sum_{j=1}^{q} a_j g(\omega_{0j} + \sum_{i=1}^{p} \omega_{ij} X_i) \quad (11)$$

where X_i represents the information (input) variables that the neuron receives from other neurons, coefficients a_j and ω_{ij} are connection weight between layers, the parameters p and q are the number of input nodes and the number of hidden nodes of the model. Furthermore the function g is activation function (transfer function) of the hidden layer. Some popular types of activation functions are: sigmoid (logistic) $g(x) = 1/(1 + e^{-x})$, Hyperbolic Tangent (Tanh) $g(x) = (e^x - e^{-x})/(e^x + e^{-x})$, and Rectified Linear Unit (ReLU) $g(x) = \max(0, x)$.

4 Data Analysis and Findings

4.1 Data Analysis

In this paper we construct a portfolio consisting 21 stocks from VNindex 30 that includes the largest amount of liquidity transaction and capitalization in the

Vietnam stock market. The weekly stock prices are collected from the beginning of 2010 up to 01/03/2023. The descriptive statistics of the stock returns are presented in the following Table 1.

Table 1. Descriptive statistics

Stocks	Mean	STD	Max	Min	Skewness	Kurtosis
ACB	0.19	3.93	18.11	−24.86	−0.29	4.37
BID	0.34	5.49	29.15	−24.01	0.03	3.98
BVH	0.13	6.35	30.60	−26.26	0.11	2.64
CTG	0.17	4.88	21.42	−19.82	0.17	1.88
FPT	0.35	3.66	21.28	−16.57	−0.08	3.47
GAS	0.28	4.96	20.28	−27.36	−0.69	4.50
HPG	0.37	5.13	24.00	−18.61	−0.02	1.77
KDH	0.12	4.80	25.94	−19.99	0.08	3.02
MBB	0.32	4.18	16.50	−19.69	−0.23	2.73
MSN	0.22	5.45	25.66	−21.93	0.13	2.59
MWG	0.36	5.40	21.78	−36.88	−0.75	7.18
NVL	−0.22	6.13	20.62	−35.86	−2.42	14.20
PDR	0.06	5.76	22.19	−35.86	−1.12	8.52
PLX	−0.01	5.10	14.56	−27.91	−0.85	3.75
SAB	0.11	4.77	33.70	−21.51	0.84	9.60
SSI	0.11	5.43	18.62	−25.75	−0.21	2.05
STB	0.13	4.91	21.21	−22.00	0.05	2.94
VCB	0.30	4.43	16.38	−14.75	0.17	1.32
VIC	0.28	4.57	22.13	−20.08	0.16	3.88
VJC	0.12	4.17	14.52	−19.71	−0.34	3.21
VNM	0.34	3.32	12.52	−14.26	0.16	2.44

We see that almost skewness are negative which means that large negative returns occur more often than large positive ones. Furthermore, most stocks have positive excess kurtosis, and, hence returns have leptokurtic distribution which indicates that stocks in the portfolio are more risky. This indicates that stable distributions are well-suited for modeling the tails of the return distribution. Additionally, it suggests that we need to address the risk minimization problem within the portfolio.

Next we use the augmented Dickey-Fuller test to check the stationary of stocks. The following results show that stock returns in the portfolio are stationary (Table 2).

Table 2. ADF test statistics

Stocks	ACB	BID	BVH	CTG	FPT	GAS	HPG	KDH
ADF	−16.5	−14.0	−19.62	−17.68	−19.38	−15.81	−17.28	−18.37
1% level	(0.00)	(0.00)	(0.00)	(0.00)	(0.00)	(0.00)	(0.00)	(0.00)
Stocks	MBB	MSN	MWG	NVL	PDR	PLX	SAB	SSI
ADF	−16.47	−19.16	−13.72	−7.35	−15.81	−12.3	−13.04	−16.52
1% level	(0.00)	(0.00)	(0.00)	(0.00)	(0.00)	(0.00)	(0.00)	(0.00)
Stocks	STB	VCB	VIC	VJC	VNM			
ADF	−17.82	−18.26	−18.66	−12.75	−17.99			
1% level	(0.00)	(0.00)	(0.00)	(0.00)	(0.00)			

4.2 Empirical Findings

We introduce a short description of the steps to implement ANN and ARMA-GARCH-Stable models for optimizing our portfolio. We use rolling window method for the out-of-sample period. Notice that all of the parameters of the models at each iteration are re-estimated. The procedure is outlined as follows. We use ARMA-GARCH model to fit conditional mean and volatility of each return, then the standardize residuals are model by stable distributions. From which, we simulate M scenarios for future values (one week) of the standardize residuals. Now using ANN with the simulated data to predict returns, and then construct the portfolio. Iterate these steps for next weeks, we obtain the cumulative portfolio returns presented in the following figure.

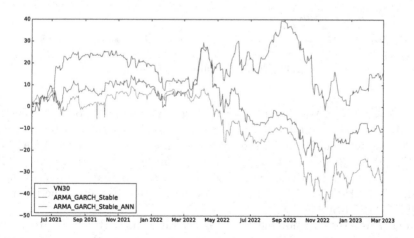

Fig. 1. The performance cumulative returns of the portfolio

The results presented in Fig. 1 show that both model ARMA-GARCH stable and ARMA-GARCH-Stable-ANN give better performances as compare with

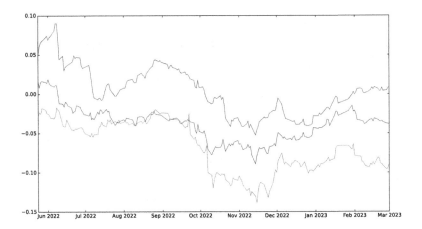

Fig. 2. The Sharpe ratios of the models

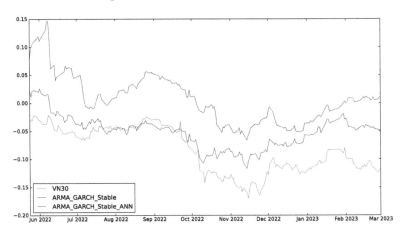

Fig. 3. The Sortio ratios of the models

market index VN30. We also see that the active return of the portfolio using the only ARMA-GARCH-Stable model is around −10%, but the benchmark drops by around −30% which implies the excess return is 15%. For the hybrid ARMA-GARCH-Stable-ANN model, the active return of the portfolio is all over +15%, hence the excess return is around 45%. This indicates that the hybrid model ARMA-GARCH-Statle-ANN are far more better than the econometric model ARMA-GARCH-Stable.

Next we use the common measures Sharpe ratio and Sortino ratio to evaluate the performance of the models. The results are presented in the following figures.

The Fig. 3 and Fig. 2 clearly illustrate that both proposed models outperform when solely employing the combination of econometric models. However, the results also demonstrate that incorporating machine learning models with econometric models yields superior performance compared to solely relying on traditional statistical and econometric models.

5   Conclusion

In this paper, we construct a portfolio from the Vietnam stock market VN30 and utilize the traditional econometric ARMA-GARCH model and statistical model stable distribution incorporating the state-of-the-art machine learning model to forecast and simulate one-week-a-head returns then solve mean-variance portfolio problem. The empirical results demonstrate that the ARMA-GARCH incorporating with Stable distribution provides effective optimization for portfolio allocation, even without a machine learning model. However, when combined with Artificial Neural Network models, the hybrid model outperforms traditional econometric models.

Acknowledgements. We would like to thank Professor Hung T. Nguyen for his enthusiasm and for introducing new directions in the field of statistics. This research is funded by International University, VNU-HCM under grant number SV2022-MA-02.

References

1. Aggarwal, C.C.: Neural Networks and Deep Learning. Springer, Cham (2018). https://doi.org/10.1007/978-3-319-94463-0
2. Berger, T.: Forecasting value-at-risk using time varying copulas and EVT return distributions. Int. Econ. **133**, 93–106 (2013)
3. Hajizadeh, E., Seifi, A., Zarandi, M.H.F., Turksen, I.B.: A hybrid modeling approach for forecasting the volatility of S&P 500 index return. Exp. Syst. Appl. **39**, 431–436 (2012)
4. Farma, E.: Mandelbrot and the stable paretian hypothesis. J. Bus. **36**(4), 420–429 (1963)
5. Huang, C.-W., Hsu, C.-P.: Portfolio optimization with GARCH-EVT-Copula-CVaR models. Bank. Finan. Rev. **7**(1), 19–31 (2015)
6. Karmakar, M.: Dependence structure and portfolio risk in Indian foreign exchange market: a GARCH-EVT-Copula approach. Q. Rev. Econ. Finan. **64**, 275–291 (2017)
7. Kim, H.Y., Won, C.H.: Forecasting the volatility of stock price index: a hybrid model integrating LSTM with multiple GARCH-type models. Exp. Syst. Appl. **103**, 25–37 (2018)
8. Krezolek, D.: The application of alpha-stable distributions in portfolio selection problem-the case of metal market. Studia Ekonomiczne **247**, 57–68 (2015)
9. Mandelbrot, B.: The variation of certain speculative prices. J. Bus. **36**(4), 394–419 (1963)
10. Markowitz, H.: Portfolio selection. J. Finan. **7**(1), 77–91 (1952)
11. Rachev, S., Han, S.: Portfolio management with stable distributions. Math. Meth. Oper. Res. **51**, 341–352 (2000)
12. Sahamkhadam, M., Stephan, A., Östermark, R.: Portfolio optimization based on GARCH-EVT-Copula forecasting models. Int. J. Forecast. **34**(3), 497–506 (2018)

13. Stay, R.: Analysis of Financial Time Series, 3rd edn. Wiley (2010)
14. Vasiukevich, A., Pinsky, E.: Constructing portfolios using stable distributions: the case of S&P 500 sectors exchange-traded funds. Mach. Learn. Appl. **10**, 10034 (2022)
15. Yu, S.L., Li, Z.: Forecasting stock price index volatility with LSTM deep neural network. In: Tavana, M., Patnaik, S. (eds.) Recent Developments in Data Science and Business Analytics. SPBE, pp. 265–272. Springer, Cham (2018). https://doi.org/10.1007/978-3-319-72745-5_29
16. Zhang, G.P.: Time series forecasting using a hybrid ARIMA and neural network model. Neurocomputing **50**, 159–175 (2003)
17. Zhong, X., Enke, D.: Predicting the daily return direction of the stock market using hybrid machine learning algorithms. Finan. Innov. **5**(24), 1–20 (2019)

Predicting Maize Yields with Satellite Information

Singh Ratna[1], Ping-Yu Hsu[1], You-Sheng Shih[1], Ming-Shien Cheng[2(✉)], and Yu-Chun Chen[1]

[1] Department of Business Administration, National Central University, No.300, Jhongda Road, Jhongli City, Taoyuan County 32001, Taiwan
984401019@cc.ncu.edu.tw

[2] Department of Industrial Engineering and Management, Ming Chi University of Technology, No.84, Gongzhuan Road, Taishan District, New Taipei City 24301, Taiwan
mscheng@mail.mcut.edu.tw

Abstract. The United States seems to have become the primary source of global corn production and export, making corn production critical to the economic activities of many countries. Many previous studies provide yield forecasts. Ground-based telemetry via satellites has recently emerged and attempts to predict vegetation indices for yield. However, except vegetation index, we should know more about vegetation area and coverage for overall consideration. Therefore, this study uses four major corn-producing areas in the United States and related data for the past nine years for model training, including multivariate linear regression, partial least squares regression, stepwise regression, and Gaussian kernel support vector regression. The experimental results show that the support vector regression with Gaussian kernel (radial basis function kernel) performs the best, and the R^2 value reaches 0.94.

Keywords: Satellite Telemetry · Telemetry Area · Vegetation Index · Regression Analysis

1 Introduction

Remote sensing data from Gao & Anderson [1] and Sakamoto & Sensing [2] have proven to help estimate crop yields in recent decades. The regular and frequent observation of the Earth's surface by satellites can effectively capture crops' developmental stages and assess crop yields' variability. However, most remote sensing studies nowadays focus on remote images with coarse spatial resolution. Zhang, Zhang, & Sensing [3] applied the coarse remote sensing images (e.g., 250 m spatial resolution remote sensing images) to a small area with limitations. In contrast, Landsat-8 provides remote sensing images with a spatial resolution of 30 m every 16 days, which can be applied to crop yield observations over a small area. The overall yield can estimate from remote sensing data over a small area. The telemetry study by Johnson [4] focused on different crops, with all studies focusing more on the telemetry of corn, soybean, and wheat, which are geographically widespread and abundant. Crops not planted in large areas are more challenging to assess by remote sensing because of the spatial resolution of the waveband. The largest crop among all crops is corn, and the United States is the largest producer of corn, so this study focuses on the observation of corn in the United States. Regardless of the crop, a common theme in remote sensing research is converting multi-spectral channels from optical sensors into vegetation indices. The Standardized Difference Vegetation Index (SDVI) is the most common vegetation index, which analyzes vegetation changes throughout the crop growth period.

Most current crop telemetry studies focus on the relationship between vegetation index and yield. Huang, Wang, Dai, Han, & Sensing [5] have pointed out that the vegetation index alone cannot apply to predict yield in all regions, so more other predictor variables require to alleviate this problem. Few studies have mentioned areas as other predictor variables compared with other remote sensing studies that used meteorological and soil characteristics factors. This study assumes that the meteorological and soil changes will reflect the vegetation index. The crop yield will increase significantly in case of the vegetation is more luxuriant or larger planting area.

This study's purpose is to extract the required band information from the satellite by remote sensing data via calculating the band information into various vegetation indices (NDVI, EVI, ARVI) and link the remote sensing area and USDA statistical information for corn production by the state and year, the developing linear and nonlinear regression models with comparing the accuracy of various regression models to achieve the purpose for the forecasting by core products in the United States.

This paper organized as follow: (1) Introduction: Research background, motivation and purpose. (2) Related work: Review of scholars' research on optical satellites, vegetation index and the differences with other crop telemetry studies. (3) Research methodology: Content of research process in this study. (4) Result analysis: Experimental results and the discussion of the test results. (5) Conclusion and future research: The contribution of the study and possible future research direction is discussed.

2 Related Work

2.1 Optical Satellites

Roy et al. [6] have proposed that the primary mission of Landsat-8 is to provide resolution measurements of visible, near-infrared, shortwave, and thermal infrared to observe the Earth's terrestrial and polar spectra. Since the launch of the first Landsat satellite, Landsat-8 has continued the Landsat record for nearly four decades. Compared to the previous Landsat satellites, Landsat-8 carries two sensors, the Operational Land Imager (OLI) and the Thermal Infrared Sensor (TIRS). Landsat-7 has equipped with the Enhanced Thematic Mapper plus (ETM +) sensor. Landsat-8 enhanced performance and new spectral bands. The new spectral bands include a short-wavelength blue band, a cirrus band, and two thermal bands. The additional short wavelength blue band improves sensitivity to suspended material for aerosol characterization, while the additional cirrus band is for cirrus cloud detection. We will use these two bands in this study, and Landsat-8 is more suitable for this study than the previous Landsat series satellites.

Justice et al. [7] have pointed out that compared to the MODIS (Moderate-resolution Imaging Spectroradiometer) sensor, Landsat-8 provides up to 36 bands with wavelengths ranging from visible light to thermal infrared light and others. It has a wide range of wavebands. This study requires more precise spatial resolution to know the satisfactory crop yield in a specific area, so Landsat-8 is suitable for our requirements. Drusch et al. [8] have pointed out that Sentinel-2 has 12 bands for users to observe, but its resolution is 60m per pixel for the short wavelength blue band and the cirrus band; this required resolution re-processing may lead to data error for this study. Therefore, we choose the data source from Landsat-8 for this study.

2.2 Vegetation Index

In 1969, Jordan [9] proposed to combine the near-infrared and red spectra into a ratio, which was highly correlated with the Leaf Area Index (LAI) using regression analysis, and this ratio was the Simple Ratio (SR), which was the cornerstone of the subsequent development of various vegetation indices. Many different vegetation indices are available today, and the three vegetation indices used in this study are described below.

Normalized difference vegetation index (NDVI) is the most commonly used vegetation index for satellite telemetry and is a vegetation index proposed by Rouse [10] based on a simple ratio. It is calculated using the physical principle that vegetation reflects visible and near-infrared light. Healthy or lush vegetation absorbs most of the visible light that hits it and reflects most of the near-infrared light, while unhealthy or sparse vegetation reflects more visible and less near-infrared light.

According to the description above, lush vegetation has the property of reflecting less visible light and more near-infrared light., the standardized vegetation index is the near-infrared light reflection minus the visible light reflection divided by the near-infrared light reflection plus the visible light reflection, as shown in Eq. (1).

$$NDVI = \frac{NIR - RED}{NIR + RED} \tag{1}$$

In Eq. (1), NIR is the near-infrared reflectance, and RED is the red light reflectance. Calculating the NDVI for a given pixel produces a number ranging from $+1$ to -1. However, the NDVI value without green foliage will be close to zero. In the case of very heavy vegetation, the NDVI will be close to $+1$.

H. Q. Liu et al. [11] have proposed that the enhanced vegetation index (EVI) is effective in reducing soil as well as aerosol impacts, as shown in Eq. (2).

$$EVI = \frac{G(NIR - RED)}{(L + NIR + C_1 * RED - C_2 * BLUE)} \tag{2}$$

L in Eq. (2) is the soil conditioning factor, while C_1 and C_2 are the aerosol scattering coefficients using the blue band to correct the red band. A. Huete, Liu, Batchily, & Van Leeuwen [12] proposed commonly used coefficients of $L = 1, C_1 = 6, C_2 = 7.5$, and $G = 2.5$. Applying the commonly used coefficients can be written as the following Equation: Eq. (3) is also used in this study.

$$EVI = \frac{2.5(NIR - RED)}{(1 + NIR + 6RED - 7.5BLUE)} \tag{3}$$

Kaufman, Tanre, & Sensing [13] mentioned that the Atmospherically Resistant Vegetation Index (ARVI) was developed to reduce the dependence of the vegetation index NDVI on atmospheric properties, using the combined reflectance of blue and red light have a self-correcting property for atmospheric effects, as shown in Eq. (4).

In Eq. (5) is for the combined reflectance, where γ is the atmospheric self-correction factor depending on the aerosol type. When γ is zero, the Equation for ARVI at this time will equal NDVI. When the vegetation cover is sparse and makes atmospheric data unknown, the value of γ usually set to 1 for better adjustment. We set γ to 1, and the resulting ARVI equation is as Eq. (6) below.

$$ARVI = \frac{NIR - RB}{NIR + RB} \tag{4}$$

$$RB = R - \gamma(B - R) \tag{5}$$

$$ARVI = \frac{NIR - (2RED - BLUE)}{NIR + (2RED - BLUE)} \tag{6}$$

2.3 Differences with Other Crop Telemetry Studies

Nordberg, Evertson, & Development [14] have pointed out that most remote sensing studies to assess crop yield focus on the effect of vegetation index on yield. As a matter of fact, the vegetation index yield prediction model is accurate in one region, but it cannot be used as a general model for other regions. The reason is that lower crop area in other regions may affect crop yield prediction or unknown variables in other regions may affect the prediction results. Therefore, some studies have added meteorological factors into the crop yield prediction models. (Prasad, Chai, Singh, Kafatos, & geoinformation [15]). Soil properties are added to predict the number of yields, such as soil organic matter, cation exchange capacity, magnesium, potassium, and soil pH (Khanal et al. [16]). Sakamoto & Sensing [2], and Chen & Jing [17] used plant canopy reflectance alone for yield prediction. Whether using only plant canopy reflectance or other predictive variables, each has its own advantages, and the aim is to eliminate unknown variables in different areas. In this study, the meteorological factors and soil characteristics are reflected in the vegetation index, and the yield prediction was made regarding the crop planting area with the vegetation index proposed by Huang et al. [5]. Therefore, we selected the planting area of corn for remote sensing, the remote sensing area as the prediction variable, and the vegetation index to complete the modeling.

3 Research Methodology

The process of this study was divided into five parts and showed as Fig. 1: remote sensing data acquisition, data pre-processing, vegetation index and remote sensing area acquisition, regression analysis, and final prediction results.

The remote sensing data were acquired on a collaborative project satellite of the National Aeronautics and Space Administration (NASA) and the United States Geological Survey (USGS), named Landsat-8. Landsat-8. The reflectance of band 1 (blue band), band 4 (infrared band), band 5 (near the infrared band), and band 9 (cirrus band) were extracted from the telemetry data using Matlab R2021a, and the vegetation index was calculated. Band 4 and 5 were mainly used to calculate the standardized vegetation index. Band 1 was mainly used to calculate the Enhanced Vegetation Index and the Atmospheric Impedance Vegetation Index. Band 9 was mainly used to filter out cirrus signals. Since the telemetry data are inevitably disturbed by clouds, the cloud signals would be filtered out by the cirrus band in Landsat-8. Then, Google Earth was used to calculate the remote sensing area, and the longitude and latitude coordinates provided by the remote sensing data were traced to calculate the required remote sensing area for each block. Then the vegetation index and remote sensing area were put into a regression model to obtain the predicted results.

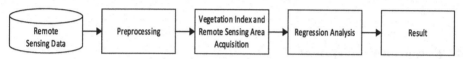

Fig. 1. Research Process

3.1 Regression Model

In this study, regarding the argument presented by Z. Y. Liu, Huang, Wu, & Dong [18], we extended the argument that there is a positive and a nonlinear correlation between different vegetation cover and all spectral indices. Based on this argument and the research motivations (Huang et al. [5]), we extended the regression model to predict corn yield using spectral index and remote sensing area. The higher the vegetation index and the larger the planted area in the predominantly corn-growing area, the higher the yield is likely to be. Therefore, a total of four regression models were used to predict corn yield in this study: multivariate linear regression, partial least squares regression, stepwise regression, and support vector regression. The order of presentation of the models is based on the study's progress.

The underlying concept of support vector regression is as follows. Smola, Schölkopf, & computing [19] and Aizerman & control [20] found a hyperplane to separate data points and made each point closest to this plane. A hyperplane is a feature that exists in the feature space, which is a multidimensional space constructed by each input variable. Each variable can be positioned in the feature space according to the value of the self-variable, and the support vector regression is to iterate through the input data to find the best hyperplane.

The following equation mathematizes the concept and finds the best solution using the objective and restriction equations. As shown in objective Eq. (7), where C is the cost parameter, the larger the value, the more intolerable the error; ε is the distance from the hyperplane to the support vector. ξ_i and ξ_i^* are the relaxation variables that measure the training sample's out-of-range deviation. The Eq. (7) is to find all data points closest to this hyperplane, but it must be achieved within the constraints of Eq. (8).

Equation (9) is the kernel function of the Gaussian kernel, whose basic concept is the feature vector of a specific input space, which is expressed as follows. The above equation $||x - x'||^2$ can be regarded as the squared Euclidean distance between two eigenvectors. Furthermore, σ is the free parameter. The Gaussian kernel shape is the bell curve, and the purpose of the free parameter is to adjust the width of the bell curve.

$$minimize\frac{1}{2}||\omega||^2 + C\sum_{i=1}^{n}(\xi_i + \xi_i^*) \tag{7}$$

$$\text{Subject to} \quad \begin{cases} y_i - f(x_i, \omega) - b \leq \varepsilon + \xi_i^* \\ f(x_i, \omega) + b - y_i \leq \varepsilon + \xi_i \\ \xi_i, \xi_i^* \geq 0 \end{cases} \tag{8}$$

$$k(x, x') = e^{-\frac{||x-x'||^2}{2\sigma^2}} \tag{9}$$

4 Research Experiment

This study used information from the U.S. Department of Agriculture (USDA) to find the major corn-producing regions in the U.S. from 2015 to 2019, which were the states with more than 10% of the total corn production in the U.S., and conducted remote sensing data capture for these states. The source of the remote sensing data was obtained from the U.S. Geological Survey website (https://earthexplorer.usgs.gov/), and the vegetation index and remote sensing area were obtained by processing the remote sensing data.

This study also captured the total annual state corn production data published on the USDA website (https://www.usda.gov/) and created a regression analysis with vegetation index and remote areas to predict corn production.

4.1 Data Collection

The USDA calculated that the significant corn-producing regions from 2015 to 2019 were the States of Iowa, the State of Illinois, the State of Minnesota, and the State of Nebraska. The U.S. Department of Agriculture has also released four counties where corn is not planted.

In this study, the part of the telemetry image that covered the red blocks was removed using Matlab R2021a, leaving only the counties where corn was planted. The USDA data shows that the main harvest month for corn in the US is September. The remote sensing data in the month prior to crop harvest can accurately predict the crop yield, so remote sensing images were selected with preference for August data. In summary, the satellite remote sensing images were selected based on the four major states and counties where corn is grown and because the Lansat-8 satellite was launched into space in 2013. Therefore, the remote sensing images were selected for the period from 2013 to 2021. If too many clouds interfere with the August remote sensing image, the remote sensing image of the end of July would be selected as a substitute. Then, depending on the size of the corn blocks in each state, the final number of blocks selected from each state would be different. Therefore, a total of 270 remote sensing files (30 blocks multiplied by nine years) were downloaded from the USGS website.

4.2 Vegetation Index Acquisition Process

The remote sensing files downloaded from the USGS website contain TIF image files for each band, as well as remote sensing text files regarding the time and date of remote sensing, the latitude and longitude of the remote sensing images, the maximum and minimum values of radiation for each band, the maximum and minimum values of reflection for each band, the size of the remote sensing images, the azimuth of the sun, the altitude of the sun, more of the same thing. In this study, the remote sensing image files of band 1 (blue band), band 4 (red band), band 5 (near the infrared band), and band 9 (cirrus band) extracting from each remote sensing data. Only four bands were selected because band 1, band 4, and band 5 could be available as calculate three vegetation indices, NDVI, EVI, and ARVI. The reason for selecting band 9 was that the remote sensing files at the end of July were selected if there was cloud interference during the data collection. Although this selection rule was applied to select the remote sensing files, the selected remote sensing files still had some cloud interference which caused the actual vegetation index to decline.

Then we applied Matlab R2021a to read the TIF files, cut the telemetry images, convert the reflectance to the Top-of-Atmosphere (TOA), filter the cirrus signal, and calculate the vegetation index. At first, Matlab R2021a was used to read the TIF files of the four bands and read out the digital number (DI) of the pixels in each band, and all the surrounding zeros were replaced with Nan values to avoid the subsequent vegetation index averaging. A telemetered map should filter out if the telemetered image is located at the state boundary or covers a county that does not produce corn. The remote sensing image should be cut in this case, leaving only the required part. After cutting the remote sensing image, the text file in the remote sensing information would be read, and the digital value of the pixels in the band would be converted to the reflectance of the top of the atmosphere by using the adjusted value of reflectance and the height of the sun.

After having the maximum atmospheric reflectance for each band, the pixels affected by the cirrus were picked out using band 9, and the positions of the pixels affected by the cirrus were marked. The NDVI value of each pixel was calculated in Band 4 and Band 5 based on the equation of standardized vegetation index, and the EVI and ARVI values of each pixel were calculated in Band 1, Band 4, and Band 5 based on the equations of enhanced vegetation index and atmospheric impedance vegetation index, respectively. Then, the pixel positions in Band 9 were applied to the corresponding NDVI, EVI, and ARVI pixel positions and set to the Nan value, leaving only the pixels that were not disturbed by the cloud signal. Referring to the method of processing vegetation indices of Huang et al. [5], we averaged the vegetation indices from the pixels that were not disturbed by cloud signals to obtain the desired vegetation indices, i.e., we completed the data pre-processing for each geographic code and averaged the vegetation indices for each block of each state (e.g., nine blocks per year for Iowa, six blocks per year for Illinois, and others.). A total of 36 vegetation indices were obtained.

4.3 Remote Measurement Area Acquisition Process

The largest corn-producing regions in the U.S. are Iowa, Illinois, Minnesota, and Nebraska, according to the U.S. Department of Agriculture. The remote sensing data

were also obtained around the counties where corn is grown in these four states. Therefore, obtaining the remote sensing area to correspond with the selected remote sensing
data would be necessary. The remote sensing area was obtained by entering a set of
latitude and longitude coordinates and closing the area into a region through the function on Google Earth (https://earth.google.com) website. The information on latitude
and longitude was obtained from the text file in the remote sensing file, and the remote
sensing area of four states was obtained by adding up each block of each state. Chen
et al. [21] wanted to avoid the discrepancy between the values of the remote sensing
area, corn yield, and vegetation index, and this study standardized the remote sensing
area and corn yield.

4.4 Experimental Results and Analysis

This study selected four regression models to find the best model to predict corn
yield, ranging from multiple linear regression, partial least squares regression, stepwise
regression, and support vector regression.

Because the results of stepwise regression were nonlinear, this study referred to
Gleason & Im [22]. In these two papers, support vector regression using the Gaussian
kernel showed good prediction results and was relatively stable. Therefore, the method
used in this study supports vector regression non-linearly, and the kernel function was
Gaussian. In this study, Matlab R2021a was used for support vector regression analysis.
At first, we used the built-in all SVMs to see which kernel function performed the
best, the Gaussian kernel performed the best, and the linear kernel performed the worst.
This is consistent with the previous results of stepwise regression, where the nonlinear
prediction was superior.

Tsamardinos, Rakhshani, & Lagani [23] have pointed out that because of the small
number of data and the need to select the best hyperparameter, this study used Nested
Cross-validation to select the hyperparameter by referring to the literature and divided
36 data into three training sets and one test set (one set has 9 data). Therefore, there were
27 data in the training set, will in the outer loop. The 9 test data were put into the inner
loop, and the 27 data were divided into eight training sets and one validation set (one set
has 3 data, so the training set has 24 data).The training data of one outer run entered into
the inner loop, and nine inner runs were made (for each inner run, the Optimizer was
Bayesian optimization, the Acquisition function was Expected improvement per, and the
number of iterations of the system was 300, and the best hyperparameter of the inner run
was selected among 300 times). Furthermore, the best hyperparameter would be selected
from the nine inner runs. The second outer run did the same thing, except the test set
would pick the 9 data initially treated as the training set. The data from the previous
test set would be added to the training set for the inner loop, and the four outer loops
would be conducted this way. Finally, the model built by the four sets of hyperparameters
would be tested with individual test sets to obtain four sets of R-squared, the best set of
hyperparameters would be applied to the test set, and the 36 data would be thrown in,
and the final result would be confirmed.

As shown in Fig. 2, the values of R-squared for the four groups of outer runs were
0.78, 0.89, 0.75, and 0.76, respectively. The hyperparameters of the best R-squared group
were Kernel scale = 7.3083 (i.e., γ, which was equal to $1/ \lfloor 2\sigma \rceil^2$), Box constraint =

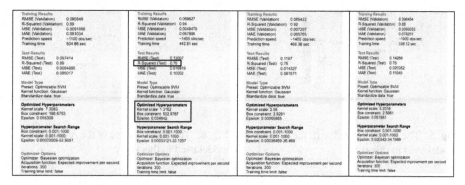

Fig. 2. The results of four sets of outer runs (the best R-squared set is marked with a red box)

198.6753, and Epsilon $= 0.055309$. After applying this set of hyperparameters and putting all 36 pieces of data into this model, the results are shown in the following figure.

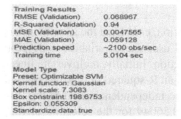

Fig. 3. The results of building the best hyperparameters

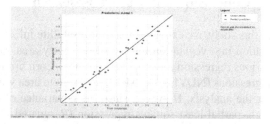

Fig. 4. The difference between the actual value and model with the predicted value of the best hyperparameter model

In Fig. 3, the result of inputting 36 data into the best model is R-squared$= 0.94$ and RMSE$=0.068967$, which was better than the previous linear regression result. In Fig. 4, showed the difference between the actual and predicted values of each data--the blue dots are the actual values, the yellow dots are the predicted value, and the errors are shown as solid orange lines.

Finally, the regression model used in this study is summarized in Figs. 5 and 6. From the R-squared values, the use of nonlinear support vector regression was significantly better than linear regression analysis. In the linear regression, when a single vegetation index was used with the remote area as the independent variable (R-squared values of 0.4426, 0.3486, and 0.3016 for NDVI + remote area, EVI + remote area, and ARVI + remote area, respectively), the result was worse than using three vegetation indices with the remote area as the independent variable (R-squared value of 0.56). Later, the partial least squares regression and stepwise regression (R-squared values of 0.56049 and 0.596, respectively) were used to solve the problem of multicollinearity, and the predictive power was better than that of the multivariate linear regression. The last nonlinear regression model was the Gaussian kernel support vector regression with an

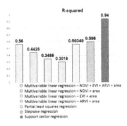

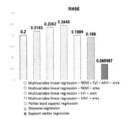

Fig. 5. The R-squared value of the regression model used in this study

Fig. 6. The RMSE value of the regression model used in this study

R-squared of 0.94, which was the best performance among the regression models. The RMSE value showed that the nonlinear regression also performed better. The experiments found that a nonlinear regression was more appropriate to predict the yield of the major corn-producing regions in the U.S.

5   Conclusion and Future Research

In this study, we extracted Landsat-8 satellite information, converted the satellite information into vegetation index and telemetry area, and used these variables to develop various regression models to predict U.S. corn production. Initially, multivariate linear regressions (NDVI, EVI, ARVI, and remote area with an R-squared of 0.56) were used for the analysis. However, since the covariance of the independent variables was too high, partial least squares regression (R-squared of 0.56049 for the selection of four principal components) and stepwise regression (R-squared of 0.596) were used to try to solve the problem of covariance of the independent variables. The results of both regressions were also better than the initial multivariate linear regressions. From the regression equation of stepwise regression analysis, we obtained the non-linearity of the relationship between the self-variable and the strain. Finally, we used the support vector regression with the Gaussian kernel (R-squared of 0.94). The results obtained were the most accurate in this study. Therefore, it can be concluded that the nonlinear regression model could help users forecast U.S. corn production. However, it is recommended that corn production forecasts for different regions should be evaluated.

For future research, Boryan, Yang, Mueller, & Craig [24] have suggested that CDL (Cropland Data Layer) can be applied to the corresponding telemetry data. This can capture the crop we want to predict, and the data is processed differently than in this study. This data processing method can be used to confirm whether the prediction effect can be improved. On the other hand, Pahlevan, Chittimalli, Balasubramanian, & Vellucci [25] have mentioned that Landsat-9 is already operating normally and the sensors on board are not much different from Landsat-8. The waveband is the same, and the data acquisition time is misaligned with Landsat-8; if combined with Landsat-8, we can have one telemetry data every eight days. If only Landsat-8 is used as the data source, there will be one telemetry data every 16 days. This can reduce the interference of the cloud layer, and more telemetry data can be used for analysis.

Acknowledgment. We would like to thank National Science and Technology Council, Taiwan for generously supporting this research through project #112-2410-H-008-017-MY2.

References

1. Johnson, D.M.: An assessment of pre- and within-season remotely sensed variables for forecasting corn and soybean yields in the United States. Remote Sens. Env. **141**, 116–128 (2014). https://doi.org/10.1016/j.rse.2013.10.027
2. Sakamoto, T.: Incorporating environmental variables into a MODIS-based crop yield estimation method for United States corn and soybeans through the use of a random forest regression algorithm. ISPRS J. Photogrammetry Remote Sens. **160**, 208–228 (2020)
3. Gao, F., Anderson, M.: Evaluating yield variability of corn and soybean using Landsat-8, Sentinel-2 and Modis in Google Earth Engine. Paper presented at the IGARSS 2019-2019 IEEE International Geoscience and Remote Sensing Symposium (2019)
4. Zhang, X., Zhang, Q.: Monitoring interannual variation in global crop yield using long-term AVHRR and MODIS observations. ISPRS J. Photogrammetry Remote Sens. **114**, 191–205 (2016)
5. Huang, J., Wang, H., Dai, Q., Han, D.: Analysis of NDVI data for crop identification and yield estimation. IEEE J. Sel. Top. Appl. Earth Observations Remote Sens. **7**(11), 4374–4384 (2014). https://doi.org/10.1109/JSTARS.2014.2334332
6. Roy, D.P., et al.: Landsat-8: Science and product vision for terrestrial global change research. Remote Sens. Environ. **145**, 154–172 (2014)
7. Justice, C.O., et al.: The moderate resolution imaging spectroradiometer (MODIS): land remote sensing for global change research. IEEE Trans. Geosci. Remote Sens. **36**(4), 1228–1249 (1998). https://doi.org/10.1109/36.701075
8. Drusch, M., et al.: Sentinel-2: ESA's optical high-resolution mission for GMES operational services. Remote Sens. Environ. **120**, 25–36 (2012). https://doi.org/10.1016/j.rse.2011.11.026
9. Jordan, C.F.: Derivation of leaf-area index from quality of light on the forest floor. Ecology **50**(4), 663–666 (1969)
10. Rouse, J.W.: Monitoring the vernal advancement of retrogradation of natural vegetation. Type III, final report, greenbelt, MD, p. (1974)
11. Liu, H.Q., Huete, A.: A feedback based modification of the NDVI to minimize canopy background and atmospheric noise. IEEE Trans. Geosci. Remote Sens. **33**(2), 457–465 (1995). https://doi.org/10.1109/TGRS.1995.8746027
12. Huete, A.: A comparison of vegetation indices over a global set of TM images for EOS-MODIS. Remote Sens. Env. **59**(3), 440–451 (1997)
13. Kaufman, Y.J., Tanre, D.: Atmospherically resistant vegetation index (ARVI) for EOS-MODIS. IEEE Trans. Geosci. Remote Sens. **30**(2), 261–270 (1992). https://doi.org/10.1109/36.134076
14. Nordberg, M.L., Evertson, J.: Vegetation index differencing and linear regression for change detection in a Swedish mountain range using Landsat TM® and ETM+® imagery. Land Degrad. Dev. **16**(2), 139–149 (2005). https://doi.org/10.1002/ldr.660
15. Prasad, A.K., Chai, L., Singh, R.P., Kafatos, M.: Crop yield estimation model for Iowa using remote sensing and surface parameters. Int. J. Appl. Earth Obs. Geoinf. **8**(1), 26–33 (2006). https://doi.org/10.1016/j.jag.2005.06.002
16. Khanal, S., Fulton, J., Klopfenstein, A., Douridas, N., Shearer, S.: Integration of high resolution remotely sensed data and machine learning techniques for spatial prediction of soil properties and corn yield. Comput. Electron. Agric. **153**, 213–225 (2018). https://doi.org/10.1016/j.compag.2018.07.016

17. Chen, P., Jing, Q.: A comparison of two adaptive multivariate analysis methods (PLSR and ANN) for winter wheat yield forecasting using Landsat-8 OLI images. Adv. Space Res. **59**(4), 987–995 (2017)
18. Liu, Z.-Y., Huang, J.-F., Wu, X.-H., Dong, Y.-P.: Comparison of vegetation indices and red-edge parameters for estimating grassland cover from canopy reflectance data. J. Integr. Plant Biol. **49**(3), 299–306 (2007)
19. Smola, A.J., Schölkopf, B.: A tutorial on support vector regression. Stat. Comput. **14**(3), 199–222 (2004)
20. Aizerman, M.A., Braverman, E.M., Rozonoer, L.I.: Theoretical foundations of the potential function method in pattern recognition learning. Translated from Avtoatika I Telemehanika **25**, 821–837 (1964)
21. Wold, S., Ruhe, A., Wold, H., Dunn, W.J.: The collinearity problem in linear regression. The partial least squares (PLS) approach to generalized inverses. SIAM J. Sci. Stat. Comput. **5**(3), 735–743 (1984)
22. Gleason, C.J., Im, J.: Forest biomass estimation from airborne LiDAR data using machine learning approaches. Remote Sens. Environ. **125**, 80–91 (2012)
23. Tsamardinos, I., Rakhshani, A., Lagani, V.: Performance-estimation properties of cross-validation-based protocols with simultaneous hyper-parameter optimization. Int. J. Artif. Intell. Tools **24**(05), 1540023 (2015)
24. Boryan, C., Yang, Z., Mueller, R., Craig, M.: Monitoring US agriculture: the US department of agriculture, national agricultural statistics service, cropland data layer program. Geocarto. Int. **26**(5), 341–358 (2011)
25. Pahlevan, N., Chittimalli, S.K., Balasubramanian, S.V., Vellucci, V.: Sentinel-2/Landsat-8 product consistency and implications for monitoring aquatic systems. Remote Sens. Env. **220**, 19–29 (2019)

Immune-Based Algorithm for the Multi-server Home Care Service Problem with Various Frequencies: An Example of Taiwan

Yi-Chih Hsieh[1]([✉]), Peng-Sheng You[2], and Ta-Cheng Chen[3]

[1] Department of Industrial Management, National Formosa University, Yunlin, Taiwan
yhsieh@nfu.edu.tw
[2] Department of Business Administration, National Chiayi University, Chiayi, Taiwan
psyuu@mail.ncyu.edu.tw
[3] Department of Information Management, National Formosa University, Yunlin, Taiwan
tchen@nfu.edu.tw

Abstract. This study investigates the scheduling of multi-server home care services in which multiple junior and senior servers can cooperatively serve a customer. We assume that different server types have different service costs and service speeds, and customers have various demand frequencies for the service. This problem aims to determine the number of the two types of servers per day and schedule the service route of the vehicle to various customers, so as to minimize the total cost of servers and vehicle travel. The studied problem is an NP-hard problem, and traditional optimization methods are time-consuming to solve when the problem size is larger. This paper presents a new decoding procedure to convert a random permutation of integers into a feasible solution to the problem. Then the decoding procedure is embedded into an immune-based algorithm to solve the problem. An example of Taiwan has been solved and analyzed by the proposed method to illustrate the effectiveness of the proposed method.

Keywords: Multi-Server · Home Care · Immune-Based Algorithm · Optimization

1 Introduction

Due to changes in medical care and lifestyles, the average life expectancy of citizens worldwide continues to increase. Today, as the social population tends to age, the ratio of the elderly population, chronic disease patients, and disabled people is increasing year by year. At present, due to the phenomenon of declining birth rate and social composition, small families have become the mainstay. However, most young and middle-aged members are the primary source of family income and usually need to go out to work, resulting in a lack of care for older adults at home. Therefore, in response to the advent of an aging society, the Ministry of the Interior of Taiwan guides the local governments to provide various home care services such as house cleaning, meal delivery, bathing, and other services, and allows the elderly to contact society to improve the quality of life of the elderly and reduce the burden of care for their families.

Over the past few decades, the so-called home health care (HHC) has received much attention due to its importance in today's society. Some HHC services require specialized skills, such as wound care for pressure sores or surgical wounds, intravenous injection, tracheal incisions, nasogastric tubes, cleaning, and replacing urinary catheters, etc. The main problems in HHC are related to vehicle routing and scheduling problem, and it is named home health care routing and scheduling problem (HHCRSP). In an HHCRSP, one has to simultaneously decide on the visit of customers and schedule the order of visiting for each day.

Liu et al. (2017) investigated an HHCRSP and developed a three-index mathematical programming model. They decomposed the mathematical programming model into a master problem and several sub-problems, and they solved these problems using a branch-and-price algorithm. Decerle et al. (2019) studied an HHCRSP with four objectives, including minimizing the total working time of the caregivers, minimizing the violation of patients' time windows, and minimizing the maximal working time difference between nurses and auxiliary nurses. Moussavi et al. (2019) studied an HHCRSP with minimizing three objectives, namely, the total distance traveled by a caregiver in a day, in all days (planning time horizon), and the total distance traveled by all caregivers in all days. They presented a matheuristic approach to solving the NP-hard problem. Liu et al. (2020) considered a periodic HHCSRP with the consideration of the medical skills of caregivers and their workload balance. They proposed a region-partition-based algorithm to solve the problem, and they designed different tabu search (TS) algorithms to solve the optimization problem for each region. Li et al. (2021) studied an HHCRSP with the consideration of the outpatient service in which a group of doctors was assigned to server outpatients. They presented a mixed integer programming model for the problem and proposed a hybrid genetic algorithm (HGA). Liu et al. (2021) considered an HHCRSP with a hard time window constraint for patients in which two caregivers must serve simultaneously for each patient. They developed four meta-heuristics to solve the problem. Related research on HHCRSP refers to the excellent survey paper by Cisse et al. (2017) and Euchi et al. (2022).

Home care services generally refer to helping the elderly and/or disabled with daily activities such as shopping, bathing, cleaning, and cooking. In this paper, we consider a multi-server home care service problem with various frequencies for customers (MHCSP-VF). In the MHCSP-VF, we assume that multiple junior and senior servers cooperate to serve each customer. Therefore, as the number of servers increases, the total service time for a customer decreases. However, more servers result in higher costs. In Taiwan, home care services are included in long-term care and are partially or fully subsidized. As a result, the demand for home care services has increased drastically.

The considered MHCSP-VF has to determine the number of junior and senior servers for each day in the planning time horizon and to schedule the route of vehicles to customers. It differs from the typical HHCRSP, because (i) in the MHCSP-VF, we assume that the service time of a customer is related to the number junior and senior servers. However, in the typical HHCRSP, the service time is usually fixed. (ii) Customers have various frequencies of demand for the home care service, and once a customer begins receiving the service, it usually lasts for a long time. Therefore, in this paper, we assume that each customer has its frequency of demand for home care service. The considered

MHCSP-VF is a variant of vehicle routing problems (VRP) and scheduling problems. Hence it is an NP-hard problem.

In past research, the main approaches to solving an HHCRSP include:

(1) Exact solution method: For example, Tanoumand and Ünlüyurt (2021) studied an HHCRSP in which home patients had two types of service needs. They presented a mathematical programming model and proposed a branch-and-price algorithm to solve it.
(2) Heuristic and hybrid method: For example, tabu search method (Rest and Hirsch (2016)), large neighborhood search heuristic (Braekers et al. (2016)), memetic algorithm (Decerle et al. (2018)), hybrid method based on ant colony system and clustering algorithm (Euchi et al. (2020)).

The exact solution method can obtain the global optimal solution for the HHCRSP, but it is time-consuming when the problem size is larger. Therefore, the current mainstream of solving HHCRSP is to use a heuristic or hybrid heuristic to quickly obtain a good/acceptable feasible solution within a reasonable time. Recently, many evolutionary algorithms have been developed to solve various optimization problems. Mirjalili et al. (2017) illustrated more than 100 evolution algorithms based on natural phenomena or animal behaviors. Among these evolutionary algorithms, the immune-based algorithm (IBA) has attracted much attention due to its numerous successful applications, for example, Li et al. (2022). Therefore, in this paper, we adopt IBA to solve MHCSP-VF. Firstly, we present a new decoding method to convert any permutation of a sequence of integers into a feasible solution of the considered MHCSP-VF, then embed it into IBA to solve it. The main advantage of the new decoding method is that any permutation of integer sequences can be transformed into a feasible solution to the problem, which directly improves the efficiency of individual evolution in IBA.

This paper is organized as follows. Section 2 introduces the MHCSP-VF, including assumptions and notations. Section 3 presents a new decoding method to convert a random permutation of integers into a feasible solution of MHCSP-VF. An example is used to illustrate the new decoding procedure. Additionally, an IBA is presented in this section. Section 4 presents a practical test problem in Taiwan. The numerical results of IBA are presented and discussed. Finally, Sect. 5 summarizes the conclusions of this paper and lists future research.

2 Problem Statement

2.1 Assumptions

(1) There are n customers needing home care service in the planning time horizon T. It is assumed $T = 2$ (days).
(2) The demand frequency of service for customer i is f_j, in T. It is assumed that $f_j = 1$, once per two days, or $f_j = 2$, twice per two days.
(3) Two types of servers, namely, junior server (e) and senior server (E), can serve cooperatively for a customer. The cost and service speed of the junior and senior servers are different.

(4) A customer is simultaneously served by a vehicle with k_1 junior servers and k_2 senior servers. The service time and service cost for customer i are set as $Sm_i/(k_1 + k_2/ssr)$ and $Y \times (k_1 + k_2 \times ssc)$, respectively, where S and Y are the given basic service time of a customer and basic service cost of a server, m_i is the number of required service units for customer i, and ssc and ssr are service speed ratios and salary cost ratios of senior and junior servers. It assumed that $S = 30, 45,$ and 60, respectively, and $Y = $ NT\$200/hr in this paper. To analyze the effect of service speed and cost of junior/senior servers, we assume three different ssr and ssc, namely, $0.5, 0.7, 0.9$ for ssr and $1.2, 1.5, 1.8$ for ssc, respectively.

(5) On each day, at most two vehicles are available. Let (v_1, v_2) be the number of vehicles for day 1 and day 2, respectively. The capacity of a vehicle is four servers, that is, $1 \le k_1 + k_2 \le 4$. Therefore, all possible 14 server combinations ($Q = 14$) in a vehicle are as follows.

  (i) 1 server: e, E, (combination 1–2)
  (ii) 2 servers: eE, ee, EE, (combination 3–5)
  (iii) 3 servers: eee, eeE, eEE, EEE, (combination 6–9)
  (iv) 4 servers: eeee, eeeE, eeEE, eEEE, EEEE.

(6) The regular working time of a server is 8 h per day. If the working time exceeds 8 h, then the following penalty of a server is added to the objective value. Penalty = (overtime) $\times w$, where w is the weight of the penalty. It is assumed $w = 40$.

(7) The vehicle departs from the long-term care center and returns to the center after serving all customers every day.

(8) Assume that the average fuel consumption of a vehicle is $f_c = 4.35$ L/hr, and the fuel price is $f_p = $ NT\$35/L.

(9) The MHCSP-VF has to determine the number of junior/senior servers in the vehicle and schedule the route of the vehicle on each day to minimize the following objective.

Objective = the working cost of junior/senior servers for customers

+ the traveling cost of vehicles
+ the penalty of overtime of servers.

2.2 Notations, Variables and Objective

Notations

C The node set of all customers, $|C| = n$.

V $= \{0\} \cup C \cup \{n + 1\}$, the set of all nodes. The starting node of the vehicle is node 0, and the end is node $n + 1$.

E $= \{(i,j) \mid i,j \in V, i \ne j\}$, the link set between node i and node j.

d_{ij} The shortest traveling time (minute) from node i to node j, $(i,j) \in E$.

T The planning time horizon. It is assumed $T = 2$ (days).

v_1, v_2 The number of vehicle available on day 1 and day 2.

f_i The demand frequency of service for customer $i, i \in C$. It is assumed $f_j = 1$ or $f_j = 2$ in two days. That is, once per two days or twice per two days.

Q The set of all possible combinations of junior and senior servers. It is assumed $Q = 14$.

S The basic service time of a customer, and we assume $S = 30, 45, 60$.

Y The basic service cost of a server and we assume $Y = \text{NT\$200}$.

m_i The required service unit for customer i, $i \in C$.

f_c The average fuel consumption of the vehicle. It is assumed $f_c = 4.35$ L/h.

f_p The unit fuel price. It is assumed $f_p = \text{NT\$24.4/L}$.

t_s, t_j The service time of senior and junior servers for a customer.

c_s, c_j The cost of senior and junior servers for a customer.

ssr $= t_s/t_j$, the service time ratio of senior/junior server. It is assumed 0.5, 0.7, and 0.9, respectively.

ssc $= c_s/c_j$, the service cost ratio of senior and junior servers. It is assumed 1.2, 1.5, and 1.8, respectively.

S_i $=Sm_i/(k_1 + k_2/ssr)$, the service time (minute) for customer i on day l, $l \in T$, if he/she is simultaneously served by k_1 junior and k_2 senior servers.

Y_l $=Y \times (k_1 + k_2 \times ssc)$, the cost (per hour) of a server in day l, $l \in T$, if k_1 junior and k_2 senior servers are assigned.

w The weight of penalty for overtime. It is assumed $w=40$.

U_l $= \sum_{(i,j) \in E} x_{ijl}(d_{ij}+S_l)/60$, the sum of the traveling time (hour) and service time of a server on day l, $l \in T$.

Decision Variables

$x_{ijl} = 1$ if the vehicle arrives at node i on day l, and serves node j, $(i,j) \in E$, $l \in T$.

Objective.

   $Min \sum_{l \in T} \sum_{(i,j) \in E} x_{ijl}(d_{ij}/60)f_p f_c + Y_l U_l + Y_l F_l$.

The objective is the sum of the traveling cost of vehicles, the working cost of servers, and the penalty of overtime of servers. Using the above notations, an integer linear programming model can be developed for the MHCSP-VF. However, it is time-consuming to solve. Thus, we present the following IBA to solve the problem.

3 Method

3.1 Immune-based Algorithm (IBA)

The main steps of IBA are as follows, and Fig. 1 illustrates the flowchart.

Step 1. Generate a random initial population of individuals (feasible solutions).

Step 2. Calculate the fitness value for each individual.

Step 3. Select the best the k individuals with the best fitness values.

Step 4. Clone the best k individuals in Step 3 using the crossover and mutation in genetic operation (Holland (1975)). The details of clonal selection and affinity maturation processes are referred to De Castro and Von Zuben (2000).

Step 5. Update the memory set of individuals. That is, replace the inferior individuals in the memory set with superior individuals, and delete the individuals with too similar structures.

Step 6. Check the stopping criterion of maximum generations. If not stop, then go to Step 2. Otherwise, go to the next step.

Step 7. Stop. Report the optimal or near-optimal solution(s) from the memory set.

Note that, in this paper, we use the following new decoding procedure to covert a permutation of integers into a feasible solution of the considered problem and use it to calculate the fitness values for a population of individuals generated in Step 1.

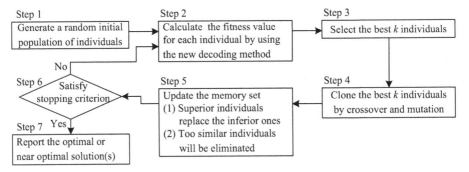

Fig. 1. The main steps of IBA.

Table 1. New decoding for the example.

Customer	1		2		3		4		5		6		7		8	
f_i	2		1		1		2		1		2		2		1	
R	15	11	4	1	12	14	16	9	10	5	8	3	6	2	7	13
Day	2	1	2	1	1	2	2	1	2	1	2	1	2	1	1	2
Car	2	2	2		1		1	2	1		1	2	1	1	1	2
Server	6	3	6		1		13	3	13		13	3	13	1	3	
type	eee	eE	eee		e		eEEE	eE	eEEE		eEEE	eE	eEEE	e	eE	

3.2 The New Decoding Procedure

We use the following example to illustrate the new decoding procedure which can be used in Step 2 of IBA in Fig. 1. Assuming that $n = 8$, $T = 2$, $f_l = 1$ for customers 2,3,5,8 and $f_l = 2$ for customers 1,4,6,7, $(v_1,v_2) = (2,2)$, then we may use a sequence of permutation of 1 to 16 ($=T \times n$) to construct a feasible solution of the problem. Assume that the random permutation is: $R = 15, 11, 4, 1, 12, 14, 16, 9, 10, 5, 8, 3, 6, 2, 7, 13$.

Step 0: Construct a matrix for customers, f_i and R as rows 1 to 3 in Table 1, in which two random numbers in R correspond to a customer. For customers with $f_l = 1$, their first cell of R is marked in grey in row 3, and for customers with $f_l = 2$, the two cells of R are marked in grey in row 3.

Step 1: (determine the service day for a customer)
(a) $k = 0$.
(b) If $R(k + 1) < R(k + 2)$, then assign days 1 and 2 to row 4 (Day) for customer $1 + (k/2)$ in Table 1.

Otherwise, assign days 2 and 1 to row 4 (day) in Table 1.

(c) $k = k + 2$. Repeat (b) until all cells in row 4 in Table 1 have been assigned.

Step 2: (determine the vehicle for a customer)

    (a) $k = 0$.

    (b) For customer $1 + (k/2)$,

        (i) Assign vehicle $\{1 + mod(R(k + 1), v)\}$ to the first cell of row 5 (Vehicle used) in Table 1.

        (ii) Assign vehicle $\{1 + mod(R(k + 2), v)\}$ to the first cell of row 5 (Vehicle used) in Table 1.

Where v is the number of vehicles available on the day.

$k = k + 2$, Repeat (b) until all cells in row 5 of Table 1 have been assigned.

Step 3: (determine the traveling route of a vehicle)

Construct the route for each vehicle on each day by sorting the corresponding random numbers R in grey, from small to large.

For example, in Table 1, $R(14) = 2$ and $R(5) = 12$ are for vehicle 1 on day 1, corresponding to customers 7 and 3, respectively. Then the route of the vehicle for the vehicle on day 1 is:

Day 1 Vehicle 1: 7-3   ($R(14) = 2, R(5) = 12$)

Similarly, we have:

Day 1 Vehicle 2: 6-8-4-1 ($R(12) = 3, R(15) = 7, R(8) = 9, R(2) = 11$)
Day 2 Vehicle 1: 7-6-5-4 ($R(13) = 6, R(11) = 8, R(9) = 10, R(7) = 16$)
Day 2 Vehicle 2: 2-1 ($R(3) = 4, R(1) = 15$)

Step 4: (determine the server combination of a vehicle on each day)

Based on the customers in the vehicle each day, sum up their corresponding random number R, say Sum_R, and use the $\{1 + mod (Sum\_R, Q)\}$th vehicle.

For example, using the results in Step 3,

(day 1, vehicle 1)$1 + mod (2 + 12,Q)$=$1 + mod (2 + 12,14) = 1$. Thus assign the 1st server combination (e) to this vehicle.

(day 1, vehicle 2)$1 + mod (3 + 7 + 9 + 11,Q) = 1 + mod (30,14) = 3$. Thus assign the 3rd server combination (eE) to this vehicle.

(day 2, vehicle 1)$1 + mod (6 + 8 + 10 + 16,Q) = 1 + mod (40,14) = 13$. Thus assign the 13th server combination (eEEE) to this vehicle.

(day 2, vehicle 2)$1+mod (4 + 15,Q) = 1 + mod (19,14) = 6$. Thus assign the 6th server combination (eee) to this vehicle.

Therefore, the objective of the problem is computed using the server combination and vehicle route on each day.

4 Numerical Results and Discussions

4.1 Test problems

(1) This research takes an area of Chung-Li in Taiwan as the test problem, which is illustrated in Fig. 2. Figure 2 contains 212 nodes, and the long-term care center is at node 209.

(2) This test problem contains 20 customers, and we can use a random sequence of 1 to 40 (=$n{\times}T$) to present a feasible solution of the problem.

(3) The demand frequency (f_i), node number, and service time units (m_i) for the 20 customers are as follows.

Once per two days (f_i=1):5(2), 34(1), 47(1), 71(1), 81(2), 112(2), 132(1), 159(2), 175(1), 187(2)

Twice per two days(f_i=2):8(1), 12(1), 39(2), 50(1), 78(2), 127(2), 135(2), 142(2), 204(1), 211(2)

4.2 Test Results and Discussions

This research uses IBA coded in MATLAB. The parameters set by IBA are::the initial population number = 100, the maximum evolutionary generation = 2000, the mating rate = 0.48, and the mutation rate = 0.08. Each test problem was executed 50 times by IBA, and the best result was recorded. For test problems with different ratios of server speed (*ssr*), ratios of service cost (*ssc*), service time combinations, and various numbers of vehicles on each day, we report the numerical results of the total cost in Tables 2 and 3, and the best server combination in Tables 4 and 5.

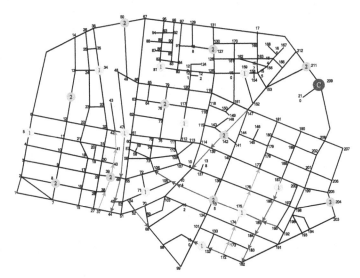

Fig. 2. The example of Chung-Li, Taiwan.

From Tables 2 and 3, we can see that:

(1) When the server service speed ratio (*ssr*) and server cost ratio (*ssc*) are fixed, with the increase of the basic service time (*S*) of each customer, the total cost increases. For example, in Table 2, the best total cost is 4723.4 when *ssr* = 0.9, *ssc* = 1.8, and *S* = 30 (min), and the best total cost is 12550.0 when *ssr* = 0.9, *ssc* = 1.8, *S* = 60 (min).

(2) When *ssc* and *S* are fixed, with the increase of *ssr*, that is, the gap between senior and junior service speed becomes smaller, the total cost increases accordingly. For example, in Table 2, if $ssc = 1.8$ and $S = 30$ (min), the total cost increases from 4393.0 to 4723.4 when *ssr* rises from 0.5 to 0.9.

(3) When *ssr* and *S* are fixed, with the increase of *ssc*, that is, the gap between senior and junior salary expenses becomes larger, the total cost increases slightly. For example, in Table 3, if $ssr=0.9$ and $S=30$ (min), the total cost increases from 4430.3 to 4511.5 when *ssc* rises from 1.2 to 1.8.

(4) When *ssr*, *ssc* and *S* are fixed, with the increase of vehicle, the total cost decreases. For example, in Table 3, if $ssc=1.8$, $ssr=0.9$, and $S=60$ (min), the total cost decreases from 12550.0 to 7782.2 when (v_1, v_2) rises from (1,1) to (2,2).

Table 2. Numerical results of $C = (1,1)$ with 50 trials of IBA.

S	ssr	0.5				0.7				0.9			
	ssc	best	μ	σ	CPU	best	μ	σ	CPU	best	μ	σ	CPU
30	1.2	3317.9	3577.4	186.4	746.8	4260.7	4508.2	121.1	743.1	4706.9	5077.8	199.1	706.0
	1.5	3987.2	4268.7	249.2	762.4	4645.6	5054.3	208.1	739.8	4766.7	5175.6	243.2	733.2
	1.8	4393.0	4749.5	261.2	742.5	4733.0	5133.6	258.0	740.8	4723.4	5172.5	243.4	722.5
45	1.2	4434.7	4662.6	114.0	735.1	5595.4	6279.6	224.7	693.8	6597.3	7088.1	284.1	695.1
	1.5	5248.7	5619.3	230.1	716.0	6594.9	7034.1	253.5	691.7	6502.1	7110.1	270.3	692.4
	1.8	6063.5	6524.2	186.7	723.9	6490.9	7107.3	233.9	707.5	6552.9	7178.3	243.9	696.4
60	1.2	5561.3	6060.5	310.6	685.6	7366.1	7865.6	282.3	668.0	8867.9	18168.0	6466.0	657.2
	1.5	6577.3	7190.5	352.6	686.5	8435.9	9163.5	343.3	666.3	10689.0	20454.0	6081.1	654.4
	1.8	7732.1	8473.3	291.4	675.9	9074.3	10080.0	575.3	669.2	12550.0	25872.0	7448.2	644.8

Table 3. Numerical results of $C = (2,2)$ with 50 trials of IBA.

S	ssr	0.5				0.7				0.9			
	ssc	best	μ	σ	CPU	best	μ	σ	CPU	best	μ	σ	CPU
30	1.2	3316.6	3650.0	147.3	924.4	3958.6	4227.9	115.1	922.9	4439.3	4767.1	156.4	957.3
	1.5	3995.4	4250.7	120.3	937.5	4303.4	4701.6	163.2	909.6	4490.2	4857.9	172.1	925.0
	1.8	4357.2	4649.8	119.4	928.3	4546.3	4853.2	175.8	910.5	4511.5	4795.4	149.0	915.7
45	1.2	4222.5	4420.5	127.7	997.2	5346.1	5807.1	194.9	1034.5	6250.9	6495.5	137.6	1043.9
	1.5	4989.4	5218.0	133.9	1020.2	6239.6	6493.3	156.4	1035.3	6256.2	6562.3	154.4	1021.4
	1.8	5787.2	6007.5	121.1	1007.4	6207.3	6535.6	143.0	1020.8	6220.6	6540.9	167.1	1025.4
60	1.2	5173.8	5572.4	242.1	1025.7	6867.3	7204.0	187.1	1013.0	7729.1	8093.5	214.7	1003.0
	1.5	6191.9	6695.4	244.7	1035.6	7897.8	8150.8	180.8	1031.7	7638.2	8199.0	189.6	1007.5
	1.8	6910.6	7519.1	197.6	1022.9	7853.0	8250.7	205.9	1020.3	7782.2	8192.6	214.5	999.3

From Tables 4 and 5, we can see that:

(1) When *ssr* and *ssc* are fixed, with the increase of *S* of each customer, the number of servers used increases accordingly. For example, in Table 4, the server allocation is

Table 4. The optimal combination of server for $C = (1,1)$ with various ssc, ssr, and S.

ssc	Day	ssr	0.5			0.7			0.9		
		S	30	45	60	30	45	60	30	45	60
1.2	day 1	1	E	EE	EEE	EE	EE	EEE	eee	eee	EEEE
	day 2	1	EE	EE	EE	EE	EEE	EEEE	ee	eeee	eEEE
1.5	day 1	1	eE	EE	EE	ee	eeee	eeeE	eee	eee	EEEE
	day 2	1	E	EE	EEE	eE	eee	eeeE	ee	eeee	eEEE
1.8	day 1	1	E	EE	EE	eee	eeee	eeeE	eee	eeee	eEEE
	day 2	1	eE	eE	eEE	ee	eee	eeeE	ee	eeee	EEEE

e = junior server, E = senior server

Table 5. The optimal combination of server for $C = (2,2)$ with various ssc, ssr, and S.

ssc	Day	ssr	0.5			0.7			0.9		
		Vehicle/S	30	45	60	30	45	60	30	45	60
1.2	day 1	1	E	E	EE	E	EE	EE	e	eE	eE
		2	E	E	E	E	E	EE	e	ee	ee
	day 2	1	E	E	E	E	E	E	E	eE	ee
		2	E	E	E	E	E	EE	e	e	ee
1.5	day 1	1	E	E	E	e	ee	ee	e	ee	ee
		2	E	E	E	e	e	ee	e	ee	ee
	day 2	1	E	E	E	e	ee	ee	e	ee	ee
		2	E	E	EE	e	E	eE	ee	ee	ee
1.8	day 1	1	e	E	E	e	ee	ee	e	ee	ee
		2	e	E	E	ee	e	ee	ee	ee	ee
	day 2	1	e	E	E	e	ee	ee	e	ee	ee
		2	E	E	E	e	ee	eee	e	e	ee

e = junior server, E = senior server

2 senior servers (EE) on days 1 and 2 when $ssr = 0.7$, $ssc = 1.2$, $S = 30$ (min), while it is 3 senior servers (EEE) on day 1 and 4 senior servers (EEEE) on day 2 when $ssr = 0.7$, $ssc = 1.2$, $S = 60$ (min).

(2) When ssc is fixed, with the increase of ssr, that is, the service speed gap between senior and junior servers becomes smaller, it tends to assign junior server. For example, in Table 4, on day 1, if $ssc = 1.2$, $S = 30$ (min), the assignment will be changed from 1 senior server (E) to 3 junior servers (eee) when ssr increases from 0.5 to 0.9.

(3) When ssr is fixed, with the increase of ssc, that is, the gap of cost between senior and junior servers becomes larger, it tends to use junior server. For example, in Table 4, if $ssr = 0.7$ and $S = 30$ (min) on day 1, the assignment will be changed from 2 senior servers (EE) to 3 junior servers (eee) when ssc increases from 1.2 to 1.8.

(4) When *ssr*, *ssc*, and *S* are fixed, with the increase in vehicles, it tends to use junior servers or few servers. For example, in Table 4 with $(v_1, v_2) = (1,1)$, when $S = 60$ and *ssr* = 0.9, each vehicle needs 4 servers, while 2 servers are enough for most of the vehicles in Table 5 with $(v_1, v_2) = (2,2)$.

5 Conclusions

In this paper,

(1) We have introduced a new MHCSP-VF with multiple vehicles in which multi-servers work cooperatively to reduce the service time for customers.
(2) We have presented a new decoding method to convert a permutation of random integers into a feasible solution of MHCSP-VF.
(3) Using the new decoding method, this study has proposed an IBA to simultaneously obtain the vehicle routes and determine the server combination of senior and junior servers for each vehicle in the planning time horizon.
(4) This study takes an area of Chung-Li in Taiwan as the test problem. The numerical results have shown that IBA can obtain reasonable solutions for this MHCSP-VF.

The future research is as follows.

(1) The proposed method in this paper can be extended to a class of more general home care service problems for multiple vehicles with various capacities and time windows of customers.
(2) Other artificial evolutionary algorithms can be developed to solve the considered problem and compare their efficiency and effectiveness with the proposed method.

Acknowledgments. The authors thank Mr. C.L. Hsu for collecting the numerical results. This research is partially supported by National Science and Technology Council, Taiwan, under grant MOST 111-2221-E-150-014.

References

Braekers, K., Hartl, R.F., Parragh, S.N., Tricoire, F.: A biobjective home care scheduling problem: analyzing the tradeoff between costs and client inconvenience. Eur. J. Oper. Res. **248**, 428–443 (2016)

Cisse, M., Yalcindag, S.Y., Kergosien, Y., Sahin, E., Lente, C., Matta, A.: OR problems related to home health care: a review of relevant routing and scheduling problems. Oper. Res. Health Care **13–14**, 1–22 (2017)

De Castro, L.N., Von Zuben, F.J.: The clonal selection algorithm with engineering applications. In: Workshop Proceedings of the GECCO 2000, pp. 36–37 (2000)

Decerle, J., Grunder, O., El Hassani, A.H., Barakat, O.: A memetic algorithm for a home health care routing and scheduling problem. Operat. Res. Health Care **16**, 59–71 (2018)

Decerle, J., Grunder, O., El Hassani, A.H., Barakat, O.: A memetic algorithm for multi-objective optimization of the home health care problem. Swarm Evol. Comput. **44**, 712–727 (2019)

Euchi, J., Masmoudi, M., Siarry, P.: Home health care routing and scheduling problems: a literature review. 4OR J. Oper. Res. **20**, 351–389 (2022)

Euchi, J., Zidi, S., Laouamer, L.: A hybrid approach to solve the vehicle routing problem with time windows and synchronized visits in-home health care. Arab. J. Sci. Eng. **45**, 10637–10652 (2020)

Holland, Y.J.H.: Adaptation in Natural and Artificial Systems. University of Michigan Press, Ann Arbor, Michigan (1975)

Li, L., Lin, Q., Ming, Z.: A survey of artificial immune algorithms for multi-objective optimization. Neurocomputing **489**, 211–229 (2022)

Li, Y., Xiang, T., Szeto, W.Y.: Home health care routing and scheduling problem with the consideration of outpatient services. Transp. Res. Part E **152**, 102420 (2021)

Liu, W., Dridi, M., Fei, H., El Hassani, A.: Hybrid metaheuristics for solving a home health care routing and scheduling problem with time windows, synchronized visits and lunch breaks. Expert Syst. Appl. **183**, 115307 (2021)

Liu, R., Yuan, B., Jiang, Z.: Mathematical model and exact algorithm for the home care worker scheduling and routing problem with lunch break requirements. Int. J. Prod. Res. **55**, 558–575 (2017)

Liu, R., Yuan, B., Jiang, Z.: The large-scale periodic home health care server assignment problem: a region-partition-based algorithm. IEEE Trans. Autom. Sci. Eng. **17**, 1543–1554 (2020)

Mirjalili, S., Gandomi, A.H., Mirjalili, S.Z., Saremi, S., Faris, H., Mirjalili, S.M.: Salp swarm algorithm: a bio-inspired optimizer for engineering design problems. Adv. Eng. Softw. **114**, 163–191 (2017)

Moussavi, S.E., Mahdjoub, M., Grunder, O.: A matheuristic approach to the integration of worker assignment and vehicle routing problems: application to home healthcare scheduling. Expert Syst. Appl. **125**, 317–332 (2019)

Rest, K., Hirsch, P.: Daily scheduling of home health care services using time dependent public transport. Flex. Serv. Manuf. J. **28**, 495–525 (2016)

Tanoumand, N., Ünlüyurt, T.: An exact algorithm for the resource constrained home health care vehicle routing problem. Ann. Oper. Res. **304**, 397–425 (2021)

Economic Applications

Variable Selection Methods-Based Analysis of Macroeconomic Factors for an Enhanced GDP Forecasting: A Case Study of Thailand

Roengchai Tansuchat[1,2], Pichayakone Rakpho[2,3], and Chaiwat Klinlampu[2,3(✉)]

[1] Faculty of Economics, Chiang Mai University, Chiang Mai 50200, Thailand
[2] Center of Excellence in Econometrics, Chiang Mai University, Chiang Mai 50200, Thailand
chaiwatklinlampu@gmail.com
[3] Office of Research Administration, Chiang Mai University, Chiang Mai 50200, Thailand

Abstract. Forecasting Gross Domestic Product (GDP) accurately is essential for understanding the behavior of an economy, but it requires considering many factors. This study aims to identify the most precise forecasting model for Thailand's GDP given such competing variable selection methods as Ridge, Lasso, Elastic Net, and SCAD. We collected data of key macroeconomic variables from 1970 to 2021 and divided them into training and testing sets at the 80:20 ratio. We compared the forecasting results using RMSE and Adjusted R-squared. Our findings indicate that the Elastic Net regression model produced the most accurate forecasts, with a 99.55% accuracy rate for predicting Thailand's GDP from the eight macroeconomic variables we examined. This study confirms that the variable selection methods can generate more accurate results than the conventional econometric forecasting methods. The Elastic Net regression model, the most accurate in this study, only uses industrial value added, exchange rate, import value, export value, and government spending to predict GDP.

Keywords: Ridge · Lasso · Elastic Net · SCAD · Macroeconomic variables · GDP

1 Introduction

The task of forecasting a country's Gross Domestic Product (GDP) is of paramount importance for policymakers, investors, and business leaders who seek to gauge the overall health and direction of the economy [1]. Forecasting provides valuable insights into the present and future circumstances that have the potential to unfold. Policymakers can leverage the results of these forecasts to devise suitable policies that facilitate economic growth, ensure financial stability, and enhance resilience in the face of crises and downturns. From the OECD's standpoint, forecasting plays a pivotal role in policy planning. In the formulation of a policy plan, it is imperative to pose pertinent questions regarding the current economic situation, factors that will impact the economy in the short, medium, and long run, and suitable measures to be implemented. The answers to these questions can be gleaned from prior forecasting models or by devising fresh predictions [2].

Recent years have seen Thailand's economic growth decline, with data from the Bank of Thailand indicating that the country's GDP contracted by 6.1% in 2020. This contraction was primarily caused by the COVID-19 pandemic, which led to lockdowns and restrictions throughout the country, resulting in a significant decline in economic activity during the second and third quarters. Nevertheless, the government has implemented a range of measures to support the economy, such as stimulus packages, low-interest loans, and debt restructuring for affected businesses. Beyond the pandemic, Thailand has faced other economic challenges, such as political instability and trade tensions, which have impacted the country's economic performance. The appreciation of the baht has also made exports less competitive. Despite these challenges, the Thai government has taken steps to address these issues and promote economic growth like making public its plans to increase investment in infrastructure and digital technology.

As previously mentioned, accurately forecasting economic growth requires considering various key macro-level determinants that can either hinder or support it. Therefore, many studies have recommended incorporating these factors into forecasting models to improve their accuracy [3, 4]. However, achieving precise predictions can be challenging since finding the right predictor to enhance economic growth forecasting is crucial. Unfortunately, traditional economic forecasting methods have typically used econometric models that require a large number of observations to avoid over-parameterization [5, 6] This means that if too many factors are included in the model, the dimensionality increases, leading to the degrees-of-freedom problem. Consequently, to preserve the degrees-of-freedom and avoid losing vital information, the number of factors included in the model should be limited [7].

Forecasting economic variables is a challenging task, especially when dealing with time series predictors. As such, this study proposes an alternative approach to the traditional forecasting methods that can accommodate a large number of predictors simultaneously. This approach involves using shrinkage estimation methods, such as penalized least squares estimation, which regularizes the model complexity and avoids over-fitting. The penalty term penalizes the size of regression coefficients, which helps in avoiding the curse of over-parametrization and conserves degrees-of-freedom.

The proposed approach maps the set of predictors into a high-dimensional space of linear functions of the predictors, allowing for the estimation of a forecast equation in an infinite-dimensional space. This approach has been confirmed to be useful in forecasting problems, as demonstrated by various studies such as [8, 9] who applied SCAD and LASSO for forecasting macroeconomic variables. Other innovative tools for forecasting include Ridge, Lasso, Elastic Net, SVM, Neural Network, Principal Component Regression (PCR), Gradient Boosting, and Random Forest models suggested by [10–12], who also confirmed that shrinkage estimation methods perform better than those conventional regression models without shrinkage.

The study aims to identify the most suitable set of predictors for forecasting Thailand's GDP by utilizing shrinkage estimations, including Ridge, Lasso, Elastic Net, and SCAD. The goal is to estimate and select the variables simultaneously and evaluate the forecasting accuracy of these models while examining their underlying mechanisms.

Through this investigation, the study aims to provide insights into the potential of shrinkage estimations in improving GDP forecasting and informing economic decision-maker not only in Thailand but also in other countries.

The rest of this paper is outlined as follows. Section 2 will cover the methodology, Sect. 3 will discuss the data used in the study, Sect. 4 will present the empirical results and analysis, and Sect. 5 will provide the conclusion.

2 Methodology

In this study, we employ shrinkage estimation methods, namely Ridge, Lasso, Elastic Net, and SCAD regression, all of which are based on OLS. The basic equation for OLS utilized in this study consists of macroeconomic indicators, with one dependent variable and independent variables, as follows

$$Y = X\beta + e$$

where Y is GDP and X is the set of macroeconomic variables. Meanwhile, β is the set of the coefficients that correspond to the macroeconomic variables, and e is the error term.

2.1 Ridge Regression

Ridge regression is a commonly used method for analyzing multicollinearity in multiple regression data, especially in cases where the number of predictor variables exceeds that of observations [13]. The first step in ridge regression is to standardize both the independent and dependent variables by subtracting their means and dividing the result by their standard deviations. Ridge regression can be employed to obtain the least squares estimate, which is close to zero. The Ridge regression coefficient estimator is formulated by adding a penalty term to the least squares objective function. This penalty term is proportional to the square of the magnitude of the coefficients and is controlled by a tuning parameter known as lambda (λ). By adjusting the value of λ, ridge regression can shrink the coefficients towards zero, thereby mitigating the impact of multicollinearity and enhancing the predictive accuracy of the model. The ridge regression coefficient estimator can be mathematically defined as:

$$minimize_\beta \quad (Y - X\beta)'(Y - X\beta)$$

$$\text{subject to} \quad \beta'\beta \leq t$$

To achieve optimization under constraints, the Lagrangian approach can be utilized as follows

$$L(Y, \beta, \lambda) = (Y - X\beta)'(Y - X\beta) - \lambda(\beta'\beta - t) \tag{1}$$

Then take the derivative equal to zero and solve the equation for the value of $\hat{\beta}$

$$-2X'Y + 2X'X\beta - 2\lambda\beta = 0 \tag{2}$$

$$(X'X - \lambda I_p)\beta = X'Y \tag{3}$$

2.2 Lasso Regression

Lasso regression is a form of linear regression that employs shrinkage, which involves shrinking data values towards a central point [14]. When adding a penalty term to the regression equation, lasso regression can effectively reduce the influence of less important variables, while promoting the retention of the most important predictors [15]. LASSO regression coefficient estimator can be defined by

$$minimize_\beta \quad (Y - X\beta)'(Y - X\beta)$$

$$\text{subject to} \quad \sum_{j=1}^{p} |\beta_j| \leq k$$

when $k \geq 0$ is a tuning parameter that controls the shrinkage size of $\hat{\beta}$ estimator. If $k_0 = \left| \sum \hat{\beta}^{ols} \right|$ when configuring $k < k_0$, the coefficient estimator will be pushed down towards zero, and some estimators are equal to zero, which can be written by the following equation:

$$\hat{\beta}^{lasso} = \arg \min_{\beta} (Y - X\beta)'(Y - X\beta) + \lambda \sum_{j=1}^{p} |\beta_j| \tag{4}$$

where λ is a nonnegative regularization parameter. The second term in (4) is the L1 penalty [16]. If λ is large enough, all coefficients are shrunk to exactly zero.

2.3 Elastic Net Regression

Elastic net linear regression is a regularization technique that combines the penalty functions of lasso and ridge regression. This approach effectively overcomes the limitations of the individual techniques by leveraging their strengths to achieve better regularization of statistical models. The method is particularly suitable for situations where the number of predictors is large, or where there is a high degree of multicollinearity between the predictors. Through careful adjustment of these parameters, the method can fine-tune the balance between variable selection and coefficient shrinkage, resulting in a more robust and accurate model [17]. Thus, the penalty functions are based on $\sum_{j=1}^{p} |\beta_j|$ and $\sum_{j=1}^{p} (\beta_j)^2$. The elastic net method becomes

$$\hat{\beta}^{elasticnet} = \arg \min_{\beta} (Y - X\beta)'(Y - X\beta) + \lambda_1 \sum_{j=1}^{p} |\beta_j| + \lambda_2 \sum_{j=1}^{p} (\beta_j)^2 \tag{5}$$

where, λ_1 and λ_2 are the tuning parameters and positive numeric values ($\lambda_1, \lambda_2 > 0$). The tuning parameters govern both the regularization strength and the selection of predictor variables [18].

2.4 SCAD Regression

SCAD is a regularization method used in regression analysis for variable selection. It aims to balance the penalties of Lasso and Ridge regression, being less severe than Lasso for small coefficients and more severe than Ridge for large coefficients. In the context of linear regression models with p nonoverlapping sets of d predictors, the SCAD method has been shown to perform well in terms of both variable selection and prediction accuracy [19]. The general objective function is

$$Q(\beta) = \frac{1}{2N}\|Y - X\beta\|_2^2 + \sum_{j=1}^{p} p_{\lambda n}(\|\beta_j\|_2), \tag{6}$$

where $X \in \mathbb{R}^{n \times p}$ is the design matrix, $Y \in \mathbb{R}^n$ is the vector of response variables, $\beta \in \mathbb{R}^p$ is the vector of coefficients, N is the number of observation and $p_\lambda(t)$ is the SCAD penalty which is defined by

$$p_\lambda'(\beta) = \lambda \left\{ I(\beta \leq \lambda) + \frac{(a\lambda - \beta)_+}{(a-1)\lambda} I(\beta > \lambda) \right\}, \tag{7}$$

where a is a constant that determines the shape of the penalty function and λ is the penalty parameter. The different values of λ can lead to different penalty function which can be shown as

$$p_{\lambda,a}'(\beta) = \begin{cases} \lambda|\beta| & \text{if } |\beta| \leq \lambda \\ \frac{a\lambda|\beta|-\beta^2-\lambda^2}{2(a-1)} & \text{if } \lambda < |\beta| \leq a\lambda \\ \frac{(a+1)\lambda^2}{2} & \text{if } |\beta| > \lambda \end{cases} \tag{8}$$

2.5 Forecast Performance Measures

In evaluating the forecasting performance of various models, it is a common practice to reserve a subset of data, denoted as n_o, for testing the out-of-sample forecasting performance. This is done using data from the first n_i observations as training and estimation data, with the remaining 20 percent of the data series (the hold-out sample) for testing. Note that $N = n_o + n_i$. . There are several ways to measure the forecasting performance of a model. Among the most commonly used criteria is the root mean squared error (RMSE), which is based on the loss function of the model and can be computed as:

$$RMSE(h) = \frac{1}{n_i + h} \sum_{h=1}^{N} \left(Y_{n_i+h} - \widehat{Y}_{n_i+1+h} \right), \tag{9}$$

where h is the forecast horizon, n_i is the number of in-sample observations. Y_{n_i+h} is the out-of-sample observations and $\widehat{Y}_{n_i+1+h}$ is the forecasting observations. In addition, as our forecasting models may have different number of selected predictors, the Adjusted R-squared is used to evaluate the goodness of fit of our model.

$$Adj - R - squared = 1 - \frac{SS_{residuals}/(n - P)}{SS_{Total}/(n - 1)}, \tag{10}$$

where $SS_{residuals}$ is $\sum_{t=1}^{N} \left(Y_t - \widehat{Y_t}\right)^2$, and SS_{Total} is $\sum_{t=1}^{N} \left(Y_t - \overline{Y}_t\right)^2$.

3 Data Description

This study focuses on Thailand, covering the time period of 1970 to 2021. Data for this study was obtained from the World Bank database. A total of nine macroeconomic variables were included in our analysis, with GDP as the dependent variable and eight independent variables: CPI, which measures the average change over time in prices paid by urban consumers for a basket of goods and services, MANUV, which represents the value-added from manufacturing, POP, which denotes the total population of Thailand, EXR, which is the official exchange rate, IM, which represents the value of imports of goods and services, EX, which denotes the value of exports of goods and services, GOVE, which is the general government final expenditure, and FOREI, which is the total foreign direct investment (sum of net inflows and outflows). In order to estimate the value of each regression model, the data was divided into training and testing sets in an 80:20 ratio. The calculated values from the training set data were then used to forecast GDP values, conditional on selected macroeconomic variables, in the testing set data. The unit of GDP, MANUV, IM, EX, GOVE, and FOREI variables are current US\$, while EXR variable is LCU per US\$ (Table 1).

Table 1. Descriptive Statistics

	GDP	CPI	MANUV	POP	EXR	IM	EX	GOVE	FOREI
Mean	25.2910	4.0040	23.9034	17.8477	3.3405	23.7193	23.6986	23.2380	23.6843
Median	25.5638	4.1699	24.2700	17.9068	3.2856	24.3298	24.2080	23.5089	24.1966
Maximum	27.0224	4.7335	25.6606	18.0866	3.7940	25.8777	25.9934	25.2446	25.9857
Minimum	22.6815	2.5476	20.8453	17.3932	3.0124	20.0756	19.8207	20.4964	19.7840
Std. Dev	1.2642	0.6557	1.4122	0.2082	0.2420	1.8480	1.9584	1.4068	1.9671
Skewness	−0.4560	−0.7511	−0.5667	−0.6501	0.1546	−0.5094	−0.4749	−0.3102	−0.4772
Kurtosis	2.1426	2.4923	2.1984	2.1841	1.8213	1.9304	1.8582	2.0707	1.8589
Jarque-Bera	3.3951	5.4473	4.1755	5.1055	3.2174	4.7277	4.7793	2.7053	4.7942
Probability	0.1831	0.0656	0.1240	0.0779	0.2001	0.0941	0.0917	0.2586	0.0910

All data in this study have been transformed into the log form to ensure that the model is more robust and can handle outliers better. The majority of the data with high average values are related to MANUV, IM, EX, FOREI, and GOVE, suggesting that these variables are likely driving Thailand's economy. From the standard deviation of the variables, we can see IM, EX, and FOREI have relatively high volatility compared to other factors. We also observe that all variables have negative skewness, indicating that the data is concentrated on the higher end of the range and has a long tail to the left. This means that for all variables, there are a few very high values that are driving the average upward.

4 Results

In the first set of results, all shrinkage estimations were compared using two metrics: RMSE and Adjusted R-squared. The findings indicate that the Ridge, Lasso, Elastic Net, SCAD, and OLS models had Adjusted R-squared values of 99.02%, 99.54%, 99.55%, 43.08%, and 98.76%, respectively. Additionally, the RMSE values used to evaluate model error were 0.1245, 0.0858, 0.0847, 0.1473, and 0.8096 for the Ridge, Lasso, Elastic Net, SCAD, and OLS models, respectively. Based on the criterion of selecting a model with the lowest RMSE and the highest Adjusted R-squared, the elastic net regression model was deemed the most accurate model for predicting Thailand's GDP from the selected macroeconomic variables, as outlined in Table 2.

Table 2. Model Accuracy

	Ridge	Lasso	Elastic Net	SCAD	OLS
RMSE	0.1245	0.0858	0.0847	0.1473	0.8096
Adjusted R-squared	0.9902	0.9954	0.9955	0.4308	0.9876

Our study's findings are consistent with prior research works that employed the Elastic Net model to forecast Gross Domestic Product (GDP). For example, [20] discovered that the Elastic Net model had the lowest error rate when compared to other models like LASSO, Ridge, and VAR. Therefore, it was deemed the most suitable model for forecasting the GDP in China's agricultural sector. Likewise, [21] demonstrated the high performance of the Elastic Net model in forecasting GDP during the financial crisis that occurred between 2007 and 2009. Another study by [22] confirmed that the Elastic Net model was useful for improving the nowcasting of GDP, particularly in situations where the number of explanatory variables is fewer than the number of samples.

The outcome of our study may not come as a surprise since one of the main advantages of the Elastic Net model is its ability to combine the strengths of Ridge and LASSO, allowing it to select the optimal performance criteria for estimating beta and producing more accurate results [23]. Elastic Net ensures that coefficients are close to each other, making the regression more stable and preventing the model from discarding significant variables compared to LASSO. The LASSO model, on the other hand, has a propensity to select only one variable with a higher value and adjust the other variables to zero when explanatory variables suffer from multicollinearity problems, resulting in important variables being lost in the model [18]. Additionally, Elastic Net is more flexible than OLS in variable selection and can prevent overfitting [24]. Our results also confirm the higher performance of the Elastic Net model in this particular application study.

Table 3. Estimated coefficients β of macroeconomic variables from the 5 models.

β	Ridge	Lasso	Elastic Net	SCAD	OLS
Intercept	−2.1234	4.7520	4.8170	4.3365	−21.5173
CPI	0.2676	-	-	-	−0.2857
MANUV	0.1649	0.3202	0.3005	0.6932	0.3570
POP	0.6680	-	-	-	1.6672
EXR	−0.3396	-	−0.0996	0.1807	−0.4158
IM	0.0890	0.0107	0.0499	−0.3293	0.6267
EX	0.0815	-	0.0168	0.1409	0.2531
GOVE	0.2195	0.5403	0.5052	0.5438	−0.0677
FOREI	0.0630	-	-	−0.1334	−0.4259

Table 3 presents the selected variables from the five regression models used to forecast GDP. As mentioned earlier, Ridge and OLS regressions include all independent variables to predict GDP. However, the effect of each variable on GDP varies across the models. For example, in Ridge regression, the exchange rate has a negative impact on GDP, while in OLS regression, variables such as CPI, exchange rate, government expenditure, and foreign direct investment negatively affect GDP. Conversely, Lasso regression uses industrial value added, import value, and government spending to predict GDP, and all variables have a positive effect on GDP. In SCAD regression, manufacturing value-added, exchange rate, export value, and government expenditure positively impact GDP, whereas import value and foreign direct investment have a negative influence. Notably, the Elastic Net regression, which is the most accurate model in this study, only employs industrial value added, exchange rate, import value, export value, and government spending to predict GDP, with the exchange rate having a negative influence.

These estimation results support the notion that machine learning techniques produce better predictions compared to OLS, going beyond the traditional evaluation metrics of RMSE and Adjusted R-squared. For instance, previous research has shown that government expenditure is critical for Thailand's economic growth in the short and the long term [25]. Government budget is also considered a significant financial tool for Thailand, and the budget allocated for research and development plays a crucial role in enhancing human resources and boosting the country's competitiveness (Jaroensathapornkul, 2010). All predicted GDP values from the five models are illustrated in Fig. 1. Figure 1 shows the out-of-sample forecasts of actual GDP versus predicted GDP from the five models used in this study.

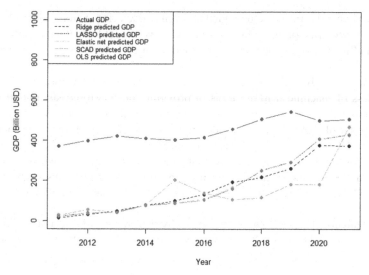

Fig. 1. Out-of-Sample forecasts: Actual GDP vs. Predicted GDP from five models.

5 Conclusion

Forecasting Gross Domestic Product (GDP) is critical for policymakers, investors, and businesses to make informed decisions about the economy. As a key indicator of a country's economic health, accurate GDP forecasting can help identify potential economic trends, opportunities, and challenges. Precise GDP forecasts can also assist policymakers in formulating effective economic policies, businesses in making informed investment decisions, and investors in making profitable trades in financial markets.

This study aimed to identify the most accurate forecasting model for Thailand's GDP, utilizing machine learning tools such as Ridge, Lasso, Elastic Net, and SCAD regressions. The study employed eight macroeconomic variables from 1970 to 2021 and divided them into training and testing sets at an 80:20 ratio. The models' accuracy was compared using RMSE and Adjusted R-squared.

The study's results revealed that the Elastic Net regression model is the most accurate in predicting Thailand's GDP, with a 99.55% accuracy rate. The model utilized only industrial value added, exchange rate, import value, export value, and government spending as independent variables. These findings demonstrate the superiority of machine learning tools over traditional econometric models in providing accurate GDP forecasts.

The implications of this study are significant for policymakers, investors, and businesses. The accuracy of the Elastic Net model can assist policymakers in identifying the key factors that influence economic growth and formulating effective economic policies. Investors can leverage the model to make profitable trades in financial markets, and businesses can make well-informed investment decisions based on its insights. Moreover, the study's findings can aid policymakers in identifying the most crucial variables that impact the country's economic growth, allowing them to develop crucial strategies for the country's development. For example, the elastic net model's largest beta is in government expenditure and value-added in the manufacturing sector. Thus, policymakers should concentrate on these factors to make government budget expenditures more efficient and to create the highest value-added from the country's key manufacturing sector.

In sum, this study highlights the significance of accurate GDP forecasting and the effectiveness of machine learning tools in providing precise forecasts. Future studies can expand on this research by incorporating additional macroeconomic variables or using other shrinkage estimations to identify the most accurate forecasting model for Thailand's GDP. The examples of macroeconomic variables include the policy interest rate, Thai household debt, the employment rate of new graduates, and the minimum wages in each year. These variables have not been investigated in this study, but they have been identified in the past as having an impact on Thailand's GDP development.

Acknowledgments. This research work was partially supported by Chiang Mai University and the authors are grateful to the Centre of Excellence in Econometrics for the suggestion support.

References

1. Bai, J., Ng, S.: Forecasting economic time series using targeted predictors. J. Econometrics **146**(2), 304–317 (2008)
2. OECD: OECD Economic Outlook, vol. 2011, Issue 1. OECD Publishing, Paris (2011). https://doi.org/10.1787/eco_outlook-v2011-1-en
3. Rakpho, P., Yamaka, W.: The forecasting power of economic policy uncertainty for energy demand and supply. Energy Rep. **7**, 338–343 (2021)
4. Goulet Coulombe, P., Leroux, M., Stevanovic, D., Surprenant, S.: How is machine learning useful for macroeconomic forecasting? J. Appl. Economet. **37**(5), 920–964 (2022)
5. Maneejuk, P.: On regularization of generalized maximum entropy for linear models. Soft. Comput. **25**(12), 7867–7875 (2021). https://doi.org/10.1007/s00500-021-05805-2
6. Yamaka, W.: Sparse estimations in kink regression model. Soft. Comput. **25**(12), 7825–7838 (2021). https://doi.org/10.1007/s00500-021-05797-z
7. Li, J., Chen, W.: Forecasting macroeconomic time series: LASSO-based approaches and their forecast combinations with dynamic factor models. Int. J. Forecast. **30**(4), 996–1015 (2014)
8. Weiye, C.: Penalized Methods and their Applications to Genetic Research and Economic Forecasting. University of Notre Dame (2015)
9. Muhammadullah, S., Urooj, A., Khan, F., Alshahrani, M.N., Alqawba, M., Al-Marzouki, S.: Comparison of weighted lag adaptive LASSO with Autometrics for covariate selection and forecasting using time-series data. Complexity **2022**, 1–10 (2022)
10. Richardson, A., van Florenstein Mulder, T., Vehbi, T.: Nowcasting GDP using machine-learning algorithms: a real-time assessment. Int. J. Forecast. **37**(2), 941–948 (2021)
11. Agu, S.C., Onu, F.U., Ezemagu, U.K., Oden, D.: Predicting gross domestic product to macroeconomic indicators. Intell. Syst. Appl. **14**, 200082 (2022)
12. Yoon, J.: Forecasting of real gdp growth using machine learning models: gradient boosting and random forest approach. Comput. Econ **57**(1), 247–265 (2020). https://doi.org/10.1007/s10614-020-10054-w
13. Hoerl, R.W.: Ridge analysis 25 years later. Am. Stat. **39**(3), 186–192 (1985)
14. Glen, S.: Lasso Regression: Simple Definition (2022). https://www.statisticshowto.com/lasso-regression/. Retrieved 19 Feb 2023 from StatisticsHowTo.com: Elementary Statistics for the rest of us!

15. Zou, H.: The adaptive lasso and its oracle properties. J. Am. Stat. Assoc. **101**(476), 1418–1429 (2006)
16. Chen, S.S., Donoho, D.L., Saunders, M.A.: Atomic decomposition by basis pursuit. SIAM Rev. **43**(1), 129–159 (2001)
17. Al-Jawarneh, A.S., Ismail, M.T., Awajan, A.M.: Elastic net regression and empirical model decomposition for enhancing the accuracy of the model selection. Int. J. Math. Eng. Manag. Sci. **6**(2), 564 (2021)
18. Zou, H., Hastie, T.: Regularization and variable selection via the elastic net. J. Royal Stat. Soc.: Ser. B (Stat. Methodol.) **67**(2), 301–320 (2005)
19. Guo, X., Zhang, H., Wang, Y., Wu, J.L.: Model selection and estimation in high dimensional regression models with group SCAD. Statist. Probab. Lett. **103**, 86–92 (2015)
20. Qiu, Z.: An elastic net based algorithm for china agriculture GDP prediction. In: 2022 International Conference on Economics, Smart Finance and Contemporary Trade (ESFCT 2022), pp. 843–849. Atlantis Press (2022)
21. Jung, J.-K., Patnam, M., Ter-Martirosyan, A.: An algorithmic crystal ball: Forecasts-based on machine learning. IMF Working Papers **18**(230), 1 (2018)
22. Nakazawa, T.: Constructing GDP Nowcasting Models Using Alternative Data (No. 22-E-9). Bank of Japan (2022)
23. Martin, L.C.: Machine learning vs. traditional forecasting methods: An application to South African GDP (No. 12/2019) (2019)
24. Tiffin, A.: Seeing in the Dark: A Machine-Learning Approach to Nowcasting in Lebanon. IMF Working Paper (2016)
25. Suanin, W.: Growth-Government Spending Nexus: the Evidence of Thailand. Thailand World Econ. **35**(3), 1–33 (2017)

Revolutionizing SET50 Stock Portfolio Management with Deep Reinforcement Learning

Sukrit Thongkairat[1,3], Donlapark Ponnoprat[2], Phimphaka Taninpong[2],
and Woraphon Yamaka[3(✉)]

[1] Program in Applied Statistics, Department of Statistics, Faculty of Science,
Chiang Mai University, Chiang Mai 50200, Thailand
[2] Department of Statistics, Faculty of Science, Chiang Mai University,
Chiang Mai 50200, Thailand
[3] Center of Excellence in Econometrics, Faculty of Economics,
Chiang Mai University, Chiang Mai 50200, Thailand
woraphon.econ@gmail.com

Abstract. In this study, we explore the potential of using Reinforcement Learning (RL) algorithms to develop a stock trading strategy that maximizes investment returns. We apply RL to monitor all 50 stocks listed in the SET 50, a stock market index in Thailand, and train deep learning agents to acquire this trading technique. Our results demonstrate that RL can successfully develop a trading strategy that outperforms the fundamental strategies, offering a promising alternative tool for investors and fund managers seeking to navigate the stock market.

Keywords: Reinforcement learning · Automated trading · Machine Learning · SET50 format

1 Introduction

The Stock Exchange of Thailand (SET) 50 Index is a benchmark index used to evaluate the performance of the top 50 publicly traded companies listed on the SET. It serves as an indicator of the Thai stock market and provides insights into the country's overall economic growth. Portfolio allocation involves dividing an investment portfolio into various asset classes, such as stocks, bonds, commodities, and others, with the aim of achieving a balance between risk diversification and return optimization.

The allocation of a SET50 stock portfolio can be performed using traditional methods such as equal weighting, market capitalization weighting, or risk-based methods, as well as more sophisticated methods such as machine learning-based techniques. Machine learning algorithms have grown in popularity in recent years in the field of automated trading due to their potential to reduce human error and provide quicker, more accurate decision-making than manual trading [5].

Supported by Chiang Mai University.

Deep reinforcement learning (DRL) is a machine learning technique that has shown the most promise in addressing complicated and rapidly shifting financial market conditions. Its usefulness in financial applications, including stock trading [19,37,41], cryptocurrency trading [3,18,30,33], and futures trading [15,27,34], has been shown by the findings of various academic studies. This has enhanced its standing as a top technique in the field of automated trading.

The COVID-19 pandemic has brought about unprecedented changes to the global economy, which has resulted in high and wild fluctuations in financial analysis. This presents significant challenges for traditional methods of financial analysis and trading, as these approaches are often based on the assumption that the trend of volatility remains stable and unchanging. This means that it has become increasingly difficult to accurately predict future outcomes in financial markets at the presence of extreme events.

Considering these challenges, we believe it is necessary to employ advanced variants of deep reinforcement learning (DRL) to improve the accuracy of financial predictions and trading outcomes. The variants of DRL that we plan to utilize include Advantage Actor Critic (A2C), Deep Deterministic Policy Gradient (DDPG), Twin Delayed DDPG (TD3), Soft Actor Critic (SAC), and Proximal Policy Optimization (PPO). These methods were developed to address financial challenges such as overestimation, performance analysis, and trading process stability.

This study aims to assess the suitability of A2C, DDPG, TD3, SAC, and PPO as an alternative to the traditional Min-Variance method in financial trading. In addition, the performance of these DRL methods will be tested using the returns data of the SET50 Index. Moreover, this study provides a unique opportunity to evaluate the performance of DRL in a highly volatile Southeast Asian stock market during the COVID-19 pandemic. Furthermore, the results will shed light on the potential benefits and limitations of using DRL in financial trading.

The paper is structured as follows: Sect. 2 provides a comprehensive review of existing literature on deep reinforcement learning. Section 3 gives a concise overview of the reinforcement learning process. Section 4 presents a comparison of the efficiency of reinforcement learning strategies with traditional techniques. Finally, the conclusion of the study can be found in Sect. 5.

2   Literature Reviews

The integration of deep neural networks with reinforcement learning gave rise to the development of Deep Reinforcement Learning (DRL) in the mid-2000s, which has been effective in such fields as computer vision, natural language processing, and speech recognition. However, DRL also faces algorithmic stability issues, and researchers have proposed several solutions to improve stability. Riedmiller [29], Mnih et al. [23,24], Van Hasselt et al. [36], and Schulman et al. [31] have provided methods to mitigate this issue. These methods adopt a common approach, assuming that RL agents will encounter observations in a variable order and that RL updates are highly correlated. To mitigate this issue, the agent's data can be processed in batches (as demonstrated in Riedmiller [29] and Schulman et al. [31]) or randomly sampled (as demonstrated in Mnih et al. [23, 24] and Van Hasselt et al. [36]) and stored in an experience replay memory. This method aggregates memories, reducing non-stability and decorrelating updates. However, it is important to note that this technique is only applicable to off-policy reinforcement learning algorithms.

In financial markets, DRL can be applied using three approaches: the critic-only approach, the actor-only approach, or the actor-critic approach, depending on the nature of the state and action spaces, which can either be discrete or continuous [11]. Continuous action spaces offer more precise control compared to those with discrete action spaces, making them a more popular choice for DRL models in financial markets.

Chen and Gao [6], Dang [8], and Jeong and Kim [17] conducted research on the most commonly utilized critic-only approach in DRL for financial markets, which involves training an agent on a single stock or asset and employing Deep Q-Learning (DQN) and its variations to tackle the challenge of discrete action spaces. However, this approach has a crucial drawback in that it can only handle finite and discrete state and action spaces, making it unsuitable for managing large stock portfolios.

The "curse of dimensionality" is another challenge that occurs when attempting to model the behavior of multiple assets in a portfolio [1]. To mitigate this challenge, various techniques such as feature selection, dimensionality reduction, and regularization can be used to reduce the number of dimensions while preserving relevant information. In the work of Moody and Saffell [26], recurrent reinforcement learning was used to address the curse of dimensionality and improve the effectiveness of trading in finance. The actor-only approach, as described by Moody and Saffell [26], Deng et al. [9], Jiang and Liang [18], and other related works, trains the agent's policy to adjust to the best action by learning the Q-value and is ideal for dealing with continuous action environments.

Recent studies, such as Bekiros [2], Li and Shi [19], Xiong et al. [39], and Zhang, Zohren, and Roberts [41] have demonstrated that the actor-critic approach is a popular and effective method in financial markets. This approach combines both the actor network, which represents the policy, and the critic network, which represents the value function, and adjusts them simultaneously for maximum performance of the policy and value function networks. The actor-critic approach's an ability to effectively manage both the actor and the critic networks that makes it a highly appealing method in financial markets.

3   Methods

This section offers a succinct overview of the main concepts utilized in this study. The objective is to give a general understanding of the concept, rather than a comprehensive explanation. For a more in-depth explanation, please refer to the resources provided as appropriate.

3.1   Reinforcement Learning and Markov Decision Process (MDP)

Understanding the components of a Markov Decision Process (MDP) is essential for comprehending the reinforcement learning process. MDP is a mathematical framework that models decision-making problems in dynamic environments with uncertain outcomes, and it is frequently used in the study of optimization problems that can be addressed through dynamic programming.

There are five essential components of an MDP: the agent, environment, state, actions, and policy. For instance, in the case of a marketing campaign for a new product, these components are:

- The agent, whose task is to optimize the marketing campaign for the new product.
- The environment, which is the market in which the product will be launched.
- The state, which is the current state of the market, such as the level of competition or the behavior of potential customers.
- The actions that can be taken, which may include adjusting pricing, modifying product features, or changing the advertising strategy.
- The policy, which is the decision-making process used to determine which action to take at each time step (Fig. 1).

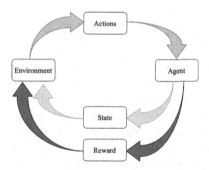

Fig. 1. Markov Decision Process

The goal of the MDP in this scenario is to maximize the sales and profits of the product by learning from the feedback obtained from the market. By carefully designing the MDP's components and the policy that drives decision-making, the agent can learn to make optimal decisions that will lead to the best possible outcome for the marketing campaign and the product itself.

3.2 Learning Strategies on Deep Reinforcement Learning

Before delving into deep reinforcement learning strategies, we must first discuss the Bellman equation, which describes the relationship between the value of a state or state-action pair and the expected future rewards that can be obtained by following a policy. The Ballman equation is consist of two functions: a state-value function $V(s_i)$ and an action-value function $Q(s_i, a_i)$. State value function $V(s_i)$:

$$V(s_i) = \mathbb{E}_\pi [\sum_{t=0}^{\infty} \gamma^t r_{t+1} | s_i] \tag{1}$$

Action value function $Q(s_i, a_i)$:

$$Q(s_i, a_i) = \mathbb{E}_\pi \left[\sum_{t=0}^{\infty} \gamma^t r_{t+1} | s_i, a_i \right] \tag{2}$$

where, $\mathbb{E}_\pi$ represents the expectation taken over all possible sequences of future states and rewards under the policy π, $\sum_{t=0}^{\infty}$ indicates the summing over all possible time steps t, and γ is the discount factor that determines the weight given to future rewards. The reward obtained after taking action a_t in state s_t and transitioning to state s_{t+1} is denoted by r_{t+1} [1].

Advantage Actor Critic (A2C). Advantage Actor-Critic (A2C) is a synchronous, online algorithm that learns a policy and a value function simultaneously. It combines the advantages of policy gradient methods and value-based methods. The policy is optimized using the advantage estimation which measures how much better the action taken in a particular state is than the expected value of the state [25]. The A2C loss function is defined as:

$$L_{A2C} = \frac{1}{N} \sum_{i=1}^{N} \left[\log \pi_\theta (a_i | s_i) \hat{A}_i \right] - c \frac{1}{N} \sum_{i=1}^{N} \left[V(s_i) - V_{\text{target}} (s_i) \right]^2 \tag{3}$$

where $\pi_\theta(a_i|s_i)$ is the policy network that outputs the probability distribution of actions given the state, $\hat{A}i$ is the advantage function, which is calculated as the difference between the estimated value function $V(s_i)$ and the discounted sum of rewards R_i received after taking action a_i in state s_i, β is the entropy coefficient that encourages exploration by maximizing the entropy of the policy distribution, $V(s_i)$ is the value function, which estimates the expected sum of rewards from state s_i, $V_{target}(s_i)$ is the target value used to train the critic network, c is the coefficient used to balance the actor and critic losses, and N is the batch size.

Proximal Policy Optimization (PPO). Proximal Policy Optimization (PPO) is another policy gradient method that updates the policy in a way that is more stable than traditional policy gradient methods. PPO uses a surrogate objective that limits the change in the policy, which helps to prevent large policy updates that can be harmful to the learning process [32].

$$L^{PPO}(\theta) = \mathbb{E}_t \left[\min \left(r_t(\theta) \hat{A}_t, \text{clip}(r_t(\theta), 1 - \varepsilon, 1 + \varepsilon) \hat{A}_t \right) \right] - \lambda S(\pi_\theta) \tag{4}$$

where $L^{PPO}(\theta)$ is the PPO loss function with respect to the policy parameters θ, $\mathbb{E}_t$ is the expectation over a batch of timesteps t, $r_t(\theta)$ is the ratio of the probability of taking the action under the new and old policies, i.e., $\hat{A}_t$ is the estimated advantage of taking action a_t in state s_t, $\text{clip}(x, a, b)$ clips x to between a and b, ε is a hyperparameter controlling the amount of clipping, λ is a hyperparameter controlling the strength of the entropy bonus term, and $S(\pi_\theta)$ is the entropy of the policy, defined as $S(\pi_\theta) = -\mathbb{E}s, a \sim \pi\theta[\log \pi_\theta(a|s)]$ and r_t is defined as follows:

$$r_t(\theta) = \frac{\pi_\theta(a_t|s_t)}{\pi_{\theta_{old}}(a_t|s_t)} \tag{5}$$

is the probability ratio between the new policy and the old policy, and ε is a hyperparameter controlling the size of the clipping range.

Deep Deterministic Policy Gradient (DDPG). Deep Deterministic Policy Gradient (DDPG) is an off-policy algorithm that learns a deterministic policy and a value function. It uses a replay buffer to store past experiences, which helps to stabilize learning. DDPG is particularly well-suited to continuous control tasks [21]. The DDPG loss function is defined as:

$$L_{DDPG} = -\frac{1}{N} \sum_{i=1}^{N} Q\left(s_i, a_i \mid \theta^Q\right) + \frac{1}{N} \sum i = 1^N \left(y_i - Q\left(s_i, a_i \mid \theta^Q\right)\right)^2 \tag{6}$$

where N is the batch size, s_i is the i state in the batch, a_i is the i action in the batch, $Q(s,a|\theta^Q)$ is the critic network with parameters θ^Q, which estimates the state-action value function, $\mu(s|\theta^\mu)$ is the actor network with parameters θ^μ, which estimates the optimal action for a given state, $y_i = r_i + \gamma Q'(s_{i+1}, \mu'(s_{i+1}|\theta^{\mu'})|\theta^{Q'})$ is the target value for the critic, where r_i is the reward obtained after taking action a_i in state s_i, γ is the discount factor, $Q'(s_{i+1}, \mu'(s_{i+1}|\theta^{\mu'})|\theta^{Q'})$ is the target value obtained from the target critic and the target actor networks (which are copies of the critic and the actor networks with slowly-changing parameters), and $\mu'(s_{i+1}|\theta^{\mu'})$ is the target actor network with parameters $\theta^{\mu'}$.

Twin Delayed DDPG (TD3). Twin Delayed Deep Deterministic Policy Gradient (TD3) is an off-policy algorithm that is similar to DDPG but uses two value networks to reduce overestimation bias. It also uses delayed updates to the policy and target networks, which helps to improve stability. TD3 has been shown to be particularly effective for tasks with high-dimensional action spaces [12].

$$L_{TD3}(\theta) = \mathbb{E}_{s_t, a_t \sim \rho_\pi} \left[\frac{1}{2} \left(Q_{\theta_1}(s_t, a_t) - y_t \right)^2 \right] \tag{7}$$

where $y_t = r_t + \gamma \min_{i=1}^{k} Q_{\theta_2}(s_{t+1}, \mu_{\phi_i}(s_{t+1}))$ is the target value, Q_{θ_1} and Q_{θ_2} are the two Q-networks, ρ_π is the behavior of policy, μ_{ϕ_i} are the target policies, and $\theta = (\theta_1, \theta_2)$ are the parameters of the two Q-networks.

Soft Actor-Critic (SAC). Soft Actor-Critic (SAC) is an off-policy algorithm that learns a stochastic policy and a value function. It maximizes an entropy regularized objective that encourages exploration and allows for better exploration of the state space. SAC has been shown to be particularly effective for tasks with sparse rewards [14].

Critic loss:

$$L_{critic}(\theta_q) = \mathbb{E}s, a, r, s' [(Q_{\theta_q}(s,a) - y(r,s'))^2] \tag{8}$$

where $y(r,s') = r + \gamma(1-d) \min_{i=1,...,n} Q_{\theta'_q}(s', a'_i)$, and $a'i \sim \pi\phi'(a'|s')$ are the n next state actions sampled from the target policy.

Actor loss:

$$L_{actor}(\phi) = \mathbb{E}s[\mathscr{H}(\pi\phi(\cdot|s)) - Q_\theta(s, \pi_\phi(s))] \tag{9}$$

where $\mathscr{H}$ is the entropy of the policy.

Entropy loss:

$$L_{entropy}(\phi) = \mathbb{E}s[-\mathscr{H}(\pi\phi(\cdot|s))] \tag{10}$$

Overall loss:

$$L_{SAC}(\theta_q, \phi) = L_{critic}(\theta_q) + \alpha L_{actor}(\phi) + \beta L_{entropy}(\phi) \tag{11}$$

where α and β are hyperparameters that weight the actor and entropy losses, respectively.

3.3   Measures for Performance Evaluation

Cumulative Return (*CR*) is a metric that evaluates the overall performance of an investment over a specific time frame, considering both capital gains and any generated income. The calculation involves taking the ending value of an investment (*EV*) minus the starting value (*SV*) to be divided by the starting value (*SV*), multiplying 100, and expressing the result as a percentage:

$$CR = \frac{EV - SV}{SV} \times 100 \tag{12}$$

Annual Revenue (*AR*) is an essential financial indicator that reflects the total amount of money a company generates annually from the sale of its products or services. This metric is crucial in determining a company's financial stability, growth, and profitability. The calculation is straightforward:

$$AR = \sum_{i=1}^{n} R_i \tag{13}$$

where n is the number of periods for which revenue is calculated, and R_i is the revenue generated in the i-th period. **Annualized Volatility** (*AV*) is a statistical measurement of the amount of variation in the returns of an investment over a given time period. This metric is commonly expressed as a percentage and is widely used to assess the risk associated with an investment. The calculation involves taking the standard deviation (*SD*) of the returns over the period for multiplication by the square root of the number of periods in a year (*N*):

$$AV = SD \times \sqrt{N} \tag{14}$$

Sharpe Ratio (*SR*) is a metric that evaluates the risk-adjusted return of an investment by comparing its expected return (*ER*) to its standard deviation (*SD*), which represents its risk. A higher Sharpe Ratio indicates a more attractive risk-return tradeoff. The calculation involves subtracting the risk-free rate (*RF*) from the expected return of the investment to be divided by the standard deviation:

$$SR = \frac{ER - RF}{SD} \tag{15}$$

Maximum Drawdown (*MD*) is a measure of the largest peak-to-trough decline of an investment over a specific time period. This metric provides a way to assess the risk of an investment by determining the largest loss that it has experienced over a given time frame. The calculation involves finding the maximum difference between a peak value (P) and any subsequent trough value (T) and expressing the result as a percentage of the peak value:

$$MD = \frac{P - T}{P} \times 100\% \tag{16}$$

4 Empirical Results

This study developed a deep reinforcement learning algorithm for trading using data of 50 stocks listed in the SET50 from 2008 to 2022. The algorithm was trained on an in-sample set and tested on an out-of-sample set, and five reinforcement learning agents (A2C, PPO, DDPG, SAC, and TD3) were used to evaluate performance using various metrics such as the Sharpe ratio, learning rate, and episode count. The study aimed to assess the algorithm's profitability and potential as a trading tool by training it to make better use of real-trading data, enhancing its ability to adjust to market developments.

The results showed that the DDPG agent was the most adaptable reinforcement learning agent, exhibiting a minimum annual volatility of 25.75% and a maximum drawdown of −37.11%. The DDPG agent's ability to adjust the investment ratio effectively in response to changes in the economic situation was demonstrated, and it generated impressive returns, with the highest annual return of 8.80% and cumulative return of 17.58%, outperforming the other reinforcement learning agents and fundamental strategies. Overall, this study demonstrates the potential of using deep reinforcement learning algorithms for trading and highlights the importance of utilizing real-trading data in training these algorithms (Table 1).

Table 1. Performance Evaluation Comparison.

(2020/01/01-2022/01/01)	SET50 average	Min-Variance	A2C	PPO	DDPG	SAC	TD3
Cumulative returns	3.60%	−6.16%	9.77%	9.66%	**17.58%**	10.50%	13.34%
Annual return	1.87%	−3.25%	4.97%	4.92%	**8.80%**	5.33%	6.74%
Annual volatility	22.79%	**20.82%**	25.75%	26.00%	25.75%	25.46%	26.08%
Sharpe ratio	0.20	−0.05	0.32	0.32	**0.46**	0.33	0.38
Max drawdown	−35.99%	−36.10%	−39.21%	−38.04%	−37.11%	−36.28%	−38.02%
Daily value at risk	−2.85%	−2.63%	−3.21%	−3.24%	−3.20%	−3.17%	**−3.25%**

Additionally, the results presented in Fig. 2 clearly demonstrate the superiority of the reinforcement learning strategies over the methods of SET50 and minimum variance portfolio allocations. The Sharpe ratio of the reinforcement learning strategy ranges from 0.32 to 0.46, which is significantly higher than the Sharpe ratio of 0.20 for SET50 and -0.05 for minimum portfolio allocation, indicating that the reinforcement learning strategy is more profitable. Furthermore, the DDPG reinforcement learning agent's effectiveness is observed in the higher yearly returns compared to the other reinforcement learning approaches. However, TD3's annual volatility is comparatively higher than the other approaches.

Overall, the study demonstrates that the proposed reinforcement learning strategy outperforms fundamental strategies and successfully develops a trading strategy that can effectively navigate the stock market. By training the algorithm to use real-trading data, the study enhances the algorithm's ability to adjust to market developments, providing an alternative tool for trading.

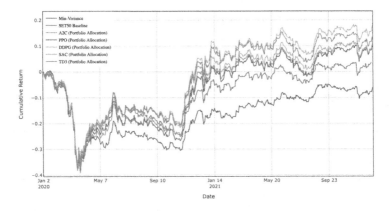

Fig. 2. Comparison of ML algorithm, min-variance portfolio, and SET50 average cumulative returns. (Initial portfolio value $1,000,000 from January 1, 2020, to January 1, 2022.)

Table 2. Comparison of Strengths and Weaknesses of RL Algorithms

Algorithm	Strengths	Weaknesses
A2C	- Adapts to changing markets effectively	- May overly focus on recent data
		- Sensitivity to hyperparameters
		- Moderate volatility performance
PPO	- Effective risk management	- Complex design can lead to implementation challenges
	- Adapts well to market changes	- Sensitive to hyperparameters
		- High volatility performance
DDPG	- High returns in certain scenarios	- Complex algorithm design
	- Adapts to economic shifts	- Sensitivity to hyperparameters
	- Supports decision-making	- Moderate volatility performance
SAC	- Good risk-adjusted returns	- Complex design may hinder ease of implementation
	- Flexible adaptation to different environments	- Sensitive to hyperparameters
TD3	- Strong returns in various economic conditions	- Complex algorithm design
	- Adapts effectively to different markets	- Sensitive to hyperparameters
		- Extreme volatility performance

5   Conclusion

This study demonstrates the potential of using deep reinforcement learning algorithms in stock trading. By subjecting five distinct reinforcement learning agents (A2C, PPO,

DDPG, SAC, and TD3) to rigorous training and evaluation using a robust dataset comprising 50 stocks listed in the SET50 index, our research underscores the remarkable adaptability demonstrated by the DDPG agent. Notably, the DDPG agent emerges as a frontrunner, boasting an impressive annual return of 8.80% and a cumulative return of 17.58%. It is imperative that we delve into the nuanced mechanisms underpinning DDPG's success, shedding light on its adaptive prowess in navigating the complexities of dynamic market conditions. However, Table 2 presents a visual representation of the distinct strengths exhibited by alternative agents. A2C, PPO, SAC, and TD3 each contribute unique attributes to create a comprehensive portfolio of strategies with varying risk-reward profiles.

A pivotal aspect of this study pertains to risk management, The compelling Sharpe ratios ranging from 0.32 to 0.46 affirm the agents' effectiveness in achieving a favorable balance between risk and reward. This becomes particularly evident when juxtaposed with the Sharpe ratio of 0.20 for the SET50 index and -0.05 for minimum portfolio allocation, highlighting the agents' ability to outperform traditional benchmarks.

In summary, this study shows that using deep reinforcement learning algorithms into stock trading yields enhanced performance compared to traditional fundamental strategies. Future work could focus on improving the reward function, optimizing transaction costs, and using better risk measurement techniques such as value-at-risk. This study also provides valuable insights into the potential for using advanced machine learning techniques in stock trading and offers a foundation for future research in this area.

Acknowledgements. This research work was partially supported by Chiang Mai University.

References

1. Bellman, R.: Dynamic programming. Science **153**(3731), 34–37 (1966)
2. Bekiros, S.D.: Heterogeneous trading strategies with adaptive fuzzy actor-critic reinforcement learning: a behavioral approach. J. Econ. Dyn. Control **34**(6), 1153–1170 (2010)
3. Borrageiro, G., Firoozye, N., Barucca, P.: The recurrent reinforcement learning crypto agent. IEEE Access **10**, 38590–38599 (2022)
4. Brockman, G., et al.: Openai gym. arXiv preprint arXiv:1606.01540 (2016)
5. Buehler, H., Gonon, L., Teichmann, J., Wood, B.: Deep hedging. Quantit. Financ. **19**(8), 1271–1291 (2019)
6. Chen, L., Gao, Q.: Application of deep reinforcement learning on automated stock trading. In: 2019 IEEE 10th International Conference on Software Engineering and Service Science (ICSESS), pp. 29–33. IEEE (2019)
7. Chong, T.T.L., Ng, W.K., Liew, V.K.S.: Revisiting the performance of MACD and RSI oscillators. J. Risk Financ. Manag. **7**(1), 1–12 (2014)
8. Dang, Q.-V.: Reinforcement learning in stock trading. In: Le Thi, H.A., Le, H.M., Pham Dinh, T., Nguyen, N.T. (eds.) ICCSAMA 2019. AISC, vol. 1121, pp. 311–322. Springer, Cham (2020). https://doi.org/10.1007/978-3-030-38364-0_28
9. Deng, Y., Bao, F., Kong, Y., Ren, Z., Dai, Q.: Deep direct reinforcement learning for financial signal representation and trading. IEEE Trans. Neural Netw. Learn. Syst. **28**(3), 653–664 (2016)
10. Dhariwal, P., et al.: Openai baselines (2017). https://github.com/openai/baselines
11. Fischer, T.G.: Reinforcement learning in financial markets-a survey (No. 12/2018). FAU Discussion Papers in Economics (2018)

12. Fujimoto, S., van Hoof, H., Meger, D.: Addressing function approximation error in actor-critic methods. In: International Conference on Machine Learning, pp. 1587–1596 (2018)

13. Gurrib, I.: Performance of the average directional Index as a market timing tool for the most actively traded USD based currency pairs. Banks Bank Syst. **13**(3), 58–70 (2018)

14. Haarnoja, T., Zhou, A., Abbeel, P., Levine, S.: Soft actor-critic: off-policy maximum entropy deep reinforcement learning with a stochastic actor. In: International Conference on Machine Learning, pp. 1861–1870 (2018)

15. Hirsa, A., Osterrieder, J., Hadji-Misheva, B., Posth, J.A.: Deep reinforcement learning on a multi-asset environment for trading. arXiv preprint arXiv:2106.08437 (2021)

16. Jagtap, R.: Understanding Markov Decision Process (MDP). Towards data science (2020). https://towardsdatascience.com/understanding-the-markov-decision-process-mdp-8f838510f150

17. Jeong, G., Kim, H.Y.: Improving financial trading decisions using deep Q-learning: predicting the number of shares, action strategies, and transfer learning. Expert Syst. Appl. **117**, 125–138 (2019)

18. Jiang, Z., Liang, J.: Cryptocurrency portfolio management with deep reinforcement learning. In: 2017 Intelligent Systems Conference (IntelliSys), pp. 905–91. IEEE (2017)

19. Li, J., Rao, R., Shi, J.: Learning to trade with deep actor critic methods. In: 2018 11th International Symposium on Computational Intelligence and Design (ISCID), vol. 2, pp. 66–71. IEEE (2018)

20. Li, Y., Ni, P., Chang, V.: Application of deep reinforcement learning in stock trading strategies and stock forecasting. Computing **102**(6), 1305–1322 (2020)

21. Lillicrap, T.P., et al.: Continuous control with deep reinforcement learning. arXiv preprint arXiv:1509.02971 (2015)

22. Maitah, M., Prochazka, P., Cermak, M., Šrédl, K.: Commodity channel index: evaluation of trading rule of agricultural commodities. Int. J. Econ. Financ. Issues **6**(1), 176–178 (2016)

23. Mnih, V., et al.: Playing atari with deep reinforcement learning. arXiv preprint arXiv:1312.5602 (2013)

24. Mnih, V., et al.: Human-level control through deep reinforcement learning. Nature **518**(7540), 529–533 (2015)

25. Mnih, V., et al.: Asynchronous methods for deep reinforcement learning. In: International Conference on Machine Learning, pp. 1928–1937. PMLR (2016)

26. Moody, J., Saffell, M.: Learning to trade via direct reinforcement. IEEE Trans. Neural Netw. **12**(4), 875–889 (2001)

27. Moriyama, K., Matsumoto, M., Fukui, K.I., Kurihara, S., Numao, M.: Reinforcement learning on a futures market simulator. J. Univers. Comput. Sci. **14**(7), 1136–1153 (2008)

28. Raffin, A., Hill, A., Ernestus, M., Gleave, A., Kanervisto, A., Dormann, N.: Stable baselines3 (2019)

29. Riedmiller, M.: Neural reinforcement learning to swing-up and balance a real pole. In: 2005 IEEE International Conference on Systems, Man and Cybernetics, vol. 4, pp. 3191–3196. IEEE (2005)

30. Sadighian, J.: Deep reinforcement learning in cryptocurrency market making. arXiv preprint arXiv:1911.08647 (2019)

31. Schulman, J., Levine, S., Abbeel, P., Jordan, M., Moritz, P.: Trust region policy optimization. In: International Conference on Machine Learning, pp. 1889–1897. PMLR (2015)

32. Schulman, J., Wolski, F., Dhariwal, P., Radford, A., Klimov, O.: Proximal policy optimization algorithms. arXiv preprint arXiv:1707.06347 (2017)

33. Shahbazi, Z., Byun, Y.C.: Improving the cryptocurrency price prediction performance based on reinforcement learning. IEEE Access **9**, 162651–162659 (2021)

34. Si, W., Li, J., Ding, P., Rao, R.: A multi-objective deep reinforcement learning approach for stock index future's intraday trading. In: 2017 10th International Symposium on Computational Intelligence and Design (ISCID), vol. 2, pp. 431–436. IEEE (2017)
35. Sutton, R.S., Barto, A.G.: Reinforcement learning. J. Cogn. Neurosci. **11**(1), 126–134 (1999)
36. Van Hasselt, H., Guez, A., Silver, D.: Deep reinforcement learning with double q-learning. In: Proceedings of the AAAI Conference on Artificial Intelligence, vol. 30, no. 1 (2016)
37. Wu, X., Chen, H., Wang, J., Troiano, L., Loia, V., Fujita, H.: Adaptive stock trading strategies with deep reinforcement learning methods. Inf. Sci. **538**, 142–158 (2020)
38. Wu, Y., Tian, Y.: Training agent for first-person shooter game with actor-critic curriculum learning (2016)
39. Xiong, Z., Liu, X.Y., Zhong, S., Yang, H., Walid, A.: Practical deep reinforcement learning approach for stock trading. arXiv preprint arXiv:1811.07522 (2018)
40. Yang, H., Liu, X.Y., Wu, Q.: A practical machine learning approach for dynamic stock recommendation. In: 2018 17th IEEE International Conference on Trust, Security and Privacy in Computing and Communications/12th IEEE International Conference on Big Data Science and Engineering (TrustCom/BigDataSE), pp. 1693–1697. IEEE (2018)
41. Zhang, Z., Zohren, S., Roberts, S.: Deep reinforcement learning for trading. J. Financ. Data Sci. **2**(2), 25–40 (2020)

Analysis of Exchange Rate Fluctuations in Japan and Thailand by Using Copula-Based Seemingly Unrelated Regression Model

Kongliang Zhu[1], Xuefeng Zhang[2(✉)] ⓘ, and Pensri Jaroenwanit[1]

[1] The Faculty of Business Administration and Accountancy, Khon Kaen University, Khon Kaen, Thailand
penjar@kku.ac.th
[2] Faculty of Economics, Chiang Mai University, Chiang Mai, Thailand
xuefengzhang999@gmail.com

Abstract. The purpose of this paper is to analyze the factors that affect exchange rate fluctuations and provide help for investors to make investment decisions in the face of uncertainty in the exchange rate market. Comparing with the Conventional SUR model, the Student-t copula SUR is applied in this study. The results show that the interest rate is negative related to the fluctuation of exchange rate for both Thailand and Japan. And the exchange rate of both countries is positively influenced by inflation rate. The empirical result also shows that the current account balance is considered weak evidence affecting the change of exchange rate. And the Market speculation is decisively affecting the fluctuation of exchange rate, while it is negatively affecting for Thailand but positively affecting for Japan.

Keywords: Exchange rate fluctuations · Student-t Copula · SUR model

1 Introduction

The outbreak of the COVID-19 epidemic and the Russian-Ukrainian war on February 24, 2022, have caused global oil prices and food prices to continue to rise, coupled with the rising cost of living in the United States, which has led to rising inflationary pressures in the United States, forcing the Federal Reserve to tighten monetary policy further. In 12 oil-producing economies, Kumeka et al. [1] investigated the trilateral interaction between the stock, exchange rate, and oil markets. They discover variables that were impacted by the dynamics of their own shocks over the time prior to COVID-19. This study applied the Copula-based Seemingly Unrelated Regression (SUR) model and used monthly data that covers the period from October 2006 to September 2022 for sets of two countries, Thailand and Japan, Japan has historically been a significant investor in Thailand. Many Japanese companies have production bases in Thailand, so fluctuations in exchange rates can have a direct impact on their operations and profitability.

By utilizing a novel multivariate dynamic conditional generalized autoregressive score model, Blazsek et al. [2] investigate nonlinear co-integration for score-driven models. The US real gross domestic product growth, US inflation rate, and effective federal funds rate are examined as empirical applications. The widening of the U.S.-Japan interest rate gap will cause a sharp depreciation of the yen. Lau and Yip [3] looked at how the Bank of Japan's four periods of unconventional monetary policies (UMPs) from 2013 to 2020 affected the Japanese financial markets. Their findings demonstrate that not all the Bank of Japan's unorthodox monetary policies have the same impact on the domestic financial markets in Japan.

The cross-spectral coherence and co-movement between the monthly return series of West Texas Intermediate (WTI) oil price and the exchange rate of the Thai baht versus the US dollar from 1986 to 2019 were examined by Kyophilavong et al. [4] in 2023. In the short, medium, and long terms, they showed adverse spill-over effects between oil prices and Thai exchange rates, indicating that the oil market offers a systemic danger to the country's foreign exchange market in the short, medium, and long terms. Lastrapes [5] attempted to distinguish empirically real versus nominal sources of fluctuations in real and nominal exchange rates. He indicates that exchange rate fluctuations lead to price fluctuations in the international market, causing people to feel insecure generally.

Exchange rates are determined by a variety of factors, including interest rates, Inflation rate, Economic growth, Current account balance, Government debt, Market speculation. Heller [6] pointed out that the exchange rates intervening regularly in the foreign exchange market can stabilize the rate. Branson [7] presented a model that integrates money, relative prices, and the current account balance as factors explaining movements in nominal (effective) exchange rates. He concluded that real exchange rates adjust to real disturbances in the current account, and time-series innovations in the current account seem to signal the need for adjustment. Exchange rate movements play a crucial role in attracting foreign investment. A stable or appreciating currency can encourage foreign investors to invest in a country's assets, such as stocks and bonds. On the other hand, a volatile or depreciating currency may raise concerns for investors about potential losses from currency fluctuations. Based on the existing findings, this article attempts to analyze the factors that affect exchange rate fluctuations between Thailand and Japan and provide help for investors to make investment decisions in the face of uncertainty in the exchange rate market.

The outline of this paper is as follows. Section 2 we do some work of literature reviews. Section 3 we discuss about the data and methodology. The result of empirical study is given in Sect. 4. Section 5 concludes.

2 Literature Reviews

According to theories and concepts in international finance, the interest rate (R), inflation rate (CPI), GDP growth rate (GDP), current account balance (CA) and market speculation (MS) index can help explain exchange rate movement. For instance, the Thai baht and The Japanese yen are influenced by global market forces, foreign investments, and the country's economic performance. The Thai government and the Japanese government issues government bonds to finance its spending and control the money supply. The central bank of the country sets interest rates to control inflation and stimulate the economy.

Higher interest rates tend to attract foreign investment, increasing demand for a country's currency and boosting its exchange rate. For example, Banchuenvijit [8] applied multiple regression by ordinary least squares shows that changes in interest rate differentials have a statistically significant effect on changes in the THB/USD exchange rate. Mirchandani [9] shows that there is a negative correlation between the inflation rate and the rupee/dollar exchange rate with a significance level of 0.01. Inflation has been a concern in many countries in recent months, as prices for goods and services have risen faster than usual. This has led some central banks to consider raising interest rates to slow down inflation. Strong economic growth can increase demand for a country's goods and services, attract foreign investment and boost the exchange rate. Investors can influence exchange rates through their perception of a country's economic prospects and expectations of future exchange rate movements.

Empirical results of Tan et al. [10] indicate that there are long-term stable and unidirectional causal relationship between the exchange rate and FDI inflow. A country with a current account deficit (imports more than it exports) typically sees its currency depreciate because it needs to borrow from other countries to finance its imports. Sawatkamon [11] studied the factors affecting the Thai baht/RMB exchange rate through multiple regression and found that imports from China have a positive impact on the exchange rate and foreign exchange reserves have a negative impact on the exchange rate.

The traditional theory of exchange rate determination based on linear model cannot reveal and explain the fluctuation law of exchange rate well. Many research show that the exchange rate fluctuation series does not obey the normal distribution. For example, Wang et al. [12] found that the yuan against the dollar and yen yield sequence does not obey normal distribution. The conventional Seemingly Unrelated Regression (SUR) model has a strong assumption of normally distributed residuals. Pastpipatkul et al. [13] suggested that the Copulas can be used appropriately to relax the assumption of normality of residuals in the conventional SUR model. Therefore, this paper use Copula-based Seemingly Unrelated Regression model which also allows for the dependence between two error components through the ability of copula.

3   Methodology

3.1   Data and Variables

This study uses monthly data that covers the period from October 2006 to September 2022 for sets of two countries, Thailand and Japan. As permitted by the data availability, the data required for this study was mainly collected from the CEIC database (https://www.ceicdata.com/en). The real exchange rate data are treated as our regressor. Furthermore, to analyze the factors that affect the exchange rate, we include the interest rate (R), inflation rate (CPI), GDP growth rate (GDP), current account balance (CA) and market speculation (MS) index of each country. We would like to note that the monthly Gross Domestic Product (GDP) and current account balance (CA) are not available, thus we interpolate quarterly time series into monthly by repeating quarterly variables two times to generate a monthly variable. The data definition is shown in Table 1.

Table 1. Data definition.

Variables	Description	Unit	Source
EX_t^i	The currency of country i against USD	*THB/USD* *JPY/USD*	CEIC database
R_t^i	Interest rate of country i	%	CEIC database
CPI_t^i	Inflation rate of country i	%	CEIC database
GDP_t^i	GDP growth rate of country i	%	CEIC database
CA_t^i	Current account balance of country i	USD million	CEIC database
MS_t^i	Market speculation of country i	Nikkei 225 Stock SET Index 50	CEIC database

According to Fig. 1, the exchange rate of the yen against the US dollar began to show a two-way fluctuation trend. In different historical periods, different factors drove the rise and fall of the yen against the US dollar. However, the exchange rate is generally within the range of 80:1 to 140:1.

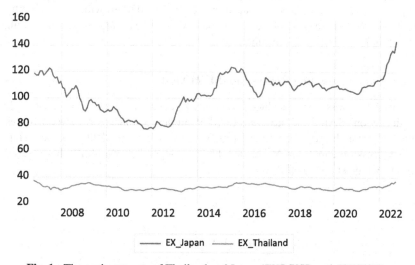

Fig. 1. The exchange rate of Thailand and Japan (*THB/USD* and *JPY/USD*).

From 2000 to 2011, the yen entered a large appreciation trend. Since Japan has basically maintained the "zero interest rate" monetary policy since 1999, under the background of the Federal Reserve's trend of interest rate cuts (only raised interest rates in 2004–2006, and finally implemented "zero interest rate" in 2008), the United States The narrowing trend of daily interest rate differentials provided support for the appreciation of the yen. In addition, the safe-haven currency at-tribute of the yen also promoted the further appreciation of the yen exchange rate in 2008–2011. The extremely loose monetary policy after 2012 pushed the yen into a depreciation trend. After the Abe government came to power at the end of 2012, it launched an extremely loose quantitative and qualitative easing (QQE) monetary policy. On the other hand, the United States has gradually promoted the normalization of monetary policy since 2013, which generally promoted the exchange rate of the yen in 2012. Continued depreciation thereafter.

In 2022, the depreciation of the yen against the U.S. dollar will be the largest since records began in 1972. The international market is highly concerned about the exchange rate of the yen breaking through 150.The U.S.-Japan interest rate gap widened as the Fed raised interest rates, pushing the dollar higher and the yen lower.

3.2 Exchange Rate Function

The exchange rate of each country is specified as a function of several key determinants, including interest rate, inflation rate (CPI), Economic growth (GDP growth rate), Current account balance and Market speculation (stock index). The exchange rate function for each country is:

$$EX_t^i = f\left(R_t^i, CPI_t^i, GDP_t^i, CA_t^i, MS_t^i\right) \tag{1}$$

where EX_t^i = exchange rate of country i at time t, R_t^i = interest rate of country i at time t, CPI_t^i = inflation rate of country i at time t, GDP_t^i = GDP growth rate of country i at time t, CA_t^i = Current account balance of country i at time t, MS_t^i = Market speculation (the stock index) of country i at time t.

3.3 Seemingly Unrelated Regression (SUR) Model

The SUR model was first introduced by Zellner [14] as a generalized system of linear regression. Considering the structure of SUR model, it consists of several regression equations, in which they are allowed to have their own dependent variables. Here we have M regression equations where each has N independent variables; the system of M equations can be shown as follows:

$$y_{t,1} = x_{t,1}\beta_1' + \varepsilon_{t,1}$$
$$\vdots \tag{2}$$
$$y_{t,M} = x_{t,M}\beta_M' + \varepsilon_{t,M}$$

The system above can be written as a vectorial form as follows:

$$Y_t = X_t\beta + \varepsilon_t \tag{3}$$

Let Y_t be a vector of dependent variables, $y_{t,i}$, $i = 1, \ldots, M$. A matrix of independent variables (regressors) is denoted as X_t, where $x_{t,ij}$, $i = 1, \ldots, M$, $j = 1, \ldots, N$. β denotes a matrix of an unknown parameters (regression coefficients), and ε_t is a vector of the error terms, where $\varepsilon_t = [\varepsilon_{t,1}, \varepsilon_{t,2}, \ldots, \varepsilon_{t,M}]'$ and $\varepsilon_{t,i} \sim N(0, \sigma_i^2)$, $i = 1, \ldots, M$. The important assumption of the SUR model which let it gain the efficiency of estimation is that the error terms are assumed to correlate across equations [15, 16]. Thus, it can be estimated jointly, $E[\varepsilon_{ia}\varepsilon_{ib}|X] = 0$; $a \neq b$ whereas $E[\varepsilon_{ia}\varepsilon_{jb}|X] = \sigma_{ij}$ [17]. The model allows non-zero covariance between the error terms of different equations in the system. Considering a variance-covariance matrix for M equations, it can be written as follows:

$$\Gamma = \Sigma \otimes I = \begin{bmatrix} \sigma_{11}I & \cdots & \sigma_{1M}I \\ \vdots & \ddots & \\ \sigma_{M1}I & & \sigma_{MM}I \end{bmatrix}, \text{ where } \Sigma\left(\varepsilon_t \varepsilon_t'\right) = \begin{bmatrix} \sigma_{11} & \cdots & \sigma_{1M} \\ \vdots & \ddots & \\ \sigma_{M1} & & \sigma_{MM} \end{bmatrix}$$

and I is an identity matrix. Hence from Eq. (3), the error terms of the SUR model are assumed to be $\varepsilon\_t \sim N(0,\Gamma)$, and this system equation can be estimated by

$$\beta_{\text{sure}} = \left(X'\Gamma^{-1}X\right)^{-1}X'\Gamma^{-1}Y \tag{4}$$

3.4 Copulas

A linkage between the marginal distributions was first introduced by Sklar in 1959 [18] called Sklar's theorem. Then, it was well described by Nelsen [19] as the dependence in Copula. Following Nelsen, this paper applies the most used copula, namely Elliptical Copulas to model the dependence structure of the SUR model. Here, the Elliptical class consists of the symmetric Gaussian and Student-t.

Gaussian Copula
Considering the case of n-dimensional, the Gaussian Copula can be defined by Schepsmeier and Stöber [20].

$$C(u_1, \ldots, u_n) = \Phi_n^{\Sigma_n}\left(\Phi_1^{-1}(u_1), \ldots, \Phi_n^{-1}(u_n)\right) \tag{5}$$

where $\Phi_d^{\Sigma_d}$ is n-dimensional standard normal cumulative distribution and Σ_n is a variance-covariance matrix. The density of the Gaussian Copula is given by

$$c(u_1, \ldots, u_n) = \left(\sqrt{\det\Sigma_n}\right)^{-1}\left(\frac{1}{2}\left(\Phi_1^{-1}(u_1)\ldots\Phi_n^{-1}(u_n)\right)\cdot\left(\Sigma_n^{-1} - I\right)\cdot\begin{pmatrix}\Phi_1^{-1}(u_1)\\ \vdots \\ \Phi_1^{-1}(u_n)\end{pmatrix}\right) \tag{6}$$

Student-t Copula
Student-t is one of the copulas in the Elliptical Copulas. It has a second parameter [20] and degree of freedom which is denoted by v. In the case of n-dimensional we can define the Student-t by

$$c(u_1, \ldots, u_n) = \int_{-\infty}^{t_v^{-1}(u_1)} \cdots \int_{-\infty}^{t_v^{-1}(u_n)} f_{t_1(v)}(x)dx \tag{7}$$

where $f_{t_1(v)}(x)$ is n-dimensional t-density function with degree of freedom v and t_v^{-1} is the quantile function of a standard univariate t_v distribution. In the estimation, the density of Student-t copula is defined by

$$c_{v,P}^t(u_1, \ldots, u_n) = \frac{f_{v,P}\left(t_v^{-1}(u_1), \ldots, t_v^{-1}(u_n)\right)}{\prod_{i=1}^n f_v\left(t_v^{-1}(u_i)\right)} \tag{8}$$

where $f_{v,P}$ is the joint density of a $t_n(v, 0, P)$ distributed random vector and P is the correlation matrix implied by the dispersion matrix Σ.

3.5 Estimation of the Copula-based SUR Model

Pastpipatkul [13] introduced the SUR model which consists of two equations ($M = 2$). Here, we use SUR to represent exchange rate of Thailand and Japan. Thus, the bivariate Copula with continuous marginal distribution is conducted in the estimation. Before estimation, we begin with checking the stationary of the data series using the Augmented Dickey-Fuller test. Next, the estimation procedures of Copula-based SUR model involve four steps. First, we estimate the conventional SUR model using a maximum likelihood technique to obtain the initial values. Second, we construct the SUR Copula likelihood using the chain rule, here we have

$$\frac{\partial^2}{\partial u_1 \partial u_2} F(u_1, u_2) = \frac{\partial^2}{\partial u_1 \partial u_2} C(F_1(u_1), F_2(u_2)) * f_1(u_1) f_2(u_2) * c(F_1(u_1), F_2(u_2)) \tag{9}$$

From Eq. (9), u_1 and u_2 are the marginal distributions which can be Guassian or Student's t distribution. $f_1(u_1)$ and $f_2(u_2)$ are normal functions of demand and supply equations, respectively. Density function of Copulas is denoted as $c(F_1(u_1), F_2(u_2))$. In this study, we are interested in the following Copula families, i.e., Gaussian and T, to construct the joint distribution function of a bivariate random variable with the univariate marginal distribution. Then, we transform Eq. (9) using a logarithm, and we get

$$\ln L = \sum_{i=1}^{T} (\ln l(\theta_1|X_{TH}) + \ln l(\theta_2|X_{JP}) + \ln f_1(u_1) + \ln f_2(u_2) + \ln c(F_1(u_1), F_2(u_2)) \tag{10}$$

where $\ln l(\theta_1|X_{TH})$ and $\ln l(\theta_2|X_{JP})$ are the logarithm of the likelihood function of exchange rate of Thailand and Japan, respectively. The logarithm of the likelihood function can be defined by

$$\ln L = -\frac{T}{2} \ln(2\pi) - \frac{T}{2} \ln(\Sigma) \left(\frac{1}{2\Gamma} (Y - X\beta)'(Y - X\beta) \right) \tag{11}$$

Moreover, the last term of (10), $\ln c(F_1(u_1), F_2(u_2))$ denotes the bivariate Copula density, which is assumed for Gaussian and T (See Sect. 3.4). Here, we employ the maximum likelihood estimation (MLE) to maximize the SUR Copula likelihood function (11) to obtain the final estimation results for the Copula-based SUR model.

4 Empirical Results and Analysis

4.1 Descriptive Statistics

The SUR model consisting of exchange rate equations of Thailand and Japan can be specified as follow:

$$REX_t^{TH} = \alpha_1 + \beta_1 R_t^{TH} + \beta_2 CPI_t^{TH} + \beta_3 GDP_t^{TH} + \beta_4 RCA_t^{TH} + \beta_5 RMS_t^{TH} + U_{1,t}$$
$$REX_t^{JP} = \alpha_2 + \delta_1 R_t^{JP} + \delta_2 CPI_t^{JP} + \delta_3 GDP_t^{JP} + \delta_4 RCA_t^{JP} + \beta_5 RMS_t^{JP} + U_{2,t} \tag{12}$$

where the error terms of the SUR model are dependent.

In our models, this study considered the log-returns, calculated as $r_t = \ln\left(\frac{X_t}{X_{t-1}}\right)$ from the exchange rate (EX), Current account balance (CA) and Market speculation (MS). We keep the interest rate (R), inflation rate (CPI) and GDP growth rate (GDP) as the original data since their unit is %. Descriptive statistics of data are presented in Table 2.

Table 2. Descriptive statistics.

		Mean	Median	Max	Min	Std.Dev	Skewness	Kurtosis	Jarque-Bera
Thailand	REX_t^{TH}	−0.001	−0.030	1.482	−1.930	0.510	−0.023	4.006	8.121**
	R_t^{TH}	2.081	1.669	5.252	0.621	1.068	0.738	3.123	17.59***
	CPI_t^{TH}	1.791	1.616	9.146	−4.357	2.270	0.490	3.952	14.96***
	GDP_t^{TH}	2.645	3.150	15.46	−12.26	4.006	−0.481	6.355	97.50***
	RCA_t^{TH}	−6.652	−24.47	3688	−4240	1393	−0.163	3.050	0.872
	RMS_t^{TH}	2.343	6.980	162.3	−140.9	41.85	−0.301	4.514	21.26***
Japan	REX_t^{JP}	0.128	0.062	8.273	−6.608	2.394	0.422	4.499	23.68***
	R_t^{JP}	0.276	0.200	0.889	0.053	0.251	1.114	3.113	39.87***
	CPI_t^{JP}	0.382	0.204	3.699	−2.558	1.132	0.632	3.952	20.05***
	GDP_t^{JP}	0.375	0.597	7.702	−9.895	2.889	−1.267	6.342	140.8***
	RCA_t^{JP}	−60.16	−116.3	9827	−1074	3568	−0.001	3.611	2.989
	RMS_t^{JP}	49.67	117.5	3456	−2682	890.6	−0.374	4.143	14.93***

Table 2 provides the summary statistics for each variable. Based on the mean growth of exchange rate, Japan has the higher fluctuation of exchange rate at 0.128, while the fluctuation of exchange rate in Thailand is negative at −0.001. it indicates that Thai Baht is more stable than Japanese Yen. Thailand has higher fluctuations in the interest rate (R), inflation rate (CPI), GDP growth rate (GDP) compared to Japan. And Japan has higher fluctuations in current account balance (CA) and market speculation (MS) index compared to Thailand. In addition, from the results of Jarque-Bera test, we may state that they do not exhibit Gaussian distribution.

4.2 Stationary Test

Before estimating the Copula-based SUR model for analyzing the determinants of exchange rate, we need to check whether the data is stationary through the unit root test. The calibrated p-value of the test statistic is obtained using the Minimum Bayes factor (MBF) method. The Goodman [21] labelled intervals are used to interpret the MBF. MBF between 1–1/3 is considered weak evidence for H_1, from 1/3 to 1/10 considered moderate evidence for H_1, 1/10 to 1/30 is considered substantial evidence, from 1/30 to 1/100 strong, from 1/100 to 1/300 very strong, and < 1/300 decisive. Table 3 shows the unit root test results, which reveal that all the data provide at least strong evidence for the stationary. Therefore, it is appropriate to further analyze the determinants of exchange rate of Thailand and Japan.

Table 3. The Augmented Dickey-Fuller (ADF) unit roots test for data.

		ADF	MBF
Thailand	REX	−9.922***	0.000
	R	−3.133**	0.007
	CPI	−3.175**	0.006
	GDP	−3.686***	0.001
	RCA	−15.233***	0.000
	RMS	−12.255***	0.000
Japan	REX	−9.9359***	0.000
	R	−4.109***	0.000
	CPI	−2.106**	0.018
	GDP	−3.506***	0.002
	RCA	−2.918**	0.014
	RMS	−12.222***	0.000

Note: MBF is Minimum Bayes Factors. $MBF(t) = \exp(-0.5 * t^2)$.

4.3 Model Comparison

In this study, the copula-based SUR models are assumed to investigate the correlation of the exchange rate between Thailand and Japan. Therefore, we need to validate our copula-based SUR models' performance by comparing it with the Conventional SUR model. In this comparison, the AIC and BIC are used as the comparison measure, and the lowest AIC/BIC indicates a more parsimonious model. Table 4 shows the model comparison of three different models for the exchange rate equations of Thailand and Japan. The result confirms that copula-based SUR models are preferable, especially Student-t copula SUR since it has the lowest AIC (-1539) /BIC (-1484) value that we highlight them with bold font.

Table 4. Model comparison.

Model	Conventional SUR	Gaussian copula SUR	Student-t copula SUR
AIC	1098	−1039	**−1539**
BIC	1157	−990.5	**−1484**

4.4 Student-t Copula SUR Empirical Results

Regarding the estimated coefficients, the inflation rate (CPI) of Japan has the highest value, reaches at 0.452, while the interest rate (R) shows the lowest value of -1.206. Both intercepts are positive which indicate that the mean of the growth of exchange rate is increasing during our empirical period (2000–2022). Moreover, the intercept term of Thailand has substantial evidence in terms of Minimum Bayes Factors. The interest rate (R) for both Thailand and Japan are negative, and the interest rate (R) of Japan is prominently negative related to the fluctuation of exchange rate. The inflation rate (CPI) of Japan (0.452) is stronger influence on exchange rate than Thailand (0.059), which coincides with the fact that the sharp depreciation of the Japanese yen pushed up the prices of imported goods. The exchange rate of both countries is positively influenced by inflation rate. The GDP growth rate (GDP) of Thailand is negatively affecting the variation of Thai baht exchange rate. Our empirical result shows that the growth of Current account balance (RCA) is considered weak evidence affecting the change of exchange rate both in Thailand and Japan. Surprisingly, the Market speculation (RMS) is decisively affecting the growth of exchange rate for both Thailand and Japan, while it is negatively affecting for Thailand (-0.048) but positively affecting for Japan (0.010) (Table 5).

Table 5. Model comparison.

Dependent variable: REX_t^i				
	Covariates	Coeff	Std. Err	MBF
Thailand	Intercept	0.169*	0.148	0.054
	R_t^{TH}	−0.105**	0.071	0.033
	CPI_t^{TH}	0.059**	0.039	0.020
	GDP_t^{TH}	−0.017*	0.020	0.071
	RCA_t^{TH}	0.038	0.007	0.974
	RMS_t^{TH}	−0.048***	0.017	0.000
Japan	Intercept	0.236	0.150	0.291
	R_t^{JP}	−1.206**	0.466	0.035
	CPI_t^{JP}	0.452***	0.088	0.000
	GDP_t^{JP}	0.022	0.031	0.777
	RCA_t^{JP}	−0.070	0.004	0.292
	RMS_t^{JP}	0.010***	0.011	0.000

Note: MBF is Minimum Bayes Factors. $MBF(t) = \exp(-0.5 * t^2)$.

5 Conclusion and Future Research Analysis

In this paper, the copula-based SUR models are applied to analyze the fluctuations exchange rate. Before estimating the Copula-based SUR model for analyzing the determinants of exchange rate, this study checks whether the data is stationary through the unit root test. The unit root test results shows that all the data provide at least strong evidence for the stationary. To validate copula-based SUR models' performance by comparing it with the Conventional SUR model, the AIC and BIC are used as the comparison measure. The result confirms that copula-based SUR models are preferable, especially Student-t copula SUR since it has the lowest AIC (-1539) /BIC (-1484) value. Therefore, this study chooses Student-t copula SUR to show empirical results.

The Estimation results of Student-t copula SUR show that the interest rate (R) is negative related to the fluctuation of exchange rate for both Thailand and Japan. The exchange rate of both countries is positively influenced by inflation rate (CPI). The GDP growth rate (GDP) of Thailand is negatively affecting the move of Thai baht exchange rate. The empirical result also shows that the change of Current account balance (RCA) is considered weak evidence affecting the change of exchange rate both in Thailand and Japan. And the Market speculation (RMS) is decisively affecting the fluctuation of exchange rate for both Thailand and Japan, while it is negatively affecting for Thailand but positively affecting for Japan. In future research, Copula approach with regime switching or other nonlinear model can be applied to analyze the fluctuations of exchange rate and it would be useful for capturing the marginal distributions as well as the dependency structure.

References

1. Kumeka, T.T., Uzoma-Nwosu, D.C., David-Wayas, M.O.: The effects of COVID-19 on the interrelationship among oil prices, stock prices and exchange rates in selected oil exporting economies. Resour. Policy **77**, 102744 (2022)
2. Blazsek, S., Escribano, A., Licht, A.: Co-integration with score-driven models: an application to US real GDP growth, US inflation rate, and effective federal funds rate. Macroecon. Dyn. **27**(1), 203–223 (2023)
3. Lau, W.-Y., Yip, T.-M.: The effect of different periods of unconventional monetary policies on Japanese financial markets. J. Financ. Econ. Policy **15**(3), 263–279 (2023). https://doi.org/10.1108/JFEP-11-2022-0275
4. Kyophilavong, P., Abakah, E.J.A., Tiwari, A.K.: Cross-spectral coherence and co-movement between WTI oil price and exchange rate of Thai Baht. Resour. Policy **80**, 103160 (2023)
5. Lastrapes, W.D.: Sources of fluctuations in real and nominal exchange rates. Rev. Econ. Stat. **74**(3), 530–539 (1992). https://doi.org/10.2307/2109498
6. Heller, H.R.: Determinants of exchange rate practices. J. Money, Credit, Bank. **10**(3), 308–321 (1978)
7. Branson, W. H.: Macroeconomic determinants of real exchange rates (No. w0801). National Bureau of Economic Research (1981)
8. Banchuenvijit, W.: Factors determining exchange rates between thai baht and US dollar as well as thai baht and chinese yuan under president donald trump's administration. J. Family Bus. Manag. Stud. **14**(2), 79–90 (2022)

9. Mirchandani, A.: Analysis of macroeconomic determinants of exchange rate volatility in India. Int. J. Econ. Financ. Issues 3(1), 172–179 (2013)

10. Tan, L., Xu, Y., Gashaw, A.: Influence of exchange rate on foreign direct investment inflows: an empirical analysis based on co-integration and granger causality test. Math. Probl. Eng. **2021**, 1–12 (2021). https://doi.org/10.1155/2021/7280879

11. Sawatkamon, P.: Analysis of the exchange rate of the Thai baht against the Chinese yuan using a support vector machine model (Doctoral dissertation, School of Mathematics Institute of Science Suranaree University of Technology) (2019)

12. Wang, B., Cao, T., Wang, S.: The researches on exchange rate risk of Chinese commercial banks based on copula-Garch model. Modern Econ. **05**(05), 541–551 (2014). https://doi.org/10.4236/me.2014.55051

13. Pastpipatkul, P., Maneejuk, P., Sriboonchitta, S.: Do Copulas improve an efficiency of seemingly unrelated regression model? Int. J. Intell. Technol. Appl. Stat. 9(2), 105–122 (2016)

14. Zellner, A.: An efficient method of estimating seemingly unrelated regressions and tests for aggregation bias. J. Am. Stat. Assoc. **57**(298), 348–368 (1962)

15. Baltagi, B.H.: The efficiency of OLS in a seemingly unrelated regressions model. Economet. Theor. **4**(3), 536–537 (1988)

16. Bilodeau, M., Duchesne, P.: Robust estimation of the SUR model. Can. J. Stat. **28**, 277–288 (2000)

17. Pastpipatkul, P., Maneejuk, P., Sriboonchitta, S.: Welfare measurement on Thai rice market: a Markov switching Bayesian seemingly unrelated regression. In: Huynh, V.-N., Inuiguchi, M., Denoeux, T. (eds.) Integrated Uncertainty in Knowledge Modelling and Decision Making, pp. 464–477. Springer, Nha Trang, Vietnam (2015)

18. Sklar, M.: Fonctions de répartition à n dimensions et leurs marges. Publications de l' Institut de Statistique de L'Université de Paris **8**, 229–231 (1959)

19. Nelsen, R.B.: An Introduction to Copulas, 2nd edn. Springer, New York (2010)

20. Schepsmeier, U., Stöber, J.: Derivatives and Fisher information of bivariate copulas. Stat. Pap. **55**, 525–542 (2014)

21. Goodman, S.N.: Toward evidence-based medical statistics. 1: the P value fallacy. Ann. Intern. Med. **130**(12), 995–1004 (1999)

The Perspective for the Economy in Cambodia, Laos, Myanmar, Vietnam and Thailand: Economic Growth, Inequality, and Environmental Considerations

Chanamart Intapan[1], Chukiat Chaiboonsri[1(✉)] (iD), and Banjaponn Thongkaw[2]

[1] Faculty of Economics, Chiang Mai University, Chiang Mai, Thailand
{chanamart.i,chukiat.chai}@cmu.ac.th
[2] Modern Quantitative Economics Research Center (MQERC), Faculty of Economics,
Chiang Mai University, Chiang Mai, Thailand

Abstract. The aim of the study was to examine the sustainability (environmental, social and economic impacts) in Cambodia, Laos, Myanmar, and Vietnam (CLMV countries) and Thailand. We evaluated sustainable development through three perspectives such as environmental indicators assessed include carbon emissions, social indicators include inequality, and the economic indicator is the growth rate of real GDP. Our theoretical model introduced Bayesian kink regression model. From the results of the study, it was found that economic development is correlated with the level of inequality. While economic development does not affect the amount of carbon emissions. This demonstrates that economic development has a greater effect on inequality than environmental problems. Economists and policymakers can use the results of empirical studies to come up with guidelines or policies that can be implemented for finding ways to develop the economy further, taking into account the impact of creating more inequality from economic development.

Keywords: CLMV · Thailand · Economic growth · Inequality · Carbon emission

1 Introduction

Over the past few years, the direction of global investment has gradually shifted towards emerging markets. One target that has continued to gain popularity is Cambodia, Laos, Myanmar and Vietnam (CLMV countries). The four countries are members of the Association of Southeast Asian Nations (ASEAN), like Thailand. For CLMV economies, these economies have maintained remarkable growth over the years despite global economic slowdowns. The educational goals of the CLMV countries are interesting. At the same time, Thailand, which is considered to be bordering the CLMV countries, is also a developing country. This is the researcher's desire to see how the economies of these two groups are similar or different. Because the researchers believe that although these two groups are similar and adjacent to each other, the effects of economic development on the environment and social are different.

© The Author(s), under exclusive license to Springer Nature Switzerland AG 2023
V.-N. Huynh et al. (Eds.): IUKM 2023, LNAI 14375, pp. 248–257, 2023.
https://doi.org/10.1007/978-3-031-46775-2_22

The economic growth based on real The GDP of CLMV countries and Thailand tends in similar direction as shown in Fig. 1. In 2020, the COVID-19 epidemic has caused all countries to suffer the same economic impact. At that time, the country CLMV and Thailand have dropped real GDP levels but after that, all 5 countries gradually recovered and improved accordingly.

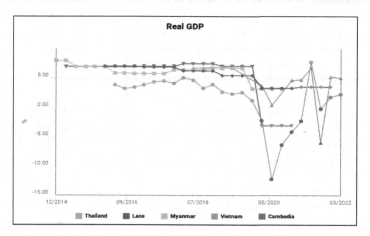

Fig. 1. The real GDP of CLMV countries and Thailand

When a country changes or develops rapidly, there will be consequences. The main impacts of economic development are environmental impacts and inequality effects. Because authors are aware of the sustainability-related issues listed in the Sustainable Development Goals (SDGs) and look to build a green economy. In this study, the researchers attempted to explore three issues to create three sustainability principles: social, economic and environmental, in accordance with the SDGs' primary objectives. The researchers believe that economic development alone cannot contribute to the sustainability of the country. The country should also develop in other areas such as the environment and society to achieve long-term sustainable development. Thus, the main objective in this study is to study how the continuous economic development of the five countries will affect the carbon emissions which is an environmental impact and affect the income inequality of the people in the country. It is an economic and social impact.

2 Literature Review

[1] stated that the big controversy that has long been the issue of the relationship between economic growth and environmental quality. One controversy is the view that greater economic activity inevitably leads to environmental degradation and ultimately to economic and ecological collapse. This has resulted in extensive studies of the relationship between the environment and sustainable economic development. Some studies agree with Shafik's statement, but others have opposite results. For instance, [2] examined the reduced-form relationship between per capita income and various environmental indicators such as urban air pollution, the state of the oxygen regime in river basins, fecal

contamination of river basins, and contamination of river basins by heavy metals. The result finds no evidence that environmental quality deteriorates steadily with economic growth. Moreover, the economic growth causes environmental deterioration initially but later improvements occur. [3] agrees that a key factor in achieving sustainable economic development is the prudent use of environmental resources. Furthermore, there are many studies of economic growth on social impacts such as poverty and unemployment rate. For instance, [4] analyzed the relationship between economic growth, unemployment and poverty in Vietnam. The research results indicate public investment has a positive effect on economic growth but negatively affects poverty. [5] analyzed the causal link between growth versus poverty. They concluded that growth is good for poverty alleviation but it is not enough. When we connect these three aspects, namely economic, social and environmental, it will lead to sustainable development. Sustainable development is gaining momentum in every region around the world in every sector. The relationship between economic, social and environmental impacts has been studied extensively in various perspectives. [6] studied the case of wildfires in Australia. They believe that wildfires often result in widespread destruction and damage to a range of economic, social and environmental assets and functions. [7] studied the impacts of cruise tourism on the economic, social, and environmental. They compare community impacts before and after the opening of a cruise ship port. The result found some negative evidences such as the ability of the local population to provide for necessities and obtain sufficient food worsened, corruption increased, and there were substantial negative environmental impacts. Large cruise tourism projects can fail to provide benefits for local populations in some ways. [8] investigated the sustainability (environmental, social and economic impacts) of tea manufacturing in Sri Lanka. The study found many issues including energy efficiency of the industry, Green House Gas (GHG) emissions, and occupational health hazards during processing stage. From most types of tea, low grown orthodox tea is the most efficient in terms of labor use, energy use and carbon emissions. [9] evaluated the potential impacts of removing energy subsidies on the Malaysian economy. The result of the study was unexpected. There are both positive and negative impacts of removing energy subsidies. It is shown that removing petroleum and gas subsidy would improve economic efficiency, increase GDP, and reduce budget deficit. However, households would be worse off due to higher price level. Recently, the issue of sustainability has been continuously studied. [10] interested in Bioeconomy because it is as a chance to focus on a sustainable mode of production and consumption. They assess its impact for implementing policies. [11] assessed the impacts of social, economic and environmental factors on the logarithm of housing prices in Hong Kong. The findings indicated that factors that are significant to house prices are economic factors and environmental factors. However, demographic factors are not as significant as expected in affecting housing prices. From the past research, we can see that the study raises the issue of various situations to look at the economic, social and environmental impact. But the difference in this study is that the researchers looked at the relationship between economic development (real GDP) and environmental impacts based on carbon emissions (carbon Emission) and social impacts in terms of inequality (Gini Index). This type of study is not widely studied, especially in emerging and developing countries such as CLMV countries and Thailand. Therefore, with all of the above, this study is an extension and doing something new that has never been studied before, especially in countries with continuous growth rates such as CLMV countries and Thailand.

3   Data and Methodology

3.1   Data

This study uses panel data of CLMV countries and Thailand, from 1997 to 2020, and uses carbon emissions (CO_2) as a proxy of environmental degradation, Gini coefficient as an indicator of inequality of income and consumption of people in the country and real GDP as an indicator of economic development. In addition, this study uses yearly time series data running from 1997 to 2020 of carbon emissions, Gini index and the growth rate of real GDP of 5 individual countries, which are member countries of the two groups, including Cambodia, Laos, Myanmar, Vietnam, and Thailand. Table 1 describes the variables employed in the study.

Table 1. Description of variables.

Variable	Description and data source	Symbol
Environmental degradation	Environmental degradation is measured using territorial CO_2 emissions, which come from the burning of fossil fuels due to human activities as well as production processes. This variable is considered a dependent variable in our analysis (unit: kt). (from CEIC data)	CO_2
Economic Development	Economic development in this analysis is measured by real GDP per capita. (from CEIC data)	GDP
Inequality	Inequality of income and consumption of people in the country (from CEIC data)	GINI

4   Methodology

4.1   Bayesian Approach for Panel Kink Regression Model

[12] utilized Bayesian kink regression to estimate unobserved thresholds, and they found that this approach worked very well to find out the unknowable threshold. In this study, we applied Bayesian kink regression in the case of the panel kink regression model with an unknown threshold (see Eq. (1)).

$$Y_{it} = \beta_1^- (X_{1.it} - \gamma)_- + \beta_1^+ (X_{1.it} - \gamma)_+ + \beta_2' X_{2,it} + \alpha_i + \varepsilon_{it}$$
$$i = 1, \ldots, N, t = 1, \ldots, T \tag{1}$$

where $Y_{it}, X_{1,it}, \varepsilon_{it}$ are the random scalars and the $X_{2,it}$ is a vector of regressor. The α_i is the unobserved heterogeneity of the i th individual which can be correlated with $X_{it} = (X_{1,it}, X'_{2,it})$. Define that the $(\alpha)_- = \min(\alpha, 0)$ and the $(\alpha)_+ = \max(\alpha, 0)$. Moreover, the β_1^- is the slope of $X_{1,it}$ and the β_1^+ is the slope of $X_{2,it}$. The γ is the unknown threshold point and $\gamma \in \Gamma$. The objective is estimating unknown parameters $(\beta_1^-, \beta_1^+, \beta_2', \gamma)$ and testing for the Kink effect of $X_{1,it}$ when N goes to infinity while T is fixed. The model (Eq. (2)) has wide potential applications. One prominent example is originally from Reinhart and Rogoff (2010, 2011). Their study is called debt-threshold effect on economic growth. To investigate the relationship across different countries, one can consider a panel regression model as follows the Eq. (2).

$$gdp_{it} = \beta_1^- (Debt_{it} - \gamma)_- + \beta_1^+ (Debt_{it} - \gamma)_+ + \beta_2' X_{2,it} + \alpha_i + \varepsilon_{it} \qquad (2)$$

where the gdp_{it} is the real GDP growth rate in t th year for the i th country. And $Debt_{it}$ is the debt to GDP ratio from the previous year. Addition, the X_{2it} is includes other variables which may affect economic growth. The result found that economic growth tends to be slow when the level of government debt relative to GDP exceeds a threshold. After that the study is recently re-examined by many researchers. The difference between them is the use of tools of threshold regression. Furthermore, the model can be adapted to investigate in many aspects such as the kink effects of income inequality on economic growth [13], the kink effects of cash flow on investment [14], and the kink effects of capital structure on firm value [15], and so on.

In this study, the authors would like to find the correlation of inequality and carbon emission to economic growth phenomena among CLMV countries and Thailand. Thus, the panel data of each group of countries are constructed and the panel kink regression model is used to fit these data. The model (Eq. (2)) can be adapted to investigate the kink effects of carbon emission and inequality on economic growth in CLMV countries and Thailand. To investigate the relationship between carbon emission and economic growth across CLMV countries and Thailand, one can consider a panel regression model as follows the Eq. (3).

$$gdp_{it} = \beta_1^- (CO2_{it} - \gamma)_- + \beta_1^+ (CO2_{it} - \gamma)_+ + \beta_2' X_{2,it} + \alpha_i + \varepsilon_{it} \qquad (3)$$

where gdp_{it} is the real GDP growth rate in t is the year for the i is the country (i = Cambodia, Laos, Myanmar, Vietnam, Thailand). And the $CO2_{it}$ is the carbon emission from the previous year. Addition, the X_{2it} is included other variables which may affect economic growth. To investigate the relationship between inequality and economic growth across CLMV countries and Thailand, one can consider a panel regression model as follows the Eq. (4).

$$gdp_{it} = \beta_1^- (GINI_{it} - \gamma)_- + \beta_1^+ (GINI_{it} - \gamma)_+ + \beta_2' X_{2,it} + \alpha_i + \varepsilon_{it} \qquad (4)$$

where the gdp_{it} is the real GDP growth rate in t is year for the i is country (i = Cambodia, Laos, Myanmar, Vietnam, Thailand). And the $GINI_{it}$ is the Gini coefficient from the previous year. Addition, the Z_{it} includes other variables which may affect economic growth.

5 Empirical Results

5.1 Data Descriptive

From Table 2 shows the descriptive statistics of all variables such as the growth rate of real GDP (**G_REAL_GDP**), the Gini coefficient (**GINI**), and the carbon emission (**CO₂**) respectively. The economic growth of CLMV and Thailand has no effect on carbon emission, but affects the level of inequality.

Table 2. Data descriptive and data visualization.

Items	G_REAL_GDP	GINI	CO₂
Mean	5.629612	36.23400	0.639652
Median	6.534114	35.80000	0.570000
Maximum	13.62797	43.10000	1.560000
Minimum	−7.634035	30.70000	0.160000
Std. Dev.	4.247341	3.038310	0.366288
Skewness	−0.595870	0.026920	0.877496
Kurtosis	3.656324	2.899716	2.759816
Normality test			
Jarque-Bera	8.869400	0.062079	15.03473
Probability	0.011859	0.969437	0.000544
Panel unit root test			
Levin, Lin & Chu t*	−1.93283	−1.87357	−1.98219
Probability	(0.0266)	(0.0305)	(0.0237)
Observations	115	115	115

* Stationary at zero level (I(0)) by significance level of 0.05
Source: Author

5.2 Model Estimation

From Table 3, it shown that the appropriated model for investigating the relationship between the economic development and inequality is the fixed effect Bayesian kink regression model. On the other hand, the appropriated model for investigating the relationship between the economic development and carbon emission is the pooled effect Bayesian kink regression model.

The R-squared for Bayesian Regression Models was created by [16] and is shown in more detail in Eq. (5).

$$Bayesian\ R_s^2 = \frac{V_{n=1}^N y_n^{pred\ s}}{V_{n=1}^N y_n^{pred\ s} + var_{res}^s}, \quad y_n^{pred\ s} = E(\tilde{y}|X_n, \theta^s) \tag{5}$$

Table 3. The value of Bayesian R^2 from model estimation

Model estimation	Bayesian R^2	
	Real GDP-Gini	Real GDP-CO_2
Pooled effect	0.03059025	**0.183269***
Fixed effect	**0.9294682***	0.04930859
Random effect	0.9270009	0.04874109
Mixed effect	0.8665936	0.08445103

* The 1st highest Bayesian R2
Source: Author

The y_n^{pred} s $= E(\tilde{y}|X_n, \theta^s)$ is the predicted value of the Bayesian regression models, and the var_{res}^s is the expected residual variance of the Bayesian regression models. The statistical value of the Bayesian R^2 is equal to 0 and 1. If the value of this statistic approaches 1, which indicates that the Bayesian regression model is appropriate, then it was suggested that this regression model is very appropriate or useful to conduct the knowledge to formulate the policy recommendation. In this case, the fixed effect model is the appropriate model because the Bayesian R^2 is very high (0.9294682*) compared with another model. Therefore, this model was utilized to explain the relationship between the Gini index and the growth rate of real GDP (see Fig. 2). From Fig. 2, we would be able to confirm that there is more economic expansion (economic growth: the growth rate of real GDP) and then more inequality (Gini coefficient) for each of the CLMV countries and also for Thailand. The pooled effect model was chosen as the appropriate model to describe the relationship between environmental protection (CO2 emissions) and economic growth. Because, when compared to other models, this model has the greatest Bayesian R^2 value (0.183269*). However, this pooled effect model still points out a weak association between environmental protection (CO2 emissions) and economic growth for the CLMV countries and Thailand (see Fig. 3).

In Fig. 3, the regression line (the red line) was conducted by the pooled effect model, which is quite bad at describing the relationship between CO2 emissions and the growth rate of real GDP for the CLMV countries and Thailand. Normally, the pooled effect model is utilized as the reference to be the baseline model when comparing another panel model. The baseline means that there is no need to design the pattern of panel data to be estimated by the fixed effect model, the random effect model, or the mixed effect model. In this economic sense, this study would confirm that the overview of CLMV countries and Thailand still maintains economic expansion or economic growth and environmental protection in balance.

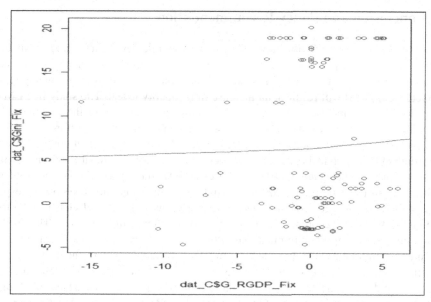

Fig. 2. The regression line (Fixed effect model) shows the relationship between the Gini index and real GDP for the CLMV countries and Thailand.

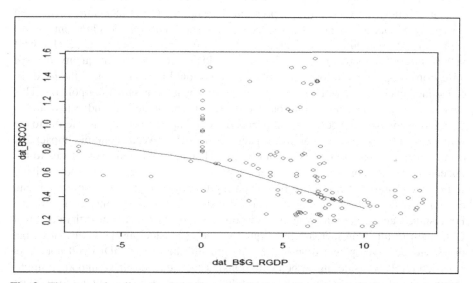

Fig. 3. The regression line (Pooled effect model) shows the relationship between the CO_2 emission and the growth rate of real GDP for the CLMV countries and Thailand.

6 Conclusion and Policy Recommendation

We summarize that the relationship between the economic development and inequality from fixed effect Bayesian kink regression model in Cambodia, Laos, Myanmar, and Vietnam (CLMV countries) and Thailand have the same direction. We can conclude that an increase in GDP will result in an increase in inequality levels. The study in CLMV countries and Thailand shows that economic development (considering the growth rate of real GDP) does not reduce inequality (considering the Gini index). Therefore, when formulating policies on economic development especially economic growth in the CLMV countries and Thailand, the negative impact of inequality should also be taken into account. On the other hand, CLMV countries and Thailand may not have a problem with emissions (considering carbon emissions) caused by economic development, especially economic balance growth (considering the growth rate of real GDP), since the finding illustrated that economic development did not affect emission levels. The results of this study show important implications: although CLMV countries and Thailand have continued economic development, at the same time, we should take into account the consequences of these economic developments. Especially on the issue of inequality arising from the rapid economic development of these countries. This is because the study found that greater economic development in terms of real GDP would result in greater inequality by using the Gini coefficient variable as a representative measure of inequality. As for the issue of environment and economic development in the CLMV countries and Thailand, it may not be much of a problem if compared to the inequality mentioned earlier. This is because the results of the study show that economic development has not yet affected carbon emissions. Therefore, in order to achieve sustainable development in the future, CLMV countries and Thailand should also consider the inequality that will arise from economic development. Policymakers should not only look at the need for economic development while ignoring these issues when implementing policies. This study makes several research contributions. First, it is one of the first studies to examine the relationship between economic impact and social impact (inequity, as referred to the welfare economics (socio-economic impacts)) [17] and between economic impact and environmental impact in CLMV countries and Thailand. Our study contributes to the understanding of economic development dynamics in CLMV countries and Thailand from a more holistic perspective by comparing the impacts of economic, social, and environmental factors. Second, the empirical results reveal various interesting effects. For instance, an increase in economic development (considering the growth rate of real GDP) will result in an increase in inequality levels (considering the Gini index), but the economic development (considering the growth rate of real GDP) will not affect emission levels (considering carbon emissions). The results thus have important implications: Current policy instruments to stimulate economic development are focused on increasing real GDP in a nation (economic factors), while little attention is paid to social or environmental factors. Our findings suggest that an increase in real GDP may be a source of inequality among the population within a country. Therefore, efforts should be made to improve economic stimulus policies while simultaneously paying attention to the issue of inequality. CLMV countries and Thailand should not focus solely on stimulating economic development based on real GDP, as perhaps economic development will lead to inequality. Therefore, in order to achieve sustainable development, these two issues should be considered together. Regarding research limitations, future research is needed to examine the economic impact beyond just the value of GDP. This is because economic growth in one country can be viewed from a different perspective than the GDP value. Moreover, social and environmental impacts can be viewed from other angles than inequality and carbon emissions.

References

1. Shafik, N.: Economic development and environmental quality: an econometric analysis. Oxford Econ. Pap. **46**, 757–773 (1994)
2. Grossman, G.M., Krueger, A.B.: Economic growth and the environment. Q. J. Econ. **110**(2), 353–377 (1995)
3. Awan, A.G.: Relationship between environment and sustainable economic development: a theoretical approach to environmental problems. Int. J. Asian Soc. Sci. **3**(3), 741–761 (2013)
4. Quy, N.H.: Relationship between economic growth, unemployment and poverty: analysis at provincial level in Vietnam. Int. J. Econ. Financ. **8**(12), 113–119 (2016)
5. Škare, M., Družeta, R.P.: Poverty and economic growth: a review. Technol. Econ. Dev. Econ. **22**(1), 156–175 (2016)
6. Stephenson, C., Handmer, J., Betts, R.: Estimating the economic, social and environmental impacts of wildfires in Australia. Environ. Hazards **12**(2), 93–111 (2013)
7. MacNeill, T., Wozniak, D.: The economic, social, and environmental impacts of cruise tourism. Tour. Manage. **66**, 387–404 (2018)
8. Munasinghe, M., Deraniyagala, Y., Dassanayake, N., Karunarathna, H.: Economic, social and environmental impacts and overall sustainability of the tea sector in Sri Lanka. Sustain. Prod. Consumption **12**, 155–169 (2017)
9. Li, Y., Shi, X., Su, B.: Economic, social and environmental impacts of fuel subsidies: a revisit of Malaysia. Energy Policy **110**, 51–61 (2017)
10. Ferreira, V., Fabregat-Aibar, L., Pié, L., Terceño, A.: Research trends and hotspots in bioeconomy impact analysis: a study of economic, social and environmental impacts. Environ. Impact Assess. Rev. **96**, 106842 (2022)
11. Li, J., Fang, W., Shi, Y., Ren, C.: Assessing economic, social and environmental impacts on housing prices in Hong Kong: a time-series study of 2006, 2011 and 2016. J. Housing Built Environ. **37**(3), 1433–1457 (2022)
12. Moeltner, K., Ramsey, A.F., Neill, C.L.: Bayesian kinked regression with unobserved thresholds: an application to the von Liebig hypothesis. Am. J. Agr. Econ. (2021). https://doi.org/10.1111/ajae.12185
13. Savvides, A., Stengos, T.: Income inequality and economic development: evidence from the threshold regression model. Econ. Lett. **69**(2), 207–212 (2000)
14. Seo, M.H., Shin, Y.: Dynamic panels with threshold effect and endogeneity. J. Econometrics **195**(2), 169–186 (2016)
15. Masulis, R.W.: The impact of capital structure change on firm value: some estimates. J. Financ. **38**(1), 107–126 (1983)
16. Gelman, A., Goodrich, B., Gabry, J., Vehtari, A.: R-squared for Bayesian regression models. Am. Stat. **73**, 1–6 (2018)
17. Berger, L., Emmerling, J.: Welfare as equity equivalents. J. Econ. Surv. **34**(4), 727–752 (2020). https://doi.org/10.1111/joes.12368

Examining the Risk Contribution of Major Stock Markets to the Global Equity Market During the COVID-19 Pandemic

Namchok Chimprang[1,2], Woraphon Yamaka[1(✉)], and Nattakit Intawong[3]

[1] Center of Excellence in Econometrics, Faculty of Economics,
Chiang Mai University, Chiang Mai 50200, Thailand
woraphon.econ@gmail.com
[2] Office of Research Administration, Chiang Mai University, Chiang Mai 50200, Thailand
[3] Faculty of Economics, Chiang Mai University, Chiang Mai 50200, Thailand

Abstract. The study aims to develop a crisis-specific investment strategy by investigating the risk contribution of stock market indexes to the global stock index during the COVID-19 pandemic. It focuses on primary capitalization indexes from each continent and utilizes the Component Expected Shortfall (CES), copula-GARCH model, VaR, and ES. The findings highlight the suitability of the ARMA (1,1)-GARCH (1,1) model with skewed Student's t distribution and Multivariate Student-t copula. The ASX index poses the highest risk, while the STOXX and BVSP indexes show no significant interaction. The study suggests considering an increased allocation to the STOXX and BVSP indexes, indicating potential benefits for risk-averse investors during the crisis.

Keywords: Risk contribution · Systemic Risk · Component Expected Shortfall · Multivariate Copula-GARCH model · Stock market index

1 Introduction

In late 2019, the world was suddenly hit by the severe outbreak of COVID-19, a global pandemic originating in China, which quickly swept across the rest of the world [1]. Consequently, numerous countries implemented stringent measures such as national lockdowns, mobility restrictions, temporary retail closures, and social isolation to curb the spread of the virus and reduce confirmed cases [2]. These preventive measures with the overall circumstances surrounding the coronavirus pandemic have caused a significant demand contraction and disruptions in global supply chains. In sum, the onset of disrupted and diminished economic activities in 2020 and 2021 resulted in a profound global recession, particularly impacting the real sector (Main Street), surpassing the severity witnessed during the global financial crisis [3]. Furthermore, the capital market (Wall Street) has been greatly influenced by these circumstances, with wild fluctuations and critical price volatility [4].

Amidst this scenario, certain investors were able to capitalize on substantial profits, while others experienced losses stemming from their lack of knowledge and inadequate preparation. According to Ilie's research findings, effective risk management strategies in addressing investor risk behavior significantly contribute to achieving success in investments [5]. Therefore, it is crucial for investors to study the volatility and risks inherent to the entire market system that play a vital role in financial risk and significantly impact investor decision-making during the crisis [6]. Furthermore, it is evident that the occurrence of systemic risks in stock indices during the pandemic has resulted in significant losses for many investors within a remarkably brief timeframe [7]. Abuzayed et al. [8] discovered that the North American and European markets are more susceptible to risks emanating from the global market as a whole, compared to the Asian market. However, Asian countries have exhibited more pronounced negative abnormal returns in comparison to their global counterparts [9]. Furthermore, Chaudhary et al. [10] ascertained that the S&P/Toronto Stock Exchange (SPTSX) Composite and the Financial Times Stock Exchange (FTSE) 100 exhibit a longer half-life for volatility compared to the Nikkei 225 and Shanghai Stock Exchange (SSE) Composite, which display the swiftest mean reversion during the crisis. Additionally, Vo et al. [11] elucidated that pandemic control measures can effectively decrease market volatility at both the country and regional levels, and the response of stock markets to the Coronavirus pandemic in the Asia-Pacific countries appears to evolve over time.

The study aims to quantitatively evaluate the influence of major stock market indexes on the overall risk of the global stock system. Specifically, it focuses on analyzing the largest capitalization stock market indexes from each continent: the New York Stock Exchange (NYSE) for North America, SSE for Asia, STOXX for Europe, Johannesburg Stock Exchange (JSE) for Africa, Australian Securities Exchange (ASX) for Australia, and Sao Paulo Bovespa (BVSP) for South America [12]. To accomplish our objective, this study adopts the Component Expected Shortfall (CES) framework introduced by Banulescu and Dumitrescu [13]. This approach can be used to accurately evaluate the individual contribution of the stock market to the overall risk within the system at a given moment in time. In a practical context, Liu et al. [14] conducted a comprehensive investigation into the systemic risk measurement, focusing on volatility and dependence employing a copula model with CES. They found that CES effectively elucidates the risk associated with financial crises. Wu [15] similarly employed the CES approach identifying risk contributions in the Chinese stock index, allowing for the measurement of systemic risk and the analysis of volatility spillover, particularly in the face of a severe health crisis. Furthermore, Liu et al. [16] innovatively combined the GARCH and CES methodologies to analyze systemic risk in global financial markets and ascertain the individual country's contributions to global systemic risk. Additionally, they developed CES portfolios based on forecasting outcomes derived from factor copula-based models, thereby enhancing the accuracy of systemic risk predictions. As a result, the CES measure holds a promise as a valuable tool for policymakers and an adequate tool for investors in identifying stock markets to monitor, thereby discouraging the accumulation of systemic risk.

Before proceeding with the risk-sharing assessment, it is essential to calculate the correlation among the stock market indexes. However, relying solely on the linear correlation and assuming normality within the conventional GARCH model are unsuitable for accurately measuring the correlation between two stock market indexes as they do not have tail dependence. Moreover, financial market index returns often exhibit heavy-tailed distributions with asymmetrical correlation and dependence [17]. Consequently, utilizing the conventional correlation model may result in an inadequate inference of the Component Expected Shortfall (CES). Numerous methodologies exist for assessing the interdependence of variables. Nevertheless, the constraints inherent in linear correlation and the underlying normality assumption have remained.

To overcome these limitations, the present study proposes a copula-GARCH model as an alternative approach. This model capitalizes on the copula methodology to construct a joint distribution of diverse marginal distributions, employing different copula structures to effectively capture the tail dependency observed in stock market indexes. Over the past decade, researchers have increasingly employed the multivariate copula to estimate dependency among large assets, as evidenced by numerous studies in the field. According to Yeap et al. [18], incorporating the modeling of the asymmetric dependence structure in financial assets has the potential to enhance the precision of risk management. However, a hedge designed under the assumption of linear correlation, which fails to conform to a multivariate normal copula due to its asymmetric and heavy nature, may not provide the desired degree of protection. Luca and Zuccolotto [19] utilize a multivariate copula approach for the hierarchical clustering of financial stock returns, enabling the construction of investment portfolios. This method offers a high degree of flexibility in modeling the interdependencies among random variables, thereby providing valuable tools for capturing joint relationships. Furthermore, Rašiová [20] suggests that to effectively address dependency in various contexts, it is imperative to emphasize the utilization of a comprehensive range of copula families, including the Elliptical and the Archimedean copulas.

This paper presents three key contributions. Firstly, it delves into an in-depth examination of different copula models and functions to determine the most suitable dependence structure for stock market indexes. Secondly, the paper explores the dependence and risk contribution among six stock market indexes, utilizing both the copula-GARCH model and the CES approach. Lastly, to identify an effective investment strategy specifically tailored to the pandemic context, the VaR and ES approaches for evaluating potential investment loss are included.

2 Methodology

2.1 ARMA-GARCH Model

To derive the marginal distribution of the log-difference of stock market index (r_t), a univariate ARMA (p, q) model with GARCH (m, n) specification is utilized, where parameter p represents the order of the autoregressive component, q denotes the order of the moving average component, m signifies the order of the term h^2 in the GARCH, and n signifies the order of the term ε^2 in the ARCH. The financial data in the study is well-established in that they exhibit heteroscedasticity and often display autocorrelation. Consequently, the GARCH (1,1) model is employed, as it effectively captures the volatility dynamics inherent in financial data, while simultaneously eliminating autocorrelation within the ARMA process. Referring back to the aforementioned notations, the model ARMA (p, q)- GARCH $(1,1)$ model can be represented as follows:

$$r_t = \mu + \sum_{i=1}^{p} \phi_i r_{t-i} + \sum_{j=1}^{q} \theta_j \varepsilon_{t-j} + \varepsilon_t \tag{1}$$

$$\varepsilon_t = h_t \eta_t \tag{2}$$

$$h_t^2 = \omega + \alpha_1 \varepsilon_{t-1}^2 + \beta_1 h_{t-1}^2 \tag{3}$$

In these equations, ϕ_i are the autoregressive component parameters of order p, θ_j are the moving average component parameters of order q. The term ε_t denotes the residual at time t consisting of the standard variance h_t and the standardized residual η_t. It is crucial to note that the standard residual term η_t adheres to the assumptions of being independent and identically distributed. In this study, the skewed-t distribution is assumed for the standard residual term. Certain standard restrictions are imposed on the GARCH parameters to ensure model validity and stability. These restrictions are commonly defined that $\omega > 0$, $\alpha_1 > 0$, $\beta_1 > 0$ and $\alpha_1 + \beta_1 < 1$.

2.2 Multivariate Copulas

The term copula refers to the multivariate distribution function that establishes the linkage between variables, ensuring that the marginal distributions are uniformly distributed. Sklar's theorem states that the n continuous marginals can be joined using a copula function $C(\cdot)$. Let $H(\cdot)$ be a joint distribution with an n dimension and marginals F_i, $i = 1, 2, ..., n$. Thus, the $H(\cdot)$ can be rewritten as

$$\begin{aligned} H(\eta_1, ..., \eta_n) &= C(F_1(\eta_1), ..., F_n(\eta_n)) \\ &= C(u_1, ..., u_n), \end{aligned} \tag{4}$$

If the marginals F_i are continuous, then the copula C associated to $H(\cdot)$ is unique, and we can further explain the Eq. (4) as follows:

$$C(u_1, ..., u_n) = H(F_1^{-1}(u_1), ..., F_n^{-1}(u_n)), \tag{5}$$

where $u \in [0, 1]^n [0, 1]^n [0,1]^n$ is the uniform distribution. The copula density distribution can be derived by performing as follows:

$$C(u_1, ..., u_n) = \frac{\partial C(u_1, ..., u_n)}{\partial u_1, ..., \partial u_n}., \tag{6}$$

In this study, we primarily investigate two fundamental copulas classes: the Elliptical copula and the Archimedean copula. Within the Elliptical class, our focus is on the symmetric Gaussian and Student-t copulas. On the other hand, within the Archimedean class, we explore the asymmetric dependence models offered by Clayton, Gumbel, Joe, and Frank copulas. The explicit mathematical expressions for these copula classes are described as follows:

Elliptical Copulas. The multivariate Elliptical copula can be expressed by

$$C(u_1, ..., u_n|\rho) = \Phi(\Phi^{-1}(u_1), ..., \Phi^{-1}(u_n); \rho) \tag{7}$$

where the term $\Phi(\cdot)$ represents either the multivariate Gaussian distribution function pertaining to the Normal copula or the multivariate Student's t distribution function incorporating degrees of freedom for the Student-t copula. ρ is the $n \times n$ correlation matrix of the dependence parameters with interval $[-1, 1]$. The function Φ^{-1} represents the inverse cumulative distribution function of the standard normal distribution for the Normal copula and the Student's t distribution for the Student-t copula [21].

Archimedean Copulas. The general form of a multivariate Archimedean copula is expressed as follows:

$$C(u_1, ..., u_n|\theta) = \Phi(\sum_{j=1}^{n} \Phi^{-1}(u_j); \theta) \tag{8}$$

where the function $\Phi(\cdot)$ denotes the Laplace transform of distributions in the univariate family, such as the Clayton, Gumbel, Joe, or Frank distribution. Additionally, the term Φ^{-1} represents an inverse cumulative distribution function of the univariate distribution. In multivariate analysis, the exchangeable dependence is employed to characterize the Elliptical copula through a symmetric positive definite matrix. As the dimensionality increases, the range of this dependence structure becomes increasingly limited. We further elaborate on the analysis concerning $n > 2$ by imposing a restriction on the dependence parameter θ to be $(0, \infty)$ for Frank, $[0, +\infty)$ for Clayton, and $[1, +\infty)$ for Gumbel and Joe copulas [22].

2.3 Component Expected Shortfall

In this section, we present the introduction to the Component Expected Shortfall (CES) concept, as proposed by Banulescu and Dumitrescu [13], to determine the weighted spillover volatility associated with the systemically risky financial stock market index. Let us assume a stock market index system composed of n indexes. r_{it} is the stock market index return i at time t and r_{mt} represents the aggregate return of all stock markets at time t.

$$r_{mt} = \sum_{i=1}^{n} w_{it} r_{it} \tag{9}$$

where w_{it} is an individual weight of stock market index $i, i = 1, ..., n$, at each time analysis t. These weights denote the relative market capitalization of stock market index i.

Let us consider the measurement of the aggregate risk of the financial system in terms of conditional Expected Shortfall (ES). In actuarial terms, the ES represents the anticipated market loss conditional on the return, contingent upon the return falling below a specified quantile. The formal representation of the conditional ES, incorporating historical information, can be expressed as follows:

$$ES_m(C) = -\mathbb{E}(r_{mt}|r_{mt} < C), \tag{10}$$

where C is the threshold or value at risk under extreme conditions. To evaluate the individual contribution of each stock market to the overall risk of the system, Acharya et al. [23] introduced the concept of the Marginal Expected Shortfall (MES). MES represents the marginal impact of a specific stock market index on the total risk of the stock market index system. This measure is quantified by assessing the change in the system's Expected Shortfall (ES) resulting from a one-unit increase in the weight of the respective stock market index within the system.

$$MES_i(C) = \frac{\partial ES_m(C)}{\partial w_i} = -\mathbb{E}(r_{it}|r_{mt} < C). \tag{11}$$

To address dependency among random variables by copula approach, the Marginal Expected Shortfall (MES) can be derived as follows:

$$MES_{it} = \left[h_{it} \cdot \kappa_{im} \frac{\sum_{t=1}^{T} \eta_{mt} \Phi\left(\frac{C-\eta_{mt}}{h_{mt}}\right)}{\sum_{t=1}^{T} \Phi\left(\frac{C-\eta_{mt}}{h_{mt}}\right)} \right] + \left[h_{it} \cdot \sqrt{1-\kappa_{im}^2} \frac{\sum_{t=1}^{T} \eta_{it} \Phi\left(\frac{C-\eta_{mt}}{h_{mt}}\right)}{\sum_{t=1}^{T} \Phi\left(\frac{C-\eta_{mt}}{h_{mt}}\right)} \right] \tag{12}$$

where η_{mt} and η_{it} are the standardized world stock market return and stock market return i. h_{mt} and h_{it} denote the variance of error at time t of the market and the individual returns, respectively. Φ is the cumulative normal distribution function and κ_i is the Kendall's tau estimated from the copula dependence parameter.

Banulescu and Dumitrescu [12] stressed the crucial significance of the scale factor in assessing the relative importance of components in total risk. They introduce the weight w_{it} based on MES and propose CES as a comprehensive measure of component contributions to overall risk.

$$CES_{it} = \frac{w_{it}\partial ES_{m,t-1}(C)}{\partial \omega_{it}} = -w_{it}E_{t-1}(r_{it}|r_{mt} < C) \tag{13}$$

A higher contribution signifies a greater level of systemic significance for the individual. However, the objective of our study is to assess the risk contribution of each individual stock market index to the global stock market index system. Therefore, it is more appropriate to measure the risk in percentage term. $CES_{it}\%(C)$ represents the proportion of systemic risk attributed by the individual i at time t. This measure is calculated by dividing the component loss by the total loss and subsequently normalizing it.

$$CES_{it}\%(C) = \frac{CES_{it}(C)}{\sum_{i=1}^{n} CES_{it}(C)} \times 100 = \frac{w_{it}E_{t-1}(r_{it}|r_{mt} < C)}{\sum_{i=1}^{n} w_{it}E_{t-1}(r_{it}|r_{mt} < C)} \times 100 \tag{14}$$

Table 1. Descriptive Statistics

	NYSE	SEE	STOXX	ASX	JSE	BVSP
Mean	0.0004	0.0001	0.0005	0.0003	−0.0004	0.0002
Median	0.0009	0.0004	0.0011	0.0011	−0.0002	0.0006
Max	0.0868	0.0755	0.0671	0.0677	0.0367	0.1302
Min	−0.1260	−0.0672	−0.1219	−0.1020	−0.0531	−0.1599
Std	0.0151	0.0111	0.0129	0.0124	0.0085	0.0192
Skewness	−0.9392	−0.1581	−1.2481	−1.0176	−0.4870	−0.9459
Kurtosis	16.8084	8.9338	17.2317	14.6106	8.6653	20.9379
JB-Test	5267.6490*	957.7859*	5662.9760*	3768.9900*	896.3269*	8825.0310*
ADF-Test						
Intercept and Trend	−1.0869*	−0.9458*	−1.0159*	−1.1584*	−0.9894*	−1.1373*

Note: * represents the p-value significant at the 1% level

3 Data

This study focuses on analyzing the highest capitalization stock market index from each continent, chosen for its representation of both the highest market value and investor popularity. The study utilizes daily log returns data from the following stock market indices: the NYSE, SSE, STOXX, JSE, ASX, and BVSP. The data spans from March 10, 2020, to March 11, 2023, comprising a total of 651 observations. All data is sourced from investing.com and the Thomson Reuters database. To provide a comprehensive overview, the descriptive statistics of the transformed data are presented in Table 1.

4 Analysis of the Results

This paper primarily endeavors to empirically investigate the risk contribution of six stock market indexes within the stock market index system. To accomplish this objective, effectively addressing the interdependence among these indexes is considerably required. We employ a robust analytical tool known as a copula, featuring a diverse range of types. The ultimate goal of the first subsection is to identify the most suitable and accurate copula that best aligns with the available data. The candidates encompass multivariate copulas falling under both the Elliptical and the Archimedean categories, as extensively discussed in Sect. 2. However, prior to meticulously selecting an appropriate copula, it is crucial to comprehensively address the initial subsection pertaining to the construction of marginal distributions. Lastly, we intricately measure the risk sharing using the Component Expected Shortfall (CES) and precisely quantify risk values by employing both ES (Expected Shortfall) and VAR (Value at Risk) metrics.

Table 2. Summary statistics results of Marginal Distribution

	NYSE	SEE	STOXX	ASX	JSE	BVSP
Mean Equation						
μ	0.0040	0.0006	0.0007^{***}	0.0029	-0.0022^{*}	0.0007
	(0.0026)	(0.0021)	(0.0002)	(0.0024)	(0.0013)	(0.0011)
AR (1)	-0.6153^{***}	0.7367^{***}	0.8748^{***}	0.4062^{**}	0.0705	0.3156^{**}
	(0.1506)	(0.0819)	(0.0904)	(0.1878)	(0.2289)	(0.1338)
MA (1)	0.5339^{***}	-0.8509^{***}	-0.9256^{***}	-0.5712^{***}	-0.2357	-0.5226^{***}
	(0.1622)	(0.0604)	(0.0687)	(0.1700)	(0.2234)	(0.1180)
Variance Equation						
ω	0.0004^{**}	0.0003^{*}	0.0000	0.0004^{*}	0.0000	0.0001^{**}
	(0.0002)	(0.0002)	(0.0000)	(0.0002)	(0.0000)	(0.0000)
α	0.2944^{***}	0.1396^{***}	0.1263^{**}	0.2253^{***}	0.1400^{***}	0.3846^{***}
	(0.0755)	(0.0416)	(0.0559)	(0.0568)	(0.0247)	(0.0412)
β	0.7046^{***}	0.8594^{***}	0.8591^{***}	0.7737^{***}	0.8590^{***}	0.6144^{***}
	(0.0501)	(0.0340)	(0.0592)	(0.0499)	(0.0236)	(0.0315)
Skewness	0.9052^{***}	0.8337^{***}	0.8101^{***}	0.7857^{***}	0.9271^{***}	0.8239^{***}
	(0.0474)	(0.0442)	(0.04901)	(0.0435)	(0.0416)	(0.0346)
Degree of freedom	4.9990^{***}	4.1377^{***}	4.4323^{***}	4.9602^{***}	4.4347^{***}	3.6531^{***}
	(0.9406)	(0.6881)	(0.7914)	(0.9280)	(0.5415)	(0.2571)
Ljung-Box Test	0.4687	0.9531	0.6417	0.3738	0.0103	0.0339
LM-ARCH Test	0.5548	0.3419	0.0824	0.7655	0.2937	0.0019

Notes: *, **, and *** denote significant at 10, 5 and 1%, respectively. The standard error is shown in the parenthesis

4.1 Modeling Marginal Distributions by ARMA-GARCH

Table 2 illustrates the estimated coefficients for the ARMA(1,1)-GARCH(1,1) model with a skewed-t distribution applied to individual stock market index return series. In this study, we judiciously adopt the ARMA(1,1)-GARCH(1,1) model to effectively characterize the underlying dynamics behavior of the time series data, as it represents a prevailing and prominently accepted approach within the field (See also, [24–27]). The significant coefficients in each equation confirm the validity of the skewed-t distribution assumption for the ARMA-GARCH model. This approach effectively captures the non-normal characteristics in stock market index returns. The LM-ARCH Test comprehensively assesses transformed marginal distribution functions of residuals. Results show no support for the null hypothesis, indicating uniform distribution within [0,1]. The Ljung-Box test checks autocorrelations of residuals, revealing no significant evidence to reject the null hypothesis, indicating no autocorrelation in the dataset.

4.2 Model Selection

In this section, we thoroughly investigate various multivariate copula models, mainly focusing on two primary classes of copula families: Elliptical and Archimedean copulas. The primary objective of this section is to explicitly indicate the optimal copula family from extensive candidates for dependency modeling. To accomplish this, we employ the Akaike Information Criterion (AIC) and the Bayesian Information Criterion (BIC) as robust measures for evaluation. The preferred copula model is identified by selecting the one that yields the minimum values of AIC and BIC, thus strengthening the validity of our findings.

Table 3. Model selection

Model type	LL	AIC	BIC
Multivariate Normal Copula	93.6774	−185.3548	−180.8763
Multivariate Student-t Copula	**112.3462**	**−220.6924**	**−211.7354**
Multivariate Clayton Copula	95.0777	−188.1553	−183.6768
Multivariate Joe Copula	36.0924	−70.18476	65.70625
Multivariate Frank Copula	84.9674	−167.9349	−163.4564
Multivariate Gumbel Copula	68.5533	−135.1066	−130.6281
Independence Copula	40.2849	−72.4390	−67.3490
Constant conditional correlation	90.8734	−181.8930	−186.2291

Note: The bold numbers indicate the lowest value of AIC and BIC, while the highest value of LL

Table 3 displays the AIC, BIC, and LL values corresponding to each copula type employed in the multivariate copula analysis. Notably, our findings reveal that the minimum AIC and BIC values are −220.6924 and −211.7354, respectively (as denoted by the bold numbers). Moreover, the maximum log-likelihood (LL) value obtained is 112.3462. Consequently, these outcomes suggest that the Student-t copula is the most suitable choice for modeling the provided dataset. Thus, we use the Student-t copula based GARCH to predict the expected return of each stock in time $t + 1$, typically around 10,000 rounds. By doing so, we gain a comprehensive view of the potential distribution of portfolio returns. This aids in refining our VaR and ES calculations, providing us with a robust understanding of risk across a wide range of conceivable scenarios. In essence, the combination of CES, VaR, and ES within the Student-t copula based GARCH framework empowers us to make well-informed decisions in portfolio management and risk assessment.

4.3 Risk Contribution Measurement

As mentioned earlier in the introduction, the main focus of this study is to assess the individual risk contributions of major stock market indexes to the global stock market.

Table 4. Summary result of Component Expected Shortfall (CES)

Component Expected Shortfall (CES)							
	NYSE	SSE	STOXX	ASX	JSE	BVSP	Sum
Contribution	0.018	0.025	0.000	0.070	0.004	0.000	0.117
Percentage (%)	15.385%	21.368%	0.000%	59.829%	3.419%	0.000%	100%

To achieve this, we adopt a direct ranking approach based on their CES% values in an equal-weighted portfolio management strategy.

Table 4 reveals the Component Expected Shortfall (CES) values, shedding light on the risk contributions of major stock markets. Among the analyzed indexes, namely ASX, SSE, NYSE, JSE, STOXX, and BVSP, the ASX index notably stands out as the primary driver of risk in global stock markets during the COVID-19 crisis. Its significant impact surpasses that of the other indexes, emphasizing the pivotal role played by the Australian market. The dominance of the ASX index can be attributed by several key factors. Australia's economy, closely tied to global trade and investments, renders it highly responsive to international economic conditions. Additionally, the composition of the ASX index, consisting of sectors such as mining, energy, and finance, exposes it to volatility stemming from fluctuations in commodity prices and market sentiment. These factors, combined with Australia's geographical location and economic interdependencies, position the ASX index as a crucial player in the global financial landscape.

4.4 Value at Risk (VaR) and Expected Shortfall (ES)

This section presents Value at Risk (VaR) and Expected Shortfall (ES) of each stock market index. VaR is a quantifiable measure representing the maximum potential loss that will not be surpassed within a specified confidence level during a predetermined period. In contrast, Expected Shortfall (ES) was formulated to align with the characteristics of coherent risk measures commonly employed when the VaR fails to adequately capture varying levels of the potential loss. The ES serves as a risk assessment technique that takes into account the tail-end of the loss distribution function. This approach computes the average of losses that exceed the VaR threshold. By incorporating both VaR and ES, it becomes possible to ascertain the loss incurred at a given significance level and estimate the mean magnitude of excess damage.

Table 5 presents the Expected Shortfall (ES) and Value-at-Risk (VaR) results for the six stock market indexes in the study. The findings align with the CES% results displayed in Table 4. The overall result indicates that three indexes, comprising ASX, NYSE, and SSE, demonstrate high ES and VaR values surpassing 19.22% and 9.43%, respectively. Notably, the ASX index exhibits the highest ES and VaR values among the six indexes. These results suggest that based on the specified confidence level, investors face the probability of losing money beyond the VaR threshold and experiencing losses equivalent to the ES value. For instance, individuals investing in the ASX index face a 1% probability of losing over 56.18% (VaR) of their investment and a 1% chance of losing 72.72% (ES). Therefore, risk-averse investors should consider allocating their

Table 5. Summary results of Value at Risk (VaR) and Expected Shortfall (ES)

	NYSE	SSE	STOXX	ASX	JSE	BVSP
At 1% confidence interval						
ES	−0.4791	−0.6481	−0.1306	−0.7272	−0.2330	−0.1561
VaR	−0.1471	−0.4195	−0.0886	−0.5618	−0.1873	−0.1207
At 5% confidence interval						
ES	−0.3009	−0.4240	−0.0688	−0.4024	−0.1408	−0.0851
VaR	−0.1583	−0.2108	−0.0322	−0.2190	−0.0795	−0.0463
At 10% confidence interval						
ES	−0.1922	−0.2550	−0.0465	−0.2845	−0.1024	−0.0621
VaR	−0.0943	−0.1300	−0.0186	−0.1295	−0.0497	−0.0330

investments to indexes with lower ES and VaR values, such as STOXX, BVSP, and JSE index.

5   Conclusion

This study undertakes a quantitative assessment to ascertain the risk contribution of six prominent stock market indexes, namely the NYSE, SEE, STOXX, JSE, ASX, and BVSP, to the aggregate global stock index system during the prevailing global COVID-19 pandemic. Consequently, the study discerns an efficacious investment strategy specifically tailored to the exigencies of the crisis. To achieve this objective, the study employs the Component Expected Shortfall (CES) approach enabling an evaluation of the individual contribution of each stock market to the overall risk inherent in the system. Unfortunately, due to limitations in relying solely on linear correlation and assuming normality within the conventional correlation model, a more sophisticated copula-GARCH framework is proposed for modeling the observed dependency in stock market returns. Lastly, the study employs risk measures such as Value at Risk (VaR) and Expected Shortfall (ES) to comprehensively assess the potential loss of an investor's investment portfolio.

Our findings reveal that the ARMA(1,1)-GARCH(1,1) model with skewed Student's t distribution, combined with the Multivariate Student-t copula, is the most suitable approach for characterizing the dependency and underlying dynamic behavior of the study's time series data. This is supported by the model's ability to minimize both the Akaike Information Criterion (AIC) and the Bayesian Information Criterion (BIC). Regarding the Component Expected Shortfall (CES), our analysis demonstrates that the ASX index presents the highest level of risk to the stock market index system, followed by the SSE, NYSE, and JSE indexes, based on their respective contributions to the system's overall risk profile. In contrast, the STOXX and BVSP indexes did not demonstrate any discernible interaction with the financial stock market index system. Furthermore, the STOXX, BVSP, and JSE indexes exhibit substantially lower Expected Shortfall (ES) and Value at Risk (VaR) measures compared to the other three indexes. These compelling results offer valuable insights for constructing investment portfolios based on investor risk preferences. For risk-averse investors, it is recommended to avoid investing in the ASX index during times of crisis and instead increase the allocation to the STOXX and BVSP indexes within their portfolios. As a secondary option, the JSE index represents a viable choice.

Acknowledgments. The authors are grateful to the Centre of Excellence in Econometrics, Chiang Mai University, for financial support. This research work was partially supported by Chiang Mai University. They are also grateful to Dr. Laxmi Worachai for her helpful comments and suggestions.

References

1. Baker, S.R., Bloom, N., Davis, S.J., Kost, K.J., Sammon, M.C., Viratyosin, T.: The unprecedented stock market impact of covid-19 (No. w26945). National Bureau of economic research (2020)
2. Huang, C., et al.: Clinical features of patients infected with 2019 novel coronavirus in Wuhan China. Lancet **395**(10223), 497–506 (2020)
3. Glocker, C., Piribauer, P.: The determinants of output losses during the covid-19 pandemic. Econ. Lett. **204**, 109923 (2021)
4. Baek, S., Mohanty, S.K., Glambosky, M.: Covid-19 and stock market volatility: an industry level analysis. Financ. Res. Lett. **37**, 101748 (2020)
5. Ilie, F.: Considerations regarding financial risk management in order to maximize earnings during the coronavirus pandemic. Sci. Bull. **25**(1), 26–32 (2020)
6. Stoja, E., Polanski, A., Nguyen, L.H., Pereverzin, A.: Does systematic tail risk matter? J. Int. Financ. Markets Institutions Money **82**, 101698 (2023)
7. Zhang, D., Hu, M., Ji, Q.: Financial markets under the global pandemic of covid-19. Financ. Res. Lett. **36**, 101528 (2020)
8. Abuzayed, B., Bouri, E., Al-Fayoumi, N., Jalkh, N.: Systemic risk spillover across global and country stock markets during the covid-19 pandemic. Econ. Anal. Policy **71**, 180–197 (2021)
9. Chien, F., Sadiq, M., Kamran, H.W., Nawaz, M.A., Hussain, M.S., Raza, M.: Co-movement of energy prices and stock market return: environmental wavelet nexus of covid-19 pandemic from the USA, Europe, and China. Environ. Sci. Pollut. Res. **28**, 32359–32373 (2021)
10. Chaudhary, R., Bakhshi, P., Gupta, H.: Volatility in international stock markets: an empirical study during covid-19. J. Risk Finan. Manag. **13**(9), 208 (2020)
11. Vo, D.H., Ho, C.M., Dang, T.H.N.: Stock market volatility from the covid-19 pandemic: new evidence from the Asia-Pacific region. Heliyon **8**(9), e10763 (2022)
12. Statista Research Department. Largest stock exchange operators worldwide. Statista. Retrieved 3 May 2023, from https://www.statista.com/statistics/270126/largest-stock-exchange-operators-by-market-capitalization-of-listed-companies/ (2023)
13. Banulescu, G.D., Dumitrescu, E.I.: Which are the Sifi? A Component Expected Shortfall (Ces) Approach To Systemic Risk. Preprint, University of Orlans (2015)
14. Liu, J., et al.: Volatility and dependence for systemic risk measurement of the international financial system. In: 4th International Symposium on Integrated Uncertainty in Knowledge Modelling and Decision Making, pp. 403–414 (2015)
15. Wu, F.: Sectoral contributions to systemic risk in the Chinese stock market. Finan. Res. Lett. **31** (2019)
16. Liu, J., Song, Q., Qi, Y., Rahman, S., Sriboonchitta, S.: Measurement of systemic risk in global financial markets and its application in forecasting trading decisions. Sustainability **12**(10), 4000 (2020)

17. Pastpipatkul, P., Maneejuk, P., Sriboonchitt, S.: The best copula modeling of dependence structure among gold, oil prices, and US currency. In: Integrated Uncertainty in Knowledge Modelling and Decision Making: 5th International Symposium, IUKM 2016, Da Nang, Vietnam, November 30-December 2, 2016, Proceedings 5, pp. 493–507. Springer International Publishing (2016)

18. Yeap, X.W., Lean, H.H., Sampid, M.G., Mohamad Hasim, H.: The dependence structure and portfolio risk of Malaysia's foreign exchange rates: the Bayesian GARCH–EVT–copula model. Int. J. Emerg. Mark. **16**(5), 952–974 (2021)

19. De Luca, G., Zuccolotto, P.: Hierarchical time series clustering on tail dependence with linkage based on a multivariate copula approach. Int. J. Approximate Reasoning **139**, 88–103 (2021)

20. Rašiová, B., Árendáš, P.: Copula approach to market volatility and technology stocks dependence. Financ. Res. Lett. **52**, 103553 (2023)

21. Joe, H., Hu, T.: Multivariate distributions from mixtures of max-infinitely divisible distributions. J. Multivar. Anal. **57**(2), 240–265 (1996)

22. Nelsen, R.B.: An Introduction to Copulas. Springer science & Business Media (2007)

23. Acharya, V., Pedersen, L.H., Philippon, T., Richardson, M.: Measuring Systemic Risk, AFA 2011 Denver Meetings Paper (2011)

24. Arashi, M., Rounaghi, M.M.: Analysis of market efficiency and fractal feature of NASDAQ stock exchange: time series modeling and forecasting of stock index using ARMA-GARCH model. Future Bus. J. **8**(1), 1–12 (2022)

25. Sharma, C., Sahni, N.: A mutual information-based R-vine copula strategy to estimate VaR in high frequency stock market data. PLoS ONE **16**(6), e0253307 (2021)

26. Wang, Y., Xiang, Y., Lei, X., Zhou, Y.: Volatility analysis based on GARCH-type models: evidence from the Chinese stock market. Econ. Res.-Ekonomska Istraživanja **35**(1), 2530–2554 (2022)

27. Lin, Z.: Modelling and forecasting the stock market volatility of SSE composite index using GARCH models. Futur. Gener. Comput. Syst. **79**, 960–972 (2018)

Dynamics of Investor Behavior and Market Interactions in the Thai Stock Market: A Regime-switching Analysis

Panisara Phochanachan[1], Supanika Leurcharusmee[2], and Nootchanat Pirabun[2,3](✉)

[1] School of Business and Communication Arts, University of Phayao, Phayao 56000, Thailand
[2] Center of Excellence in Econometrics, Faculty of Economics,
Chiang Mai University, Chiang Mai 50200, Thailand
Nootchanat_pi@cmu.ac.th
[3] Office of Research Administration, Chiang Mai University, Chiang Mai 50200, Thailand

Abstract. This paper investigates the correlation between trading activities of four different investor groups, namely individual, foreign, institutional, and proprietary traders, and the stock price index in the Stock Exchange of Thailand. Using a Markov-switching VAR (MS-VAR) model and variance decomposition analysis, we analyze the dynamics and spillover effects among investor types and their impact on the Thai stock market. The findings reveal two distinct trading value regimes: low trading value and high trading value. Stronger integration among investor types and the market is observed during the high trading value regime. The variance decomposition analysis highlights limited impact of investor trading activities on the stock market, with the stock price index playing a dominant role, especially in the low trading value regime. Foreign trading emerges as a key volatility transmitter during the high trading value regime, impacting both the market and other investor groups. These findings have implications for investors and market participants, aiding in informed decision-making and risk management.

Keywords: Spillover measure · Thai stock market · MS-VAR · Investor groups

1 Introduction

The influence of investor types on the stock market is a pivotal and multifaceted concern within the field of finance, particularly in emerging markets like Thailand. The Thai stock market, characterized by a significant presence of individual investors with limited expertise, finds itself juxtaposed against institutional and foreign investors who wield considerable power due to their extensive exposure, substantial capital, and seasoned experience. This paper endeavors to provide comprehensive insights into the intricate and nuanced effects of diverse investor types including individual, foreign, institutional, and proprietary traders on stock market performance. Furthermore, our research delves into the intricate domain of investor behavior, exploring the interconnections among different investor segments and their profound impact on the Thai stock market. Building upon the fundamental premise that emerging markets exhibit distinctive dynamics resulting from irrational behavior, particularly prevalent among individual investors, our analysis strives to unravel the idiosyncratic contributions of each investor type to stock returns while also untangling their intricate interactions with one another.

These four investor types are officially classified by regulatory bodies such as the Bank of Thailand (BOT) and the Securities and Exchange Commission of Thailand (SEC). Each category of investors exhibits distinct behaviors in financial markets, driven by diverse motivations, resources, and investment strategies. Local institutions, including banks and mutual funds, prioritize stability and risk management, emphasizing a long-term investment approach. On the other hand, proprietary trading firms employ strategies focused on short-term profit generation, leveraging their expertise and resources to exploit market inefficiencies [1]. Foreign investors bring a unique perspective to the financial landscape [2], as their investment decisions are influenced by factors such as economic conditions, political stability, and regulatory frameworks. Their objectives, risk appetites, and trading styles may differ from those of local participants [3]. Local retail investors, comprising individual investors, exhibit a wide range of behaviors influenced by such factors as sentiment, personal experiences, and long-term investment goals [4]. Their investment choices are often driven by psychological biases and sentiment rather than extensive financial analysis. Retail investors are more susceptible to short-term market movements and are prone to herd behavior [5]. It is important to note that while these observations are generally applicable, individual behavior within each investor category can vary significantly. Factors such as personal preferences, risk tolerance, and financial knowledge contribute to behavioral differences even among investors in the same category.

Several studies have conducted extensive research on the correlation between investor types and stock indices, providing valuable insights into investor behaviors and their influence on stock prices. For instance, Li [6] discovered an inverted U-shaped relationship between the shareholding ratio of institutional investors and stock price synchronicity, indicating a nuanced connection between institutional investors' stock holdings and stock price synchronization. Similarly, Zhang et al. [7] found a similar relationship between institutional investors and stock price synchronicity. In a study by Brown and Cliff [8], it was revealed that investor sentiment holds additional explanatory power in predicting co-movements in stock returns, suggesting that bullish or bearish sentiment can impact excess returns. Moreover, Sun et al. [9] uncovered a robust bidirectional Granger causality between stock prices and trading relationship indices, implying that the price of a stock is influenced by the trading behavior of its investors. They further highlighted that significant market movements often stem from the trading activities of large investors. Exploring the motivations of investors and their impact on stock prices, Liu et al. [10] identified three dimensions of motivations: epistemic motivation, achievement motivation, and experience motivation. Each dimension may lead to the adoption of different investment strategies, thereby yielding distinct effects on the stock markets.

To comprehend the relationship among different investor types, two theoretical approaches can be employed. Firstly, herding behavior manifests as the tendency of investors to imitate the actions of others rather than make independent decisions. This behavior can significantly impact trading volume as investor groups collectively engage in buying or selling activities, leading to heightened trading volume [11]. Secondly, investor sentiment captures the overall attitude and perception of market participants towards the market and specific stocks. It possesses the power to influence trading volume as different investor groups react to positive or negative sentiments by adjusting their trading activity accordingly [12]. A promising avenue for gaining valuable insights is the monitoring of sentiment indicators and exploring their correlation with trading volume among different investor groups. By doing so, a deeper understanding of the relationship between investor behavior and market dynamics can be attained.

The existing body of literature delves into the relationship between different investor types, offering valuable insights. Kacperczyk et al. [13] emphasized that foreign investors' trading activity can have a dominant influence over local investors in the market. This influence becomes particularly significant in emerging markets where a substantial number of individual investors with limited expertise are present. Foreign investors, with their extensive exposure, capital allocation, and investment experience, are capable of exerting considerable control over market dynamics, potentially overshadowing the impact of local investors. Additionally, Ferreira et al. [14] discovered that foreign institutions perform similarly to local institutions in terms of average investment performance. However, it is the domestic institutions that exhibit a trading pattern indicative of an information advantage. This finding highlights that individual investors often exhibit poor performance and contribute to increased stock return volatility. These observations align with the notion that the psychological biases and irrational decision-making of individual investors can have a destabilizing effect on stock prices and overall market volatility [15].

As we mentioned above, the existing body of literature underscores the profound impacts of investors' trading activities on stock markets and their intricate interactions with each other. Nevertheless, there remains a critical knowledge gap regarding the specific investor types that exert a more pronounced influence on stock return volatility and the magnitude of their impact. Thus, a comprehensive understanding of the dynamic relationship between investor types and the stock market is of paramount importance. Furthermore, the empirical literature highlights the presence of structural changes within the stock market, as identified by Pastpipatkul et al. [16, 17], accentuating the need to recognize the varying roles played by different investor types across distinct market states, such as upturns and downturns. This paper aims to address this crucial gap in the literature by conducting a meticulous analysis of investor trading activities and assessing their impacts on, as well as reactions to, the Thai stock market within each market state. To achieve this ambitious goal, our research employs an innovative and sophisticated methodological approach known as the Markov-switching Vector Autoregressive (MS-VAR) model. This advanced tool, as described by Diebold and Yilmaz [18], is tailor-made for financial time series analysis, allowing us to capture discrete changes within the data and discern the intricate relationships between investor types and the market. Moreover, the MS-VAR-typed models facilitate a comprehensive examination of forecast error variance decomposition, which plays a vital role in quantifying directional and net spillovers across investor types and markets. By utilizing this methodology, we can provide robust and reliable insights into the intricate dynamics of the Thai stock market and shed light on the nuanced impacts of different investor types.

The subsequent sections of the paper are structured as follows. Section 2 introduces and elaborates on the regime-switching vector autoregressive model, while also explaining the variance decomposition methodology employed in the analysis. In Sect. 3, the data used in the study is presented and discussed. Section 4 presents and interprets the results obtained from the analysis. Finally, in Sect. 5, the paper concludes by summarizing the key findings and their implications.

2 Methodology

In this study, we employ the regime-switching variance decomposition method to calculate spillover measures among the trading activities of different investor types and the Thai stock market. This technique is an extension of the estimation of the Markov-switching Vector Autoregressive (MS-VAR) model [19]. By applying this approach, we are able to analyze the dynamic interactions and transmission between investor types and the stock market in a regime-dependent manner, thereby providing a more comprehensive understanding of the interconnectedness and impact of different market participants.

2.1 Markov-Switching VAR (MS-VAR) Model

To capture the dynamic relationships and account for structural changes, we employ the two-regime Markov-switching Vector Autoregressive (MS-VAR) model. The MS-VAR model is presented as follows:

$$Y_t = v(s_t) + A_1(s_t)Y_{t-1} + \ldots + A_p(s_t)Y_{t-p} + u_t(s_t) \tag{1}$$

where Y_t denotes trading values growth of four investors types (individual, foreign, institutional, and proprietary trading) and Thai stock market return $\{FINVE_t, LIST_t, PTRAD_t, LINDI_t, SET_t\}$. $v(s_t)$ and $A_1(s_t), \ldots, A_p(s_t)$ are, respectively, mean, and autoregressive coefficient matrices for state or regime $s_t = 1, 2$ (which represents the Low and High Trading regimes, respectively). $u_t(s_t)$ are the error that follows the multivariate normal distribution with zero mean and variance $\sum(S_t)$. p is the number of autoregressive terms of order. To provide a clearer view of the full model, let's consider a lag order of 1. We can rewrite Eq. (1) as follows:

$$
\begin{bmatrix} FINVE_t \\ LIST_t \\ PTRAD_t \\ LINDI_t \\ SET_t \end{bmatrix} = \begin{bmatrix} v_1(s_t) \\ v_2(s_t) \\ v_3(s_t) \\ v_4(s_t) \\ v_5(s_t) \end{bmatrix} + \begin{bmatrix} a_{11}(s_t) \ a_{12}(s_t) \ a_{13}(s_t) \ a_{14}(s_t) \ a_{15}(s_t) \\ a_{21}(s_t) & \ddots & \\ a_{31}(s_t) & & \ddots & \\ a_{41}(s_t) & & & \ddots \\ a_{51}(s_t) & \cdots & \cdots & \cdots & a_{55}(s_t) \end{bmatrix} \begin{bmatrix} FINVE_{t-1} \\ LIST_{t-1} \\ PTRAD_{t-1} \\ LINDI_{t-1} \\ SET_{t-1} \end{bmatrix} + \begin{bmatrix} u_1(s_t) \\ u_2(s_t) \\ u_3(s_t) \\ u_4(s_t) \\ u_5(s_t) \end{bmatrix} \tag{2}
$$

The underlying concept of the MS-VAR model is that all parameters are dependent on the regime and switch based on the unobserved state variables denoted as $s_t = 1, 2$. These state variables are determined by the first-order Markov stochastic process [20], which is characterized by transition probabilities.

$$p_{ij} = \Pr(s_{t+1} = j | s_t = i), \quad \sum_{j=1}^{2} p_{ij} \ \forall i, j \in \{1, 2\}) \tag{3}$$

More precisely, it is assumed that s_t follows the irreducible ergodic *two state Markov process* with the transition matrix

$$P = \begin{bmatrix} p_{11} & p_{12} \\ p_{21} & p_{22} \end{bmatrix} \tag{4}$$

To estimate all unknown parameters in the MS-VAR model, the maximum likelihood estimation of Krolzig [19] is used.

2.2 Variance Decomposition

In this study, we extend the concept from Diebold and Yilmaz [21] to develop a regime-switching spillover measure based on the forecast error variance decomposition from the MS-VAR model. This measure quantifies the influence of one variable on the forecast error variance of another variable in each regime. Additionally, we incorporate techniques from Pesaran and Shin [22] and Diebold and Yilmaz [18] to construct directional indices and ensure invariance to ordering without relying on orthogonalization by Cholesky decomposition. These advanced methods allow us to analyze the dynamic spillover effects and directional transmissions among investor types and the Thai stock market.

Formally, consider the covariance-stationary model proposed in Eq. (1). The vector moving average (VMA) representation of that model is:

$$Y_t|s_t = \omega_k + \sum_{j=0}^{\infty} A_{k,j} u_{s_t, t-j} \tag{5}$$

where $A_{k,j}$ are $(n \times n)$ matrices of regime k that are obtained from.

$$A_{k,j} = \sum_{j=0}^{\infty} \Phi_{k,j} A_{k,j-i}$$

Note that $A_{k,0} = I_n$, and $A_{k,j} = 0$ for $j < 0$, we then establish the vector ω_k by applying the infinite-order inverse autoregressive lag-operator to v_k, namely $\omega_k = \left(I_n - \sum_{i=1}^{p} \Phi_{k,i}\right)^{-1} v_k$. Typically, the infinite-order VMA representation in Eq. (5) is truncated at h-step ahead in order to forecast the error variance.

The definition of the generalized h-step-ahead forecast error variance decomposition for each regime k is as follows:

$$\theta_{k,ij}^g(h) = \frac{\sigma_{k,jj}^{-1} \sum_{l=0}^{h-1} \left(e_i' A_{k,l} \Sigma_k e_j\right)^2}{\sum_{l=0}^{h-1} \left(e_i' A_{k,l} \Sigma_k A_{k,l}' e_i\right)} \tag{6}$$

The variance decomposition is computed based on the generalized impulse response functions, where $\sigma_{k,jj}$ represents the standard deviation of the error term for the j th equation in regime k, and e_i is a selection column vector with the i th element equal to one and zeros elsewhere.

To ensure that each row in the variance decomposition table sums up to one, we normalize the variance shares. This normalization is achieved by dividing each element in the variance decomposition table by the sum of all elements in its respective row. This normalization procedure guarantees that the sum of each row, denoted as $\sum_{j=1}^{n} \theta_{k,ij}^{g}(h)$, does not equal one.

$$\tilde{\theta}_{k,ij}^{g}(h) = \frac{\theta_{k,ij}^{g}(h)}{\sum_{j=1}^{n} \theta_{k,ij}^{g}(h)} \tag{7}$$

To quantify the contribution of volatility shock spillovers among variables to the overall forecast error variance, we introduce the concept of the total spillover index in regime k, which serves as a measure to assess the extent of volatility shock transmission among variables and its impact on the total forecast error variance.

$$S_{k}^{g}(h) = \frac{1}{n} \sum_{\substack{i,j=1 \\ i \neq j}}^{n} \tilde{\theta}_{k,ij}^{g}(h) \tag{8}$$

On the other hand, we can analyze directional spillovers to understand how each investor type or market contributes to the spillovers affecting other investor types or market and the directions in which these spillovers occur. We differentiate between the spillovers received by investor/market i from all other investor types or market and the spillovers transmitted from investor/market i to all other investor types or market.

$$S_{k}^{g}(h) = \frac{1}{n} \sum_{\substack{j=1 \\ i \neq j}}^{n} \tilde{\theta}_{k,ij}^{g}(h) \tag{9}$$
$$\underset{i \rightarrow all}{}$$

$$S_{k}^{g}(h) = \frac{1}{n} \sum_{\substack{i=1 \\ i \neq j}}^{n} \tilde{\theta}_{k,ij}^{g}(h) \tag{10}$$
$$\underset{all \rightarrow i}{}$$

In Eq. (9), the contributions from other investor types or market can be calculated as the sum of the off-diagonal elements in the row of the normalized generalized variance decomposition shares from Eq. (7). It is important to note that these contributions cannot exceed 100% since each row of the matrix sums up to one.

Similarly, in Eq. (10), the contributions to other markets can be calculated as the sum of the off-diagonal elements in the column of the normalized generalized variance decomposition shares from Eq. (7). It is possible for these contributions to exceed 100% if the market in question has a significant impact on transmitting shocks to other markets.

Furthermore, we can define the net volatility spillovers from investor/market i to all others in Eq. (11) as the difference between the gross volatility shocks transmitted from investor/market i to others and the gross volatility shocks received by investor/market i from others.

$$S_{k,i}^{g}(h) = \underset{i \rightarrow all}{S_{k}^{g}(h)} - \underset{all \rightarrow i}{S_{k}^{g}(h)} \tag{11}$$

3 Data

This study uses daily data on the Thai Stock Exchange's trading values, classified by investor type: local institutions (LINST), proprietary trading (PTRAD), foreign investors (FINVE), and local individuals (LINDI). All data are collected for the period from 15/5/2017 to 27/4/2022 and transformed into growth rate for the analysis.

It can be observed from the descriptive statistics in Table 1 that the daily average trading values range from −1.5703 for local institutions to 0.9992 for SET100. Notably, foreign investors have the highest average daily trading value of 0.7156, followed by local individuals with 0.3908 and proprietary trading with −1.2310. The standard deviation (STD) of local individuals, which is 71.4871, is the highest among all investor types, indicating that this group is the most volatile.

To determine the stationarity of the trading values, an Augmented Dickey-Fuller test with intercept is conducted. The results indicate that the trading values of all investor types and market return are stationary at the 0.01 significance level and do not exhibit unit root.

Table 1. Descriptive statistics

	FINVE	LINST	PTRAD	LINDI	SET
Mean	0.7156	−1.5703	−1.2310	0.3908	0.9992
Minimum	−315.0804	−693.0075	−412.6218	−1028.6760	0.0000
Maximum	2039.6920	957.8505	298.4000	2149.5760	1.0120
Std. Dev.	61.8000	43.2073	19.4528	71.4871	0.0288
Skewness	30.1985	3.1850	−5.6427	20.2056	−34.5791
Kurtosis	1002.0020	291.2715	232.7172	727.8222	1199.1139
ADF Test	−10.394***	−10.207***	−10.359***	−10.519***	−10.881***
No. Obs.	1186	1186	1186	1186	1186

Note: ***, **, * represent the significance levels at 0.01, 0.05 and 0.1, respectively

4 Empirical Results

4.1 States of the Market

To illustrate the changes in the trading value regime for all four investor types and the SET return, we present Fig. 1, which displays the smoothed probabilities of regime 2 (representing the high trading value regime) obtained from the MS-VAR model. The data covers the period from May 2017 to April 2022. The probability values range from 0 to 1, where a value close to 1 indicates that the market state is classified as regime 2, characterized by high trading value. Conversely, a value close to 0 suggests that the market state is classified as regime 1, indicating low trading value. By examining the smoothed probabilities in Fig. 1, we observe significant fluctuations and that periods of increased trading value are unlikely to be persistent.

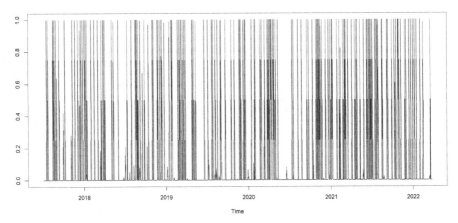

Fig. 1. Smoothed probabilities of high trading value (regime 2)

The analysis of the Thai stock market using the MS-VAR model reveals a fluctuation between zero and one in the probability of a high trading value regime. This indicates that the occurrence of such a regime is not constant during the period of 2017–2022, and there is a lack of persistence in the probability of remaining in this regime. These observations can be attributable to two key factors: external influences and investor behavior. Firstly, the Thai stock market is interconnected with global financial markets, and changes in international markets, such as fluctuations in major stock indices or shifts in global investor sentiment, can have a spill-over effect on the probability of a high trading value regime in the Thai market. These external influences may be transient in nature, thereby contributing to the observed lack of persistence in the probability estimates. Secondly, investor behavior, encompassing trading strategies and risk appetite, plays a significant role in influencing the likelihood of high trading values. Shifts in investor sentiment, market participation, and changes in trading patterns can introduce fluctuations in the probability of a high trading value regime. Behavioral factors, such as herding behavior or sudden shifts in market sentiment, could further contribute to the observed volatility in the probability estimates.

4.2 Estimation Results

In our analysis, we utilize the MS-VAR(1) model to find the relationship among investor types and the stock market in each state of trading value. The results are reported in Table 2.

Before discussing the estimation of values from the MS-VAR model, we examine the suitability of the model by comparing the BIC between VAR(1) and MS-VAR(-1). The comparison results shown in Table 2 indicate that the MS-VAR model has a lower BIC, which means that this model is better than VAR.

According to the findings presented in Table 2, there are notable differences in the interaction between investor types within the two regimes. In regime 2, the interaction between investor types is more pronounced, with both positive and negative significant effects observed among them. However, in regime 1, only a positive impact of foreign

Table 2. MS-VAR(1) estimation results

State		SET	Foreign	Institutions	Proprietary Trading	Individuals
Regime 1	Constant	−1.0906***	1.8997***	−3.4877***	7.5917**	−1.8503***
		(0.0008)	(0.1933)	0.9176)	(2.3445)	(0.4120)
	SET(-1)	−0.0905***	−1.3668***	2.9045***	−9.1022***	−1.9117***
		(0.0008)	(0.0210)	(1.9192)	(2.3465)	(0.4141)
	Foreign(-1)	0.0000**	−0.0034**	0.0028**	−0.0006	−0.0011
		(0.0000)	(0.0013)	(0.0013)	(0.0016)	(0.0016)
	Institutions(-1)	0.0000	−0.0005	0.0000	0.0009	−0.0041
		(0.0000)	(0.0019)	(0.0018)	(0.0022)	(0.0023)
	Proprietary Trading(-1)	0.0000	0.0024	0.0051	0.0104**	−0.0046
		(0.0000)	(0.0042)	(0.0041)	(0.0050)	(0.0051)
	Individuals(-1)	0.0000	−0.0006	0.0005	0.0000	0.0000
		(0.0000)	(0.0012)	(0.0011)	(0.0013)	(0.0014)
	Expected duration	6.37 days				
Regime 2	Constant	0.8580***	1.9400***	−2.8870***	7.8287***	−5.6690***
		(0.0008)	(0.3993)	(0.9176)	(2.3445)	(1.4120)
	SET(-1)	−0.0936***	−5.6230***	8.3044***	6.5448**	1.9300***
		(0.0008)	(2.0010)	(1.9192)	(2.3465)	(0.4141)
	Foreign(-1)	0.0000***	−0.1533**	0.0401	0.2927***	−0.1190
		(0.0000)	(0.0013)	(0.0013)	(0.0016)	(0.0016)
	Institutions(-1)	0.0000	−0.4723***	−0.0039***	0.0230	0.0535
		(0.0000)	(0.0019)	(0.0018)	(0.0022)	(0.0023)
	Proprietary Trading (-1)	−0.0001***	0.7656**	3.0040***	0.0619	1.0657***
		(0.0000)	(0.0042)	(0.0041)	(0.0050)	(0.0051)
	Individuals(-1)	0.0000***	−0.3442***	−0.0980**	0.1647***	−0.0318
		(0.0000)	(0.0012)	(0.0011)	(0.0013)	(0.0014)
	Expected duration	1.08 day				
Transition probability matrix	P_{11}	0.8431				
	P_{22}	0.0721				
BIC of MS-VAR(1)		17,883.77				
BIC of VAR(1)		37,586.90				

Notes: (1) Standard errors are enclosed in parentheses. (2) Significance levels are represented by ***, **, and *, which correspond to 0.01, 0.05, and 0.1, respectively.

trading on institutional trading is evident. Furthermore, the analysis highlights the significant role of the Thai stock market (SET) in influencing trading behavior across all investor types in both regimes. The Thai stock market serves as a crucial factor influencing the trading activities of investors in all categories. In addition, the regime switching analysis in the Thai stock market reveals interesting dynamics. The high trading value regime (regime 2) demonstrates a lower probability of persistence, while the low trading value regime (regime 1) has a higher likelihood of remaining. This suggests that the high trading value regime is less stable and more prone to transitions, while the low trading value regime exhibits greater stability. Moreover, the high trading value regime has a shorter duration, indicating a relatively stable period characterized by sustained high trading values. On the other hand, the low trading value regime has longer stayed, reflecting longer durations spent in this regime.

We conducted an estimation of regime-switching variance decomposition and reported the spillover measure results in Table 3. Overall, the spillover contribution among investor groups and the SET return is relatively low. This suggests that the observed volatility of each variable primarily originates from its own dynamics, accounting for almost 100% of the variability. A particularly interesting aspect in the spillover literature is comparing the spillovers during high and low trading value periods. Our findings indicate pronounced spillovers during the high trading value regime (Regime 2) compared to the low trading value regime.

In regime 1, our analysis reveals significant spillover effects from the SET index to the trading value of all four investor groups. The SET index contributes to 0.25% of the volatility of all investor groups. Specifically, foreign investors, local institutions, and proprietary trading experience spillover effects of 0.39%, 0.16%, and 0.14%, respectively, originating from the SET index. Meanwhile, the trading value of local individuals exhibits the highest spillover effect from the SET index, explaining 0.51% of their volatility. Based on these findings, it can be inferred that the SET index acts as the dominant transmitter of spillovers, influencing the trading activities of foreign investors, local institutions, proprietary trading, and local individuals. Furthermore, when examining the net directional spillover values from individual investors, which measure the net beta spillovers from one investor group to all other groups, we find that institutions and proprietary trading emerge as the dominant transmitters of spillovers. They have a positive net spillover effect on other investor groups. On the other hand, foreign investors and local individuals are net receivers of spillovers, indicating that they are more influenced by the trading activities of institutions and proprietary trading.

In contrast to regime 1, in the high trading value regime, foreign investors emerge as the top dominant transmitter of spillovers, accounting for 1.95% of the net spillover. The SET index also plays a significant role as a transmitter of spillovers in this regime. However, institutions, proprietary trading, and individuals act as net receivers of spillovers, indicating that they are more influenced by the trading activities of foreign investors and the SET index. This shift in the dominant transmitter of spillovers between regimes highlights the dynamic nature of market dynamics and the varying influence of different investor groups during different market conditions. Foreign investors take on a more prominent role in transmitting spillovers during periods of high trading value, while

Table 3. Directional detailed spillovers under regime-switching (percentage)

		SET	Foreign	Institutions	Proprietary Trading	Individuals	*FROM* others
Regime 1	SET	99.75	0.01	0.01	0.22	0.01	0.25
	Foreign	0.39	99.50	0.10	0.01	0.00	0.50
	Institutions	0.16	0.01	99.80	0.02	0.01	0.20
	Proprietary Trading	0.14	0.00	0.02	99.84	0.00	0.16
	Individuals	0.51	0.00	0.31	0.23	98.95	1.05
	TO others	1.20	0.02	0.45	0.48	0.02	
	NET spillovers	0.95	−0.49	0.25	0.32	−1.03	
	Total	100.95	99.52	100.25	100.32	98.97	0.43
Regime 2	SET	96.69	2.33	0.83	0.03	0.12	3.31
	Foreign	0.99	98.92	0.01	0.08	0.00	1.08
	Institutions	0.75	0.18	98.8	0.00	0.27	1.20
	Proprietary Trading	0.08	0.2	0.00	98.46	1.26	1.54
	Individuals	3.09	0.32	0.05	0.14	96.41	3.60
	TO others	4.91	3.03	0.89	0.25	0.02	
	NET spillovers	1.60	1.95	−0.31	−1.29	−3.58	
	Total	101.60	101.95	99.69	98.71	98.06	2.15

Notes: This table presents the percentage (%) of volatility directional spillovers under the regime-switching framework, as discussed in Sect. 2.2. It is based on the generalized variance decompositions using 10-day ahead volatility forecast errors. TO others: contributions to others; FROM others: contributions from others; Net spillovers: (TO-FROM)-a positive value indicates the first market is the spillover leader; whereas a negative value indicates the second market is the spillover leader.

institutions, proprietary trading, and individuals become recipients of these spillover effects.

Figure 2 displays the total volatility spillovers obtained from the regime-switching model. Notably, it reveals a significant disparity between regime 1 and regime 2 in terms of trading value fluctuations in the Thai stock market. The plot clearly demonstrates higher spillover effects during the high trading regime compared to the low trading state. This gap between the two regimes highlights the distinct levels of interconnectedness and transmission of volatility between different market segments or investor groups. The findings emphasize the importance of understanding and accounting for regime-specific dynamics when analyzing volatility spillovers in the Thai stock market.

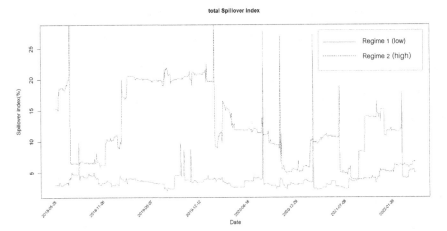

Fig. 2. Total volatility spillovers

5 Conclusions

This study aimed to examine the correlation between trading activities of various investor groups and the stock price index in the Stock Exchange of Thailand. To achieve this, we utilized a Markov-switching VAR (MS-VAR) model and conducted a variance decomposition analysis. Our objective was to gain insights into the dynamics and spillover effects among different investor types, as well as their impact on the overall Thai stock market. By employing these advanced techniques, we aimed to provide a comprehensive understanding of the interplay between investor behavior and stock market performance in Thailand.

The empirical findings of this study have provided valuable insights into several important aspects. Firstly, the analysis has revealed the presence of two distinct trading value regimes in the Thai stock market: the low trading value regime and the high trading value regime. These regimes signify periods characterized by varying levels of trading activities among investor categories. By employing the MS-VAR model, we were able to effectively capture the structural shifts and dynamics across these regimes, leading to a more comprehensive understanding of market behavior. Secondly, our analysis has demonstrated that the integration among investor types and the market is more pronounced during the high trading value regime. This suggests that during periods of heightened trading activity, there is a greater interplay and interconnectedness among different investor groups. Thirdly, the regime-switching variance decomposition analysis has provided valuable insights into the spillover effects among investor groups and the SET Index. The results indicate that trading activities of investors have a relatively smaller impact on the stock market volatility. On the other hand, the SET Index plays a dominant role in influencing the behavior of investors, particularly during the low trading value regime.

The findings of this study carry significant implications for investors and market participants in the Stock Exchange of Thailand. By gaining insights into the dynamics and interactions among different investor types, market participants can make well-informed investment decisions and assess market risks more effectively. Understanding the impact of investor behavior on stock market volatility enables the development of strategies to mitigate risks and capitalize on potential opportunities.

However, it is important to acknowledge certain limitations of this study. The analysis focused exclusively on the Thai stock market and the trading activities of four specific investor groups. Therefore, the generalizability of the findings to other markets or investor types may be limited. Additionally, the analysis was based on daily data, and conducting future research using higher-frequency data could provide deeper insights into short-term dynamics. Further studies could expand the analysis to include a broader range of markets and investor types, while considering additional factors that may influence investor behavior and market dynamics. Incorporating sentiment analysis or other qualitative measures could also enhance our understanding of investor motivations and sentiments.

Acknowledgments. The paper is supported by School of Business and Communication Arts, University of Phayao, and this research work was partially supported by Chiang Mai University.

References

1. Menkveld, A.J.: High frequency trading and the new-market makers. J. Fin. Mark. **16**(4), 712–740 (2013)
2. Bekaert, G., Harvey, C.R., Lundblad, C.: Liquidity and expected returns: lessons from emerging markets. Rev. Fin. Stud. **20**(6), 1783–1831 (2007)
3. Kacperczyk, M., Sialm, C., Zheng, L.: Unobserved actions of mutual funds. The Rev. Fin. Stud. **21**(6), 2379–2416 (2008)
4. Barber, B.M., Odean, T.: All that glitters: the effect of attention and news on the buying behavior of individual and institutional investors. Rev. Fin. Stud. **21**(2), 785–818 (2008)
5. Deaves, R., Lüders, E., Schröder, M.: The dynamics of overconfidence: evidence from stock market forecasters. J. Econ. Behav. Organ. **75**(3), 402–412 (2010)
6. Li, X.: A research on relationship between the stock holdings of institutional investors and the stock price synchronicity of SME board market. Technol. Invest. **8**(1), 1–10 (2017)
7. Zhang, Y., Wang, W., Zhang, L.: Research on the Influence of institutional investors' shareholding and transaction on stock price synchronicity. In: 5th Annual International Conference on Social Science and Contemporary Humanity Development (SSCHD 2019). Atlantis Press (2019)
8. Brown, G.W., Cliff, M.T.: Investor sentiment and asset valuation. The Journal of Business **78**(2), 405–440 (2005)
9. Sun, X.Q., Shen, H.W., Cheng, X.Q.: Trading network predicts stock price. Sci. Rep. **4**(1), 3711 (2014)
10. Liu, W., Xin, X., Wang, W.: An empirical study on the relationship between different motivations of stock investors and the investment decision behavior. In: 2013 International Conference on Applied Social Science Research (ICASSR-2013), pp. 194–197. Atlantis Press (2013, August)
11. Kamesaka, A., Nofsinger, J.R., Kawakita, H.: Investment patterns and performance of investor groups in Japan. Pac. Basin Financ. J. **11**(1), 1–22 (2003)

12. Ryu, D., Kim, H., Yang, H.: Investor sentiment, trading behavior and stock returns. Appl. Econ. Lett. **24**(12), 826–830 (2017)
13. Kacperczyk, M., Sundaresan, S., Wang, T.: Do foreign institutional investors improve price efficiency? The Review of Financial Studies **34**(3), 1317–1367 (2021)
14. Ferreira, M.A., Matos, P., Pereira, J.P., Pires, P.: Do locals know better? a comparison of the performance of local and foreign institutional investors. J. Bank. Finance **82**, 151–164 (2017)
15. Grinblatt, M., Keloharju, M.: The investment behavior and performance of various investor types: a study of Finland's unique data set. J. Financ. Econ. **55**(1), 43–67 (2000)
16. Pastpipatkul, P., Yamaka, W., Wiboonpongse, A., Sriboonchitta, S.: Spillovers of quantitative easing on financial markets of Thailand, Indonesia, and the Philippines. In: Integrated Uncertainty in Knowledge Modelling and Decision Making: 4th International Symposium, IUKM 2015, Nha Trang, Vietnam, October 15–17, 2015, Proceedings 4, pp. 374–388. Springer International Publishing (2015)
17. Pastpipatkul, P., Yamaka, W., Sriboonchitta, S.: Effect of quantitative easing on ASEAN-5 financial markets. Causal Inference in Econometrics, 525–543 (2016)
18. Diebold, F.X., Yilmaz, K.: Better to give than to receive: Predictive directional measurement of volatility spillovers. Int. J. Forecast. **28**(1), 57–66 (2012)
19. Krolzig, H.M.: The markov-switching vector autoregressive model. In: Markov-Switching Vector Autoregressions, pp. 6–28. Springer, Berlin, Heidelberg (1997). https://doi.org/10.1007/978-3-642-51684-9_2
20. Hamilton, J.D.: A new approach to the economic analysis of nonstationary time series and the business cycle. Econometrica: J. Econ. Soc. 357–384 (1989)
21. Diebold, F.X., Yilmaz, K.: Measuring financial asset return and volatility spillovers, with application to global equity markets. Econ. J. **119**(534), 158–171 (2009)
22. Pesaran, H.H., Shin, Y.: Generalized impulse response analysis in linear multivariate models. Econ. Lett. **58**(1), 17–29 (1998)

Does Cryptocurrency Improve Forecasting Performance of Exchange Rate Returns?

Chatchai Khiewngamdee[1] and Somsak Chanaim[2,3]([✉])

[1] Faculty of Economics, Chiang Mai University, Chiang Mai 50200, Thailand
[2] International College of Digital Innovation, Chiang Mai University, Chiang Mai 50200, Thailand
somsak.chanaim@cmu.ac.th
[3] Center of Excellence in Econometrics, Chiang Mai University, Chiang Mai 50200, Thailand

Abstract. The aim of this paper is to investigate the predictive capacity of cryptocurrency on exchange rate returns. The train-test split technique is applied in estimating the ARIMA and ARIMAX models. We also use an ensemble technique to improve our prediction accuracy. The results indicate there exists only a linkage between cryptocurrencies and US dollar. There found no evidence that cryptocurrencies affect the returns of Chinese Yuan and Japanese Yen. According to the RMSE criterion, the ARIMAX with cryptocurrency forecasting model produces superior results over the ARIMA model. This indicates that cryptocurrency can improve forecasting accuracy for exchange rate returns. Furthermore, we found that the ensemble model outperforms the conventional ARIMA and ARIMAX forecasting models.

Keywords: Bitcoin ARIMA · ARIMAX · Machine Learning · Ensemble Technique

1 Introduction

Exchange rates play a vital role in a country's trade performance. Change in relative valuations of currencies and their volatilities often have important repercussions on international trade, the balance of payments and overall economic performance of the country. The exchange-rate dynamics also affect the real return of an investor's portfolio, exporters, and importers. Even a small change in value of exchange rate can entitle some involved individuals and entities to loss a huge amount of money. For these reasons, forecasting exchange rate is a crucial endeavor for international economists and policymakers to make informed decisions as well as for investors and multinational corporations worldwide to mitigate risks and explore investment opportunities. Since the early development of financial market up to now for several decades, the number of studies has grown investigating the interrelationship between foreign exchange rates and financial variables. For instance, Wong [1] found a negative relationship between

V.-N. Huynh et al. (Eds.): IUKM 2023, LNAI 14375, pp. 285–294, 2023.
https://doi.org/10.1007/978-3-031-46775-2_25

286 C. Khiewngamdee and S. Chanaim

real exchange rate return and real stock price return in Malaysia, Singapore, Korea and the UK. Xie et al. [2] and Ding [3] also found that stock prices are useful to predict exchange rates. Furthermore, some previous studies suggested the causality between oil prices and exchange rates (Dai et al. [4]; Chkir et al. [5]; Wu et al. [6]) and the ability of commodity price to predict exchange rate (Ferraro et al. [7]; Zhang et al. [8]; Liu et al. [9]). Apergis [10] found that gold prices do matter in forecasting the Australian dollar.

Cryptocurrencies have gained huge attention of not only investors but also academics and researchers around the globe in the past few years. A wide range of studies concentrate on forecasting returns of cryptocurrencies. Nonetheless, few literatures investigated the predictive capacity of cryptocurrency in relation to other financial assets. For example, Rehman and Apergis [11] investigated the predictive power of cryptocurrencies on commodity futures and found there is a causality running from cryptocurrencies to commodity futures both in terms of mean and volatility. Isah and Raheem [12] suggested that Bitcoin can predict the return of S&P500. Some literatures also found the predictive power of cryptocurrency uncertainty on precious metal returns (Shang et al. [13]; Wei et al. [14]). According to the asset pricing theory, the present-value model suggests that Bitcoin prices should predict the relevant fundamentals, including exchange rates. This theory was first examined by Feng and Zhang [15] who studied the predictive power of Bitcoin on currency exchange rates. Using the autoregressive distributed lag approach, they found that Bitcoin has a capacity to predict exchange rates. Moreover, some previous papers found a linkage between Bitcoin and US dollar (Baur et al. [16]; Mokni and Ajmi [17]; Wang et al. [18]). However, there remains a paucity of studies that investigate the predictive capacity of Bitcoin or other cryptocurrencies on foreign exchange rates. Therefore, our paper seeks to address this gap by investigating the forecasting performance of cryptocurrencies on exchange rate. Additionally, we propose a machine learning approach, in particular the ensemble model, which has proven more suitable as a forecasting model than the traditional methods.

An ensemble model is one that takes a machine learning approach to combine multiple other models and data types in its prediction process. This model overcomes several technical problems of a single model, namely, high variance, low accuracy, and bias. Numerous works suggest that ensemble model dominates individual model and enhance forecasting capability (He et al. [19]; Hajek et al. [20]; Wang et al. [21]; Sun et al. [22]). In this paper, we propose an ensemble forecasting model based on the ARIMA and the ARIMAX models.

The objectives of this paper are to investigate the predictive ability of cryptocurrencies on exchange rate and compare the forecasting accuracy between our proposed ensemble model and single model. Firstly, we use, as a benchmark, the ARIMA model, and the ARIMAX model which includes Bitcoin or Ethereum as an exogenous variable, to estimate exchange rates. We also conduct a model averaging ensemble model. Secondly, we employ the train-test split technique to evaluate the exchange rate forecasting performance of each model. Finally, we compare the forecasting performance using root mean squared error (RMSE).

The rest of this paper is organized as follows. Section 2 describes research methodology. Section 3 provides data and descriptive statistics. Section 4 presents the empirical results and discussion. Section 5 concludes the paper.

2 Methodology

2.1 ARIMAX Model

ARIMAX is the Autoregressive Integrated Moving Average with Exogenous Variables. It is a time series forecasting model extended from the ARIMA model by incorporating exogenous or external variables that can potentially influence the time series being forecasted. We define it as

$$r_t = \alpha_0 + \sum_{i=1}^{p} \alpha_i r_{t-i} + \beta r_t^x + \theta r_t^y + \varepsilon_t \tag{1}$$

where

r_t is the time series of asset return at time $t, t = 1, 2, \cdots, n$
r_t^x and r_t^y are exogenous variables of asset return time series at time t
ε_t is the random variable of error term with $N\left(0, \sigma^2\right)$.
$\alpha, \alpha_1, \ldots, \alpha_p$ are the parameters in the ARIMA part and β and θ are the parameters for exogenous variables.

2.2 Ensemble Model

Ensemble model is applied to time series forecasting to improve the accuracy and robustness of predictions and we use weighted averaging method where predictions from individual models are combined by assigning different weights to each model's prediction. The weights can be determined based on the performance of the models on validation data or through optimization techniques. Weighted averaging is straightforward to implement and can be effective in reducing bias and improving accuracy. The model is defined as

$$r_t = \sum_{i=}^{m} w_i E\left(f_i\right) + \varepsilon_t \tag{2}$$

where $0 \leq w_i \leq 1, i = 1, 2, \ldots, m$ and $\sum_{i=1}^{m} w_i = 1, w_i$ is the weight for the prediction value from model i. $E\left(f_i\right)$ is the predicting value from model $i, i = 1, 2, \ldots m$.

In this research, we choose 3 models from ARIMAX as

$$r_t = \alpha + \alpha_1 r_{t-1} + \varepsilon_{t,1} \text{ as model 1 or } f_1$$
$$r_t = \alpha + \alpha_1 r_{t-1} + \beta r_t^x + \varepsilon_{t,2} \text{ as model 2 or } f_2$$
$$r_t = \alpha + \alpha_1 r_{t-1} + \theta r_t^y + \varepsilon_{t,3} \text{ as model 3 or } f_3$$

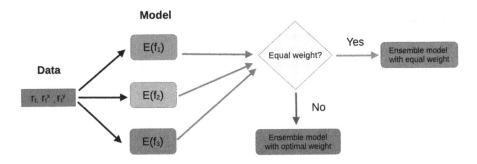

Fig. 1. The flowchart of the ensemble model

and the process for ensemble modelling is defined as shown in Fig. 1.

The Fig. 1 shows the process of the ensemble method. First we estimate parameter in the model f_1, f_2 and f_3 from the time series data r_t, r_t^x, and r_t^y respectively. Second we can calculate the prediction value from each model by the expected value denote by $E(f_i)$, $i = 1, 2, 3$. Third we use equal weight and optimal weight for the ensemble model. So that, the ensemble model with equal weight is defined by

$$r_t = \frac{1}{3}E\left(f_1\right) + \frac{1}{3}E\left(f_2\right) + \frac{1}{3}E\left(f_3\right) + \varepsilon_t$$

and the ensemble model with optimal weight can be defined by minimizing RMSE with the constraint in the form

$$\min_{w_1, w_2, w_3} \sqrt{\frac{\sum_t^n \left(w_1 \cdot \varepsilon_{t,1} + w_2 \cdot \varepsilon_{t,2} + w_3 \cdot \varepsilon_{t,3}\right)^2}{n}}$$

Subject to $w_1 + w_2 + w_3 = 1$, $w_i \geq 0$, $i = 1, 2, 3$.

2.3   Forecasting Comparison Criteria

In this study, our forecasting models use the Root Mean Squared Error (RMSE) as a loss function. It calculates the difference between actual and anticipated volatility. The RMSE formula is

$$RMSE = \sqrt{\frac{\sum_{t=1}^n \left(\varepsilon_t - \overline{\varepsilon_t}\right)^2}{n}}$$

where ε_t is the error value from the difference between actual value and predicted value and n is the number of forecasting data.

3   Data

This paper focuses on investigating the forecasting accuracy of four major foreign exchange rates, namely Chinese Yuan (CNY/USD), US Dollor (US-INDEX),

Euro (EUR/USD), and Japanese Yen (JPY/USD), under the influence of two main cryptocurrencies, in particular, Bitcoin (BTC) and Ethereum (ETH). To achieve this, ARIMAX models are employed, and external variables in the form of cryptocurrencies (Bitcoin and Ethereum) are incorporated. The dataset comprises weekly data spanning from January 1st, 2019 to April 25th, 2023, totally 226 observations. The foreign exchange rates and cryptocurrency data were obtained from the Thomson Reuters database, and log-returns were calculated for analysis.

Table 1 provides a descriptive overview of the data. Notably, all return series exhibit kurtosis values below 3, suggesting a departure from normal distribution. Additionally, the Jarque-Bera test strongly rejects the null hypothesis of normality for all return series, as evidenced by the Probability JB test. Furthermore, the augmented Dickey-Fuller (ADF) unit root test confirms the stationary condition of the return series for all variables. Price and return movement of the data are presented in Fig. 2.

Table 1. Descriptive statistics

	CNY	USD	EUR	JPY	BTC	ETH
Mean	0.0047	0.033	−0.0052	0.1077	1.3749	1.8922
Median	0.0015	−0.0627	0.0182	0.1135	0.2961	1.6672
Maximum	2.1264	5.0896	4.1582	3.5374	37.747	43.8813
Minimum	−2.0445	−3.8348	−4.1323	−4.9632	−36.715	−45.2409
Skewness	0.2366	0.5801	0.0109	−0.5102	0.2441	0.1538
Kurtosis	1.6846	4.0504	2.2843	2.1186	1.783	1.4866
Jarque-Bera	30.2015*	172.165*	51.1877*	54.0438*	33.6592*	22.8468*
ADF test	−4.8728*	−5.1927*	−5.1987*	−5.3846*	−4.5576*	−5.7213*
Observations	226	226	226	226	226	226

Note: * indicates the significance level at 1%.

4   Empirical Results and Discussions

4.1   Estimation Results from the ARIMA and ARIMAX Models

This section provides parameter estimation from the ARIMA and the ARIMAX models. The Autoregressive Integrated Moving Average (ARIMA) model is denoted as the AR model. In this paper, we propose two ARIMAX models, one with Bitcoin and the other with Ethereum as an exogenous variable, which are denoted as the BTC model and the ETH model, respectively. The parameters are estimated from training dataset. The results are presented in Table 2. It can be seen that returns of Bitcoin and Ethereum both have a negative relationship with return of US dollar. An increase in returns of cryptocurrencies, in particular,

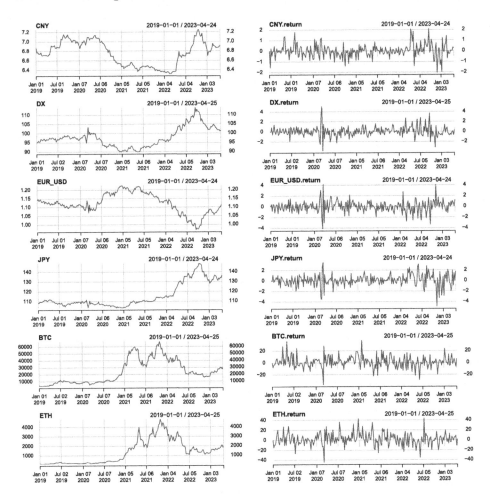

Fig. 2. Price and return movement

Bitcoin and Ethereum, tend to weaken the US dollar and vice versa. One possible explanation is because cryptocurrencies are considered as risky asset; when returns of cryptocurrencies increase, investors tend to increase long position on cryptocurrencies and its derivatives. As a results, the demand for US dollar will decrease as some investors may prefer to hold their wealth in cryptocurrencies and hence US dollar depreciates. On the other hand, when cryptocurrencies' return fall, investors tend to shift their investments towards safer assets like US dollar. This result is consistent with Baur et al. [16], Mokni and Ajmi [17] and Wang et al. [18]. Also, we found that the return of Ethereum positively affects Euro exchange rate. It means an increase in the return of Ethereum leads to an appreciation of Euro currency. This might be in consequence of the depreciation of US dollar that we mentioned earlier.

Table 2. Parameter estimation from training dataset.

CNY	AR Coef	AR SE	BTC Coef	BTC SE	ETH Coef	ETH SE
Intercept	−0.01045	0.061709	−0.014063	0.063922	−0.009067	0.060943
MA(1)	−0.792644**	0.1833	−0.802029**	0.17033	−0.789031**	0.189424
AR(1)	0.861119**	0.155694	0.871958**	0.143053	0.856298**	0.162569
BTC.return			0.002896	0.004038		
ETH.return					−0.001005	0.003314
USD	**Coef**	**SD**	**Coef**	**SD**	**Coef**	**SD**
Intercept	0.054618	0.053274	0.075654	0.052147	0.085505	0.049306
MA(1)	−0.548946	0.200778	−0.562192	0.198359	−0.583346	0.172556
AR(1)	0.395603	0.226392	0.397554	0.225329	0.396245	0.200568
BTC.return			−0.014053*	0.006474		
ETH.return					−0.015383**	0.00519
EURO	**Coef**	**SD**	**Coef**	**SD**	**Coef**	**SD**
Intercept	−0.040262	0.056438	−0.054893	0.057212	−0.062949	0.055062
MA(1)	−0.325093	0.29528	−0.276497	0.32062	−0.324257	0.299027
AR(1)	0.092552	0.310308	0.029396	0.332207	0.060393	0.312822
BTC.return			0.009779	0.006986		
ETH.return					0.011395*	0.005503
JPY	**Coef**	**SD**	**Coef**	**SD**	**Coef**	**SD**
Intercept	0.12595	0.068482	0.129797	0.068785	0.127248	0.069249
MA(1)	−0.462312	0.396017	−0.472888	0.382321	−0.465962	0.391117
AR(1)	0.410886	0.410912	0.418656	0.397333	0.413568	0.405579
BTC.return			−0.002679	0.007166		
ETH.return					−0.000673	0.005857

Note: *, ** indicate the significance levels at 5% and 1% respectively.

Although we find strong evidence that cryptocurrencies have a statistically negative effect on US dollar, the empirical results suggest that there is no relationship between cryptocurrencies and Chinese Yuan and Japanese Yen. The main reason is the relatively small shares of market value in cryptocurrencies for Chinese Yuan and Japanese Yen. Thus, change in return of cryptocurrencies does not affect Yuan and Yen foreign exchange markets and brings about no significant change in the exchange rate returns. According to the 2022 Geography of Cryptocurrency Report by Chainanlysis, cryptocurrency value in the USA accounted for 18% of the world market while China and Japan only accounted for 4.56% and 2.85%, respectively. In fact, Chinese Bitcoin market in the past had around 95% share of global trading until Chinese regulators enforced a policy which required local Bitcoin exchanges to implement trading fees in late 2016 and hence the traded value reduced to 8% in the following year (Young [23]).

4.2 Training and Testing the Forecasting Models

The forecasting performances of various models on future exchange rates are presented in Table 3. In this study, we use root mean squared error (RMSE) as a criterion to identify the most suitable forecasting model. To perform our forecast, we employ the autoregressive (AR) model as a benchmark model, and the ARIMAX denoted by the BTC and the ETH models that include Bitcoin and Ethereum as a predictor, respectively. Furthermore, we conduct model averaging using the ensemble technique which aggregates the prediction of each model and results in one final prediction. We provide both equal weigh and optimal weight that minimize RMSE. To conduct a forecasting procedure, we employ the train-test split technique to evaluate the performance of each machine learning algorithm. Firstly, we divide dataset into two subsets, namely, 80% for training dataset and 20% for test dataset. The training dataset is used to fit the model and the test dataset is used to provide an input for predictions and evaluate the fit machine learning model.

The empirical results show that, from training dataset, the ARIMAX forecasting models with one including Bitcoin and the other including Ethereum perform better than the benchmark autoregressive model according to the RMSE values. This implies that Bitcoin and Ethereum do improve the forecasting performance for US dollar, Chinese Yuan, Euro, and Japanese Yen. Furthermore, the optimal weight ensemble model provides the lowest value of RMSE to all models based on the training dataset for each exchange rate. This indicates that the ensemble model with optimal weight is superior to all models for exchange rate prediction.

For forecasting using the test dataset, it is found that the BTC model outperforms the others in forecasting US dollar and Euro while the suitable forecasting model for Chinese Yuan and Japanese Yen are the ETH model and the AR model, respectively. In other words, our optimal weight ensemble model is no longer the best forecasting model, but it does not appear to be the worst one. The reason maybe because change in the return of Bitcoin and that in Ethereum have a similar pattern. Since the base models are not diverse and independent, the prediction error of the model increases when the ensemble approach is used. Another possible reason is that financial time series typically have no fixed pattern overtime. The pattern of the data might change when the variable is affected by an external factor. Thus, during the assigned period for the test dataset, the

Table 3. RMSE values for forecasting model selection

		AR	BTC	ETH	Ensemble model with equal weight	Ensemble model with optimal weight
CNY	Training	0.562215	0.561403	0.562074	0.561629	**0.561294**
	Test	0.797596	0.801791	**0.796617**	0.798563	0.800267
USD	Training	0.957124	0.944962	0.935121	0.940792	**0.9351020**
	Test	1.227704	**1.219042**	1.227683	1.222259	1.227333
EUR	Training	1.018672	1.01319	1.007002	1.010356	**1.006984**
	Test	1.245761	**1.24107**	1.258004	1.24685	1.257292
JPY	Training	1.008171	1.007781	1.008134	1.007954	**1.007781**
	Test	**1.637899**	1.638795	1.638726	1.63845	1.638765

Note: The bold numbers indicate the lowest value of RMSE.

variable might be influenced by short-term intervening factors that affect the pattern and hence increase the forecasting error.

5 Conclusion

This study aims to examine whether Bitcoin and Ethereum can improve forecasting accuracy for major exchange rate returns, namely, US dollar, Chinese Yuan, Euro, and Japanese Yen. We employ the ARIMA and the ARIMAX model including the ensemble model to achieve our objective. Firstly, we divide the dataset into two subsets, training dataset and test dataset. Secondly, we apply the ARIMA model and the ARIMAX model for parameter estimation on the training dataset. Then, we conduct forecasting procedure on both training and test data sets using the ARIMA, ARIMAX and ensemble model. Finally, we compare the model forecasting performance using the RMSE criterion.

The empirical results suggest that an increase in the return of Bitcoin as well as Ethereum leads to a depreciation of US dollar and vice versa. We also found that an increase in the return of Ethereum tends to appreciate Euro. Nonetheless, in this study, we did not find any relationship between cryptocurrencies and Chinese Yuan and Japanese Yen. For forecasting model selections, the results show that adding Bitcoin or Ethereum can improve forecasting accuracy for the exchange rate returns compared with our benchmark ARIMA model. Moreover, the ensemble model outperforms a single ARIMA and ARIMAX model in predicting exchange rate returns. Last but not least, there are rooms to develop our ensemble model to better forecast foreign exchange rates.

Acknowledgements. This research work was partially supported by Chiang Mai University and Center of Excellence in Econometrics, Faculty of Economics, Chiang Mai University

References

1. Wong, H.T.: Real exchange rate returns and real stock price returns. Int. Rev. Econ. Finance **49**, 340–352 (2017)
2. Xie, Z., Chen, S.W., Wu, A.C.: The foreign exchange and stock market nexus: new international evidence. Int. Rev. Econ. Finance **67**, 240–266 (2020)
3. Ding, L.: Conditional correlation between exchange rates and stock prices. Q. Rev. Econ. Finance **80**, 452–463 (2021)
4. Dai, X., Wang, Q., Zha, D., Zhou, D.: Multi-scale dependence structure and risk contagion between oil, gold, and US exchange rate: a wavelet-based vine-copula approach. Energy Econ. **88**, 104774 (2020)
5. Chkir, I., Guesmi, K., Brayek, A.B., Naoui, K.: Modelling the nonlinear relationship between oil prices, stock markets, and exchange rates in oil-exporting and oil-importing countries. Res. Int. Bus. Financ. **54**, 101274 (2020)
6. Wu, T., An, F., Gao, X., Wang, Z.: Hidden causality between oil prices and exchange rates. Resour. Policy **82**, 103512 (2023)

7. Ferraro, D., Rogoff, K., Rossi, B.: Can oil prices forecast exchange rates? An empirical analysis of the relationship between commodity prices and exchange rates. J. Int. Money Financ. **54**, 116–141 (2015)
8. Zhang, H.J., Dufour, J.M., Galbraith, J.W.: Exchange rates and commodity prices: measuring causality at multiple horizons. J. Empir. Financ. **36**, 100–120 (2016)
9. Liu, L., Tan, S., Wang, Y.: Can commodity prices forecast exchange rates? Energy Econ. **87**, 104719 (2020)
10. Apergis, N.: Can gold prices forecast the Australian dollar movements? Int. Rev. Econ. Finance **29**, 75–82 (2014)
11. Rehman, M.U., Apergis, N.: Determining the predictive power between cryptocurrencies and real time commodity futures: evidence from quantile causality tests. Resour. Policy **61**, 603–616 (2019)
12. Isah, K.O., Raheem, I.D.: The hidden predictive power of cryptocurrencies and QE: evidence from US stock market. Phys. A **536**, 121032 (2019)
13. Shang, Y., Wei, Y., Chen, Y.: Cryptocurrency policy uncertainty and gold return forecasting: a dynamic Occam's window approach. Financ. Res. Lett. **50**, 103251 (2022)
14. Wei, Y., Wang, Y., Lucey, B.M., Vigne, S.A.: Cryptocurrency uncertainty and volatility forecasting of precious metal futures markets. J. Commod. Mark. **29**, 100305 (2023)
15. Feng, W., Zhang, Z.: Currency exchange rate predictability: the new power of Bitcoin prices. J. Int. Money Financ. **132**, 102811 (2023)
16. Baur, D.G., Dimpfl, T., Kuck, K.: Bitcoin, gold and the US dollar-A replication and extension. Financ. Res. Lett. **25**, 103–110 (2018)
17. Mokni, K., Ajmi, A.N.: Cryptocurrencies vs. US dollar: evidence from causality in quantiles analysis. Econ. Anal. Policy **69**, 238–252 (2021)
18. Wang, P., Liu, X., Wu, S.: Dynamic linkage between bitcoin and traditional financial assets: a comparative analysis of different time frequencies. Entropy **24**(11), 1565 (2022)
19. He, K., Yang, Q., Ji, L., Pan, J., Zou, Y.: Financial time series forecasting with the deep learning ensemble model. Mathematics **11**(4), 1054 (2023)
20. Hajek, P., Hikkerova, L., Sahut, J.M.: How well do investor sentiment and ensemble learning predict Bitcoin prices? Res. Int. Bus. Financ. **64**, 101836 (2023)
21. Wang, G., Tao, T., Ma, J., Li, H., Fu, H., Chu, Y.: An improved ensemble learning method for exchange rate forecasting based on complementary effect of shallow and deep features. Expert Syst. Appl. **184**, 115569 (2021)
22. Sun, S., Wang, S., Wei, Y.: A new ensemble deep learning approach for exchange rates forecasting and trading. Adv. Eng. Inform. **46**, 101160 (2020)
23. Young, J.: No Direct Correlation Between Chinese Yuan & Bitcoin Price, Here's Why. Cointelegraph, 17 April 2017. https://cointelegraph.com/news/no-direct-correlation-between-chinese-yuan-bitcoin-price-heres-why. Accessed 31 May 2023

The Perspective of the Creative Economy Stimulus on the Thai Economy: Explication by BSTS Mixed with the CGE Model

Kanchana Chokethaworn[1], Chukiat Chaiboonsri[1(✉)] [iD], Paponsun Eakkapun[2], and Banjaponn Thongkaw[2]

[1] Faculty of Economics, Chiang Mai University, Chiang Mai, Thailand
{kanchana.ch,chukiat.chai}@cmu.ac.th
[2] Modern Quantitative Economics Research Center (MQERC), Faculty of Economics, Chiang Mai University, Chiang Mai, Thailand

Abstract. The global creative economy has lately grown to play a significant role economic development, as recommended by UNCTAD in 2004. The research study aims to support the assertion that Thailand's creative economy provides a substantial contribution to the nation's long-term development through sustainable economic growth. This study used the BSTS model to predict the long-run effects of Thailand's SAM table applied in the CGE model, predicting the Thailand's economy in both a non-promote creative economy and a promote creative economy rather than using econometric data to estimate the effects of the development of Thailand's creative economy. The results of this study, based on both the BSTS model and the CGE model, have already suggested that the long-term development of Thailand's economy depends significantly on the creative economy. If this result is useful and it can be applied to Thailand's economic fortunes, then Thailand's economy or any other country will have its long-term sustainable development goal achieved as soon as possible based on creative economy stimuli.

Keywords: Creative Economy · CGE model · BSTS · SAM table · Stimulus · Thailand

1 Introduction

In recent years, the global creative economy has grown to play a significant role in economic expansion and development. It includes a broad range of businesses and sectors, including design, music, cinema, fashion, advertising, and others, that create and offer creative products and services. Additionally, it supports diversity and cultural identity. In the global economy, the creative economy has grown in significance. More and more nations are attempting to take advantage of the potential for both social and economic advancement. One of the major forces driving the economy in the twenty-first century has been characterized as the creative economy, which has a big impact on trade, investment, and job growth. Its potential to promote innovation and push sustainable development has also been acknowledged. The creative economy is one of the most dynamic and rapidly

© The Author(s), under exclusive license to Springer Nature Switzerland AG 2023
V.-N. Huynh et al. (Eds.): IUKM 2023, LNAI 14375, pp. 295–305, 2023.
https://doi.org/10.1007/978-3-031-46775-2_26

expanding areas of the global economy, according to the United Nations Conference on Trade and Development (UNCTAD). Particularly for decision-makers, business owners, and professionals who wish to promote economic growth and open up fresh prospects for wealth and growth. The creative economy is significant around the world, and many nations have put policies and initiatives in place to assist it. United Kingdom, South Korea, Australia, Singapore, and Canada are a few examples of nations that prioritize the creative economy. These nations have put in place financial schemes. Spend money on research, education, and tax breaks for creative enterprises. To help the sector flourish, they have also established incubators and creative hubs. Crafts and handicrafts, music, performing arts, visual arts, movies, broadcasting and broadcasting software, printing, advertising, design, architecture services, fashion, Thai food, traditional Thai medicine, and cultural travel are among the sectors that make up the creative economy. In recent years, Thailand's economy has seen significant growth in the creative economy sector. There are other sources of cultural and aesthetic value besides the creative economy. Additionally, it is a significant generator of cash and employment. The growth of the creative economy has been found to be an important contributor to GDP in Thailand, with a 9.1% share of GDP in 2015, indicating the potential of the creative economy to promote creative economy industries. The industry's ability to spur innovation is also acknowledged (GDP). Raise living standards and draw in international investment. For Thailand's creative economy to continue to expand and contribute to the nation's economy and culture, there must be constant investment in it (Fig. 1).

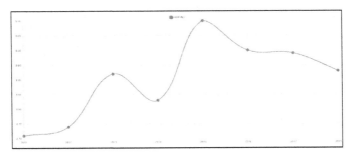

Fig. 1. The ratio of creative industries value to national GDP (%) is calculated from industry value data. According to business code (TSIC) 4 digits (excluding duplicates)

In analyzing creative economy data, the Computable General Equilibrium (CGE) model as a tool to address a variety of economic, social, and environmental issues. [1, 2] Because it incorporates systematic linkages between key economic activities to replicate the behavior of key economic activities including production, consumption, investment, exports, imports, taxation, government spending, etc.

Additionally, they should be able to provide numerical solutions that make it obvious which way the influence will be felt. Additionally, because the CGE model encompasses the entire economy, it can examine how policies affect macro variables including economic growth, employment, inflation, consumption, exports, trade balance, and more. At the same time, a thorough analysis of the effect on different economic system subunits (Micro Variables) is possible. Both in terms of winners and losers, such as the effects on different industries (for instance, level of production, price, employment, consumption, and exports), household groupings, the government sector, etc. Therefore, the CGE model is a crucial tool for developing and make policies for Stimulus Creative Economy on Thai Economy. The strength of the CGE model is that it does not require a lot of time series data to be estimated [3]. This point can be overcome by using the calibration for the calculation of the SAM table (Social Accounting Metrix) applied in the CGE model. This research study anticipated how the creative economy will affect Thailand's GDP over the long run using the Bayesian Structural Time Series (BSTS) to forecast the calibration and effect of the SAM table that is applied in the CGE model.

2   The Conceptual Framework and Methodology

2.1   The Conceptual Frame of Standard CGE Model

The flows of products and factors are then described at each point when they are merged for either production or consumption. In Fig. 2, the flows are displayed from bottom to top using the example of the t sector. First step using the composite factor production function, the capital and labor factors $F_{CAP,PRODUCT}$ and $F_{LAB,PRODUCT}$ are combined to form the composite factor $Y_{PRODUCT}$. And using the gross domestic output production function, this composite factor $Y_{PRODUCT}$ is coupled with the intermediate inputs Service and Product to form the gross domestic output $Z_{PRODUCT}$ and The gross domestic output transformation function converts the gross domestic product $Z_{PRODUCT}$ into the exports $E_{PRODUCT}$ and the domestic good $D_{PRODUCT}$ To create the composite good $Q_{PRODUCT}$ with the composite good production function, the domestic good $D_{PRODUCT}$ and imports $M_{PRODUCT}$ are merged so the Product and service sectors $\sum_j X_{PRODUCT,j}$ distribute the composite good $Q_{PRODUCT}$ among home consumption $X_{PRODUCT}^p$, government consumption $X_{PRODUCT}^g$, investment $X_{PRODUCT}^v$, and intermediary uses. As the utility function suggests, consumption $X_{PRODUCT}^p$ and $X_{SERVICE}^p$ produce household utility UU.

2.2   The Methodology for Standard CGE Model

Analysis of the CGE model in this research will be used to analyze the creative economy. It uses the equations related to Domestic production Government, Investment, Household, Export and Import prices, Substitution between import and domestic goods Transformation between export and domestic goods Market-Clearing conditions, Stimulating the creative economy to the economy of Thailand. The CGE model will be used to describe the model, which is the following equation.

Domestic production:

$$Y_j = b_j \prod_h F_{h,j}^{\beta_{h,j}} \quad \forall j \tag{1}$$

where:

Y_j = composite factor.
b_j = scale parameter in production function.

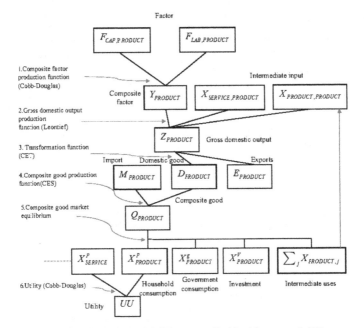

Fig. 2. The Standard CGE was applied in this research [5].

$F_{h,j}$ = the h-th factor input by the j-th firm
$\beta_{h,j}$ = share parameter in production function.

$$F_{h,j} = \frac{\beta_{h,j}p_j^y}{P_h^f}Y_j \quad \forall h,j \tag{2}$$

$$X_{i,j} = ax_{i,j}Z_j \quad \forall i,j \tag{3}$$

$$Y_j = ay_jZ_j \quad \forall j \tag{4}$$

$$P_j^z = ay_jp_j^y + \sum_i ax_{i,j}p_i^q \quad \forall j \tag{5}$$

Government:

$$T^d = \tau^d \sum_h p_h^f FF_h \tag{6}$$

where:

T^d = direct tax
T_j^z = production tax
$T_j^z T_i^m$ = import tariff
X_i^g = government consumption

$$T_j^z = \tau_j^z p_j^z Z_j \quad \forall j \tag{7}$$

$$T_i^m = \tau_j^m p_j^m M_i \quad \forall i \tag{8}$$

$$X_i^g = \frac{\mu_i}{p_i^q}(T^d + \sum_j T_j^z + \sum_j T_j^m - S^g) \quad \forall i \tag{9}$$

Investment:

$$X_i^v = \frac{\lambda_i}{p_i^q}(S^p + S^g + \varepsilon S^f) \quad \forall i \tag{10}$$

where:
X_i^v = investment demand
S^p = private saving
S^g = government saving

$$S^p = ss^p \sum_h p_h^f FF_h \tag{11}$$

$$S^g = ss^g (T^d + \sum_j T_j^z + \sum_j T_j^m) \tag{12}$$

Household:

$$X_j^p = \frac{\alpha_i}{p_i^q}(\sum_h p_h^f FF_h - S^p - T^d) \quad \forall i \tag{13}$$

where:
X_j^p = the household consumption of the i-th good

Export and Import prices and the balance of payments constraint:

$$P_i^e = \varepsilon p_i^{We} \quad \forall i \tag{14}$$

$$P_i^m = \varepsilon p_i^{Wm} \quad \forall i \tag{15}$$

where:
P_i^e = export price in US dollars
P_i^m import price in US dollars

$$\sum_i p_i^{We} E_i + S^f = \sum_i p_i^{Wm} M_i \tag{16}$$

Substitution between import and domestic goods (Armington composite):

$$Q_i = \gamma_i(\delta m_i M_i^{\eta_i} + \delta d_i D_i^{\eta_i})^{\frac{1}{\eta_i}} \quad \forall i \tag{17}$$

where:
Q_i = Armington's composite good

M_i = the imports
D_i = the domestic good

$$M_i = \left[\frac{y_i^{\eta_i} \delta m_i p_i^q}{(1 + \tau_i^m) p_i^m} \right]^{\frac{1}{1-\eta_i}} Q_i \quad \forall i \tag{18}$$

$$D_i = \left[\frac{y_i^{\eta_i} \delta d_i p_i^q}{p_i^d} \right]^{\frac{1}{1-\eta_i}} Q_i \quad \forall i \tag{19}$$

Transformation between export and domestic goods:

$$Z_i = \theta_i (\xi e_i E_i^{\Phi_i} + \xi d_i D_i^{\Phi_i})^{\frac{1}{\Phi_i}} \quad \forall i \tag{20}$$

$$E_i = \left[\frac{\theta_i^{\Phi_i} \xi e_i (1 + \tau_i^z) p_i^z}{p_i^e} \right]^{\frac{1}{1-\Phi_i}} Z_i \quad \forall i \tag{21}$$

$$D_i = \left[\frac{\theta_i^{\Phi_i} \xi d_i (1 + \tau_i^z) p_i^z}{p_i^d} \right]^{\frac{1}{1-\Phi_i}} Z_i \quad \forall i \tag{22}$$

Market-Clearing conditions:

$$Q_i = X_i^p + X_i^g + X_i^v + \sum_j X_{i,j} \quad \forall i \tag{23}$$

$$\sum_j F_{h,j} = FF_h \quad \forall h \tag{24}$$

2.3 Bayesian Structural Time Series Forecasting Model

Time series models have many applications in economics, business, science, and engineering. The first goal of a time series model is forecasting. And it describes the model structure, for which there are several ways to analyze time series. The Bayesian structural time series models are one of them that were utilized to predict the time series data, especially the time series data that are found to have high levels of uncertainty or unclear information [4] To deal with this problem, Bayesian forecasting must be used instead of classical statistical forecasting. This method relies on predicting random parameters instead of fixed parameters (the classical forecasting methods) to predict more precisely in the case of uncertainty in time series data, especially for the prediction of the creative economy's contribution to Thailand's economy for the short- and long-run periods. This creates a predictive model by structuring the elements of the time series. in a structured time, series model the observations at times of d y_t are defined according to the following observation equations.

$$y_t = z_t^T \alpha_t + \varepsilon_t \tag{25}$$

where α_t is a latent variable vector and Z_t is a vector of model parameters. Where the error t follows the Gaussian distribution with $\mu = 0$ and $\sigma^2 = H_t$ Furthermore, α_t is expressed as the following transformation equation.

$$\alpha_{t+1} = T_t \alpha_t + R_t \eta_t \tag{26}$$

where ηt has a Gaussian distribution with $\mu = 0$ and $\sigma^2 = Q_t$ This equation shows the improvement of the unobserved latent variable α_t over time. T_t And R_t are the transformation matrices and structural parameters. Respectively, Z_t, T_t and R_t have values 0 and 1, indicating their relevance for structural calculations.

Equations (25) and (26) describe the state space of the observed data. Using this model, we can model time series for short-term and long-term forecasts. Bayesian statistics is applied to many statistical fields such as regression, classification, clustering, and time series analysis. Bayesian statistics are based on the Bayes Theorem as follows:

$$P(\theta|x) = \frac{P(x|\theta)P(\theta)}{P(x)}$$

where x is the observed data and θ is the model parameter. P(x) is calculated as follows: $P(x) = \sum_\theta P(x, \theta) = \sum_\theta (P(x|\theta)P(\theta)$ P(θ) and P(X|θ) represent the preceding function and the probability, respectively. In addition, P(θ|x) is the latter function. This is enhanced by learning the observed data x provided θ (probability) in Bayesian statistics. We can get various analysis results. According to various selections of the BSTS prior distribution is also based on the general structure time series model in Eqs. (25) and (26). Up from 0 to infinity [0,1), the following represents the basic structure of BSTS:

$$yt = \mu t + \tau t + \beta^T xt + \varepsilon t$$

where $\mu_t = \mu_{t-1} + \delta t - 1 + \mu t$, $\delta t = \delta t - 1 + vt$, and $\tau_t = -\sum_{S=S}^{S-1} \tau_{t-s} + \omega_t$ where τ_t is distributed to the Gaussian.

3   The Result of the Creative Economy Prediction

Figure 3 confirms that the creative economy has a high potential positive trend growth rate of more than 50% (R-squared = 0.53). And this creative economy has still contributed to Thailand's GDP from 2011 to 2018 at a statistically significant 99% (P-value = 0.040). According to a study from a Bayesian structural time series forecasting model, the creative economy will eventually contribute around 8–10% to Thailand's GDP or economy in the period between 2019 and 2025 (see Fig. 4).

From the detail of prediction results at level 25%, it will be described from the perspective that the creative economy will be derived to contribute to Thailand's GDP on average, approximately 8.82% per year between 2019 and 2025 (see Table 1). In this case, it was implied that if Thailand's policy does not concern itself with the creative economy driving economic and social development based on creativity or reduces the significant level of promotion of the creative economy in Thailand, then Thailand's economy will receive less and less contribution from the creative economy, which drives both economic and social development.

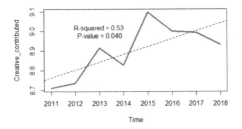

Fig. 3. The creative economy contributed on GDP of Thailand since 2011–2018.

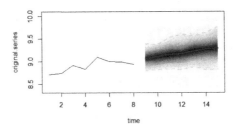

Fig. 4. The result of a Bayesian structural time series forecasting model to predict the contribution of the creative economy to Thailand's economy between 2019 and 2025

Table 1. Average creative economy contribution to GDP since 2019–2025 based on a Bayesian structural time series forecasting model estimation

Years	2.5%	Mean	Median	97.5%
2019	8.78	9.06	9.05	9.38
2020	8.81	9.10	9.10	9.37
2021	8.83	9.15	9.15	9.53
2022	8.81	9.17	9.17	9.59
2023	8.88	9.23	9.23	9.57
2024	8.80	9.26	9.27	9.64
2025	8.82	9.30	9.29	9.76
Average	8.82	9.18	9.18	9.55

The creative economy will also play a key role at the level of normal or general stimulus to push Thailand's GDP by about 50% if Thailand's policy continues to prioritize it. It was assumed that the public and private sectors would work together to advance and expand Thailand's creative economy by more than 50% from its current level and that the creative economy would then contribute roughly 9.18% of Thailand's GDP in the future (see Table 1). Finally, if those of them, such as the public sector, private organizations, and international organizations, work together to push and promote the policy driving the creative economy up to its maximum level (97.5%) (see Table 1), then Thailand's economy will receive a long-term contribution from the creative economy of up to 9.5% or 10% of GDP, which may be substantially more. Because the creative economy for products and services is the main core to improving the quality of life for every person in Thailand or every country around the world in the long run, especially the creative economy, it contributes to and improves societies' well-being [5].

4 The Result of the CGE Model Based on the Calibration and Computation from Bayesian Structural Time Series Forecasting Model (BSTS)

Because the standard econometric model requires a large sample size of data to measure the creative economy's contribution to Thailand's economy, the creative economy of Thailand is limited in terms of data to study; this can be overcome by using both calibration and computation applied in the CGE model [6]. From the results of the prediction based on BSTS model to predict the creative economy in Thailand in Table 2, it was found that creative economy contribution in general will have an effect on GDP equal to 8.93%, referring to the year 2018. From the analysis results, household consumption of products was 3,875,900 million bath and that of services was 3,598,500 million baths as well.

Table 2. Creative Economy Contribution to GDP based on CGE model estimation (Unit: Million Bath).

Order	Agents of Economic	Creative Economy Contribution 8.93% (GDP) (2018)		Creative Economy Contribution 8.82% (GDP)		Creative Economy Contribution 9.18% (GDP)		Creative Economy Contribution 9.55% (GDP)	
		Products	Services	Products	Services	Products	Services	Products	Services
1	Household Consumption	3,875,900	3,598,500	3,826,300	3,552,500	3,984,400	3,699,300	4,144,900	3,848,300
2	Government Consumption	192,140	1,457,900	189,680	1,439,300	197,520	1,498,800	205,470	1,559,100
3	Investment Demand	3,825,800	5,155,900	3,776,900	5,089,900	3,932,900	5,300,300	4,091,300	5,513,700
4	Export	6,126,900	103,930,000	6,048,500	102,600,000	6,298,500	106,840,000	6,552,100	111,140,000
5	Import	3,861,200	107,540,000	3,811,800	106,170,000	3,969,300	110,550,000	4,129,200	115,010,000
6	Domestic goods	19,165,000	1,411,300	18,920,000	1,393,300	19,701,000	1,450,900	20,495,000	1,509,300
7	Gross Domestic Output	19,016,000	10,770,000	18,772,000	10,632,000	19,548,000	11,071,000	20,336,000	11,517,000
8	Production function(Labor/Capital)	7,755,600	7,004,600	7,656,300	6,915,000	7,972,700	7,200,800	8,293,800	7,490,800

However, if the creative economy contributes to Thailand's GDP only 8.82% per year for the future (during period 2019–2025 on Thailand's GDP), which means the normal creative economy. If there is no stimulus or push, it will be found that in terms of products and services, there will be a continuous downward trend, not only household but also including the agent of economics in every order, which shows that if Thailand does not have stimulation or promotion in terms of the creative economy, it will result in a significant decrease in income from the creative economy. Whether to export or import will be affected in a negative way. But if Thailand starts to stimulate the creative economy, notice that it will affect 9.18% of Thailand's GDP, with every agent of economic order tending to increase. And more than that. If the government or related agencies give full support, it will be to contribute by the creative economy to GDP is 9.55%, which shows the distribution of the creative economy within Thailand. It also has a significant impact on employees and vendors. Finally, in terms of long-term development, the prediction made by the CGE model confirmed that if those countries try to promote and push their economies to be creative economies through future planning and development, an increase in economic welfare will be received by every person in Thailand or every person in every country around the world [7].

5   The Result of the CGE Model Prediction for the Effect of the Creative Economy on Thailand's Economic Growth

According to the analysis of the CGE model [8, 9] projection, this model suggested that if no more support or reductions are made to promote the creative economy in Thailand in the future, then the economic growth of this country will trend to slow down on average by -1.28% per year (see Table 3). It can be implied that in the long run, there is no mission for promoting the growth of creative industries at the regional level and connecting to all country regions, and there is also a lack of creative knowledge management support. As a consequence, all agents in Thailand's economy, such as household consumption, government consumption, investment demand, the export of the country, the import of the country, domestic goods and services of the country, gross domestic output, and production of the country, will have a negative growth rate.

Table 3. The creative economy contributes to Thailand's economic growth.

Order	Agents of Economic	Creative Economy Contribution 8.93% (GDP) (2018)		Creative Economy Contribution 8.82% (GDP)		Creative Economy Contribution 9.18% (GDP)		Creative Economy Contribution 9.55% (GDP)	
		Products	Services	Products (%)	Services (%)	Products	Services	Products	Services
1	Household Consumption	3,875,900	3,598,500	- 1.28	- 1.28	2.80	2.80	6.94	6.94
2	Government Consumption	192,140	1,457,900	- 1.28	- 1.28	2.80	2.81	6.94	6.94
3	Investment Demand	3,825,800	5,155,900	- 1.28	- 1.28	2.80	2.80	6.94	6.94
4	Export	6,126,900	103,930,000	- 1.28	- 1.28	2.80	2.80	6.94	6.94
5	Import	3,861,200	107,540,000	- 1.28	- 1.27	2.80	2.80	6.94	6.95
6	Domestic goods	19,165,000	1,411,300	- 1.28	- 1.28	2.80	2.81	6.94	6.94
7	Gross Domestic Output	19,016,000	10,770,000	- 1.28	- 1.28	2.80	2.79	6.94	6.94
8	Production function(Labor/Capital)	7,755,600	7,004,600	- 1.28	- 1.28	2.80	2.80	6.94	6.94

In contrast, if the public sector, private organizations, and international organizations work together to push and promote the policy driving the creative economy up to its maximum level for the future (9.55%), then the Thailand economy will grow on average by 6–7% per year (see Table 3)[10]. It can be implied that in the long run, the mission of promoting the growth of creative industries at the regional level and connecting to all country regions is an important priority that all organizations must be aware of, and also that all organizations must have enough knowledge support for creative knowledge management.

6   Conclusion

This study aims to support the idea that Thailand's economy could be driven by the creative economy, particularly in the wake of the COVID-19's decline. Since 2004, the UNCTAD has made an effort to assist each nation in learning how to support and utilize the creative economy to help their economies shift from traditional development to creative economy development. The long-term development of Thailand's economy depends significantly on the creative economy, according to the research study's findings from the BSTS and CGE models. Moreover, this study discovered that if the Thai government energetically supports or promotes the creative economy, Thailand's economy, comprising both aggregate demand and supply, will increase gradually. In contrast, if the Thai government stops promoting or supporting the creative economy in the future, Thailand's economy will suffer from both aggregate supply and demand. To advance and promote all aspects of the creative industries, including trade, labor, and production, all organizations and communities, including the public sector, private organizations, and international organizations, are advised in the policy recommendations resulting from this study. If the creative economy has been emphasized more in the growth of Thailand's economy or the economies of all other nations, then these nations still have a strong potential to achieve the sustainable long-term development goal.

References

1. Anantsuksomsri, S., Tontisirin, N.: Computable general equilibrium of real estate and financial crisis vulnerability. Int. J. Build. Urban Interior Landscape Technol. **11**, 29–42 (2018)
2. Alcamo, J., Thompson, J., Alexander, A., et al.: Analysing interactions among the sustainable development goals: findings and emerging issues from local and global studies. Sustain. Sci. **15**, 1561–1572 (2020). https://doi.org/10.1007/s11625-020-00875-x
3. Almarashi, A., Khan, K.: Bayesian structural time series. Nanosci. Nanotechnol. Lett. **12**, 54–61 (2020). https://doi.org/10.1166/nnl.2020.3083
4. Jun, S.: Bayesian structural time series and regression modeling for sustainable technology management. Sustainability **11**, 4945 (2019). https://doi.org/10.3390/su11184945
5. Creative Economy Agency (Public Organization), Creative Industries Foresight (2023)
6. Hosoe, N., Gasawa, K., Hashimoto, H.: Handbook of Computable General Equilibrium Modeling. University of Tokyo Press, Tokyo, Japan (2004)
7. Lofgren, H., Harris, R.L., Robinson, S.: A Standard Computable General Equilibrium (CGE) Model in GAMS (2002)
8. Kabir, K., Dudu, H.: Using Computable General Equilibrium Models to Analyze Economic Benefits of Gender-Inclusive Policies. MTI Practice Notes © World Bank, Washington, DC (2020)
9. An, K., Zhang, S., Zhou, J., Wang, C.: How can computable general equilibrium models serve low-carbon policy? A systematic review. Environ. Res. Lett. **18**(3) (2023)
10. Navas Thorakkattle, M., Farhin, S., Khan, A.A.: Forecasting the trends of Covid-19 and causal impact of vaccines using Bayesian structural time series and ARIMA. Ann. Data. Sci. **9**, 1025–1047 (2022). https://doi.org/10.1007/s40745-022-00418-4

Spatial Spillover Effect of Provincial Institutional Quality in Vietnam

Viet Quoc Nguyen[1] and Chon Van Le[2(✉)] [iD]

[1] University of Economics Ho Chi Minh City, Vietnam-Netherlands Program,
59C Nguyen Dinh Chieu, District 3, Ho Chi Minh City, Vietnam
`viet.nq@vnp.edu.vn`
[2] International University, Vietnam National University - Ho Chi Minh City,
Quarter 6, Linh Trung, Thu Duc, Ho Chi Minh City, Vietnam
`lvchon@hcmiu.edu.vn`

Abstract. This paper investigates the determinants of institutional quality at province level and its potential spillover effects across provinces in Vietnam. Three spatial econometric models are employed with three spatial weight matrices and a data set of all sixty three provinces collected from the General Statistics Office and the Vietnam Chamber of Commerce and Industry, spanning from 2013 to 2021. Substantial evidence is found to support that an improvement in the institutional quality of one province helps leverage that of adjacent provinces. GRDP per capita, industrial production, and trained labor force have both positive direct and spillover effects. However, FDI and profit before tax have no influence on institutional quality. The impact of local government spending is felt in neighboring provinces, though not in the host province.

Keywords: Institutional quality · Spillover effects · Spatial econometrics

1 Introduction

Efficient allocation of resources has been the cornerstone of sustained economic growth. Laissez-faire economists believe that the interaction of free and self-directed market forces without government intervention would maximize the size of the economic pie. However, a market economy needs a sound legal framework and a capable, responsive public service to operate effectively. The government is expected to ensure property rights, to enforce contracts, and to correct market failures. North (1991) claimed that institutions have been devised to reduce uncertainty in exchange, and to establish the incentive structure. As institutions evolve throughout history, that structure determines the direction and speed of economic growth. Acemoglu and Robinson (2012) insisted that development differences across countries were exclusively attributed to differences in political and economic institutions, not culture, geography, weather, or knowledge base. Institutional adjustment that can reallocate resources to more productive activities would create more wealth and raise social welfare (Hasan et al., 2009).

V.-N. Huynh et al. (Eds.): IUKM 2023, LNAI 14375, pp. 306–319, 2023.
https://doi.org/10.1007/978-3-031-46775-2_27

There have been several researches on the role of institutional quality in economic development and social security in Vietnam. Institutional transformation is considered the quickest and most effective way to generate internal resources, which are essential for Vietnam to expand and develop comprehensively and sustainably (Ho et al., 2011). Higher-quality institutions can promote economic growth by attracting foreign direct investment, fostering trade liberalization, and curbing corruption (Nguyen, 2019; Su et al., 2019; Nguyen et al., 2016).

The important role of institutions in economic development naturally leads to the question what determines institutional quality. At the country level, institutional change may be induced by strong economic performance. MacFarlan et al. (2003) argued that as countries grew and prospered, they might find that they needed and could afford to strengthen their legal and regulatory frameworks underpinning economic activity. Institutional quality also depends on income distribution (Alesina and Rodrik, 1993), trade openness, freedom of the press, and checks and balances in the political system (Islam and Montenegro, 2002), etc. In addition, institutional quality spills over to neighboring countries due to competition and multilateral treaties (Qian and Roland, 1998; Casella, 1996).

Our paper is to examine the determinants of institutional quality and its spillover effect among sixty three provinces in Vietnam. Focusing on Vietnam allows us to eliminate country-level heterogeneity which might not have been properly accounted for and to highlight possible spillover effect because of closer proximity and easier movement of people and goods across provinces. We use spatial econometric models with a panel data set of all provinces in Vietnam spanning from 2013 to 2021. It is found that the institutional quality of a province is enhanced by improvement in that of its bordering provinces. GRDP per capita, share of trained work force, and industrial production tend to positively affect the institutional quality of the host province and that of adjacent localities. Provincial government spending influences the latter but not the former. However, FDI and profit before tax make no difference to institutional quality possibly due to weak linkages to domestic enterprises and fraudulent financial reporting.

The rest of the paper is structured as follows. Section 2 gives a brief overview of the literature on spillover effect and underpinning factors of institutional quality. Section 3 outlines econometric models and spatial weight matrices. Section 4 presents data and empirical findings. Conclusions follow in Sect. 5.

2   Literature Review

Lin and Nugent (1995) broadly defined institutions as human-created rules that guided interactions and shaped expectations among individuals. Because markets are not able to self-create, self-regulate, self-stabilize, or self-legitimize, they need at least five institutions to perform competently, i.e., secure property rights, regulatory bodies to prevent misconduct and reduce transaction costs, fiscal and monetary frameworks for macroeconomic stability, social insurance mechanisms to mitigate income risks, and incentive systems to promote mutually beneficial projects among diverse social groups (Rodrik, 2007).

In the last several decades, a large number of institutional quality indicators have been introduced by academic institutions, risk-rating agencies, nongovernmental, and multilateral organizations. These indicators define institutional quality in terms of static efficiency, legitimacy, predictability, and dynamic efficiency (Alonso and Garcimartín, 2013). Possible determinants of institutional quality should be considered on the basis of their relation to these four criteria[1].

One of the most important explanatory variables is development level, related to the static efficiency of institutions. It affects institutional quality through both supply and demand. Wealthier countries have more resources to build good institutions. They also have a greater demand for quality institutions. Rigobon and Rodrik (2005) found that higher income improved the quality of rule of law and democratic institutions, though the second effect was statistically insignificant. Richer countries protect property rights and regulate better in spite of their higher marginal tax rates (La Porta et al., 1999). They also have more competent governments, better supply of public goods, and a larger public sector as Wagner's law suggests. Javaid et al. (2017) noted that higher income per capita, bigger tax revenues, and lower military spending would enhance institutional quality in several Asian countries. Economic development level is typically reflected by gross domestic product (GDP) or income per capita, and by industrial production index in developing countries.

The second factor is foreign direct investment (FDI), related to the dynamic efficiency of institutions. FDI by multinational corporations is believed by developing countries to be a major channel for their access to state-of-the-art technologies. It improves technical progress and acquisition of human capital in the host country (Borensztein et al., 1998). It encourages competition, learning processes and good practices imitation from developed countries. In addition, FDI can restrain corruption because foreign investors are more likely to leave if corruption isn't controlled. Wei (2000a) found that American and other OECD investors are averse to corruption in host countries. Given the importance of FDI to local economies, foreign integrity standards affect local officials and their behavior (Kimberley, 1997). Fukumi and Nishijima (2010) recognized that Latin America and the Caribbean region experienced a virtuous circle where FDI improved the quality of institutions which in turn attracted more foreign capital.

The third determinant is education which is related to dynamic efficiency of institutions and can be proxied by trained/skilled labor force. A more educated population requires more transparent and accountable governance and is able to build it. Well-trained employees motivate enterprises to quickly adapt to new technologies and innovate work processes, hence enhancing the ability of enterprises to compete in global markets. On the other hand, high skills boost job mobility, the labor market's flexibility, efficiency, and productivity, and encourage institutional quality improvement (Campbell and Pedersen, 2007).

---

[1] Because data for several determinants are available at country level, but not at provincial level in Vietnam, our review here is confined to those variables having data at provincial level.

The fourth variable is taxes or government expenditures which affect both the static efficiency and the legitimacy of institutions. While decent public expenditures provide necessary resources to establish high quality institutions, an associated sound tax system consolidates a social contract that leads to a more demanding state-people relationship (Alonso and Garcimartín, 2013). High government spending may reflect the citizens' willingness to pay high taxes because they consent to what their government does. Therefore, institutions will have more transparency and accountability. This may not hold if government spending is financed mostly by other sources such as state-owned enterprises or natural resources. However, Goel and Nelson (1998) showed that spending by state governments had a strong positive impact on corruption.

Institutional quality is shaped by internal factors of organizations and external factors part of which is named spillover effect. As early as the 11th century, merchants formed many trade rules which were gradually incorporated into official laws. Private initiatives to change commercial laws have harmonized institutions across countries (Rosett, 1992). Recently, multinational companies participate in a policy-making process, aiming for host government policies to be consistent with common principles (Leebron, 1996). World Trade Organization (WTO) and regional blocks require their acceding members to offer uniform treatment under the principles of non-discrimination and reciprocity (Christensen, 2015). It requires member countries to reconcile domestic laws and policies.

Institutional adoption is more receptive among closer countries than farther countries. Mukand and Rodrik (2005) showed that "following" countries chose to mimic a successful policy in a "leading" neighboring country to quickly reap the policy benefits without experimentation costs. Distant countries or those having different characteristics performed worse with greater variance. Institutional duplication could also take place through negotiations between governments. For example, countries with stricter environmental standards might require their partners to comply, which leads to changes in the policies of the partner countries. Persson and Tabellini (2009) discovered that the experience with democracy in neighboring countries could nurture greater domestic appreciation of democracy and greater willingness to endorse these values.

The spillover effect of institutional quality occurs not only at country level but also at province level. Poel (1976) indicated that there existed a diffusion of 25 selected public policies among Canadian provinces. Economic liberalization in a German state could be influenced by that in the neighboring states (Potrafke, 2013). Cutts and Webber (2010) suggested that there was strong spatial autocorrelation in voting patterns across constituencies in England and Wales. In China, there is a spatial spillover effect of environmental regulation on air pollution. The $PM_{2.5}$ concentration of Beijing-Tianjin-Hebei, Yangtze River Delta, and Pearl River Delta rose by 0.76, 0.147 and 0.109 for each unit increase in environmental regulation of surrounding cities, respectively (Feng et al., 2020).

3 Econometric Models

To account for the spillover effect of institutional quality among provinces in Vietnam, we use the spatial autoregressive model (SAR), spatial Durbin model (SDM), and spatial error model (SEM) (LeSage and Pace, 2009). The spatial autoregressive model takes the form

$$\mathbf{y} = \rho\mathbf{W}\mathbf{y} + \mathbf{X}\boldsymbol{\beta} + \boldsymbol{\varepsilon}, \qquad (1)$$
$$\boldsymbol{\varepsilon} \sim N(\mathbf{0}, \sigma^2\mathbf{I}_n),$$

where the dependent variable vector $\mathbf{y}$ of dimension n by 1 contains provincial institutional quality measures, the n by k matrix $\mathbf{X}$ contains explanatory variables including a constant term, the n by n matrix $\mathbf{W}$ quantifies the connections between provinces and is called the spatial weight matrix, the n by 1 disturbance vector $\boldsymbol{\varepsilon}$ is assumed to contain independent, normally distributed error terms, each with zero mean and variance σ^2. The n by 1 spatial lag vector $\mathbf{W}\mathbf{y}$ shows an average of institutional quality of neighboring provinces, and the associated scalar parameter ρ represents the strength of spatial dependence. When ρ is zero, the model in (1) becomes a linear regression model. The SAR model states that the institutional quality of each province is related to the average institutional quality of nearby provinces.

The spatial Durbin model allows explanatory variables that determine institutional quality in neighboring provinces to affect institutional quality of that province. This is done by adding the matrix product $\mathbf{W}\mathbf{X}$ which reflects an average of explanatory variables from neighboring provinces. The SDM model is

$$\mathbf{y} = \rho\mathbf{W}\mathbf{y} + \alpha\mathbf{1} + \mathbf{X}\boldsymbol{\beta} + \mathbf{W}\mathbf{X}\boldsymbol{\theta} + \boldsymbol{\varepsilon}, \qquad (2)$$
$$\boldsymbol{\varepsilon} \sim N(\mathbf{0}, \sigma^2\mathbf{I}_n),$$

where the constant term vector $\mathbf{1}$ is eliminated from the explanatory variable matrix $\mathbf{X}$.

The spatial error model considers spatial dependence in the disturbance process $\boldsymbol{\varepsilon}$,

$$\mathbf{y} = \mathbf{X}\boldsymbol{\beta} + \mathbf{u}, \qquad (3)$$
$$\mathbf{u} = \lambda\mathbf{W}\mathbf{u} + \boldsymbol{\varepsilon},$$
$$\boldsymbol{\varepsilon} \sim N(\mathbf{0}, \sigma^2\mathbf{I}_n),$$

In the SAR model, a change in an explanatory variable in province i leads to a change in institutional quality of that province which in turn affects institutional quality of neighboring provinces which causes further change in institutional quality of province i. Dependence relations among provinces result in an infinite series of feedback effects which can be captured after some manipulation of Eq. (1)

$$\mathbf{y} = (\mathbf{I}_n - \rho\mathbf{W})^{-1}\mathbf{X}\boldsymbol{\beta} + (\mathbf{I}_n - \rho\mathbf{W})^{-1}\boldsymbol{\varepsilon}, \qquad (4)$$

where $(\mathbf{I}_n - \rho\mathbf{W})^{-1} = \mathbf{I}_n + \rho\mathbf{W} + \rho^2\mathbf{W}^2 + \rho^3\mathbf{W}^3 + \dots$. The direct impact of explanatory variable l in province i on institutional quality of province i is $(\mathbf{I}_n - \rho\mathbf{W})_{ii}^{-1}\beta_l$ where $(\mathbf{I}_n - \rho\mathbf{W})_{ii}^{-1}$ is the iith element in $(\mathbf{I}_n - \rho\mathbf{W})^{-1}$. The indirect impact of explanatory variable l in province i on institutional quality of province j is $(\mathbf{I}_n - \rho\mathbf{W})_{ji}^{-1}\beta_l$.

In the SDM model, a similar arrangement gives

$$\mathbf{y} = (\mathbf{I}_n - \rho\mathbf{W})^{-1}\alpha\mathbf{1} + (\mathbf{I}_n - \rho\mathbf{W})^{-1}\mathbf{X}\beta + (\mathbf{I}_n - \rho\mathbf{W})^{-1}\mathbf{W}\mathbf{X}\theta + (\mathbf{I}_n - \rho\mathbf{W})^{-1}\varepsilon. \quad (5)$$

The direct impact of characteristic l in province i on institutional quality of province i is $(\mathbf{I}_n - \rho\mathbf{W})_{ii}^{-1}\beta_l + (\mathbf{I}_n - \rho\mathbf{W})_{i.}^{-1}\mathbf{W}_{.i}\theta_l$ where $(\mathbf{I}_n - \rho\mathbf{W})_{i.}^{-1}$ is the ith row of $(\mathbf{I}_n - \rho\mathbf{W})^{-1}$ and $\mathbf{W}_{.i}$ is the ith column of $\mathbf{W}$. The indirect impact of characteristic l in province i on institutional quality of province j is $(\mathbf{I}_n - \rho\mathbf{W})_{ji}^{-1}\beta_l + (\mathbf{I}_n - \rho\mathbf{W})_{j.}^{-1}\mathbf{W}_{.i}\theta_l$ where $(\mathbf{I}_n - \rho\mathbf{W})_{j.}^{-1}$ is the jth row of $(\mathbf{I}_n - \rho\mathbf{W})^{-1}$.

We employ three kinds of spatial weight matrices. In a spatial contiguity matrix $(\mathbf{W}^C)$, if two provinces are adjacent, their w is set as 1. If they have no adjacency, it is 0.

$$w_{ij}^C = \begin{cases} 1, \text{ provinces } i \text{ and } j \text{ are neighbors,} \\ 0, \text{ provinces } i \text{ and } j \text{ are not neighbors or } i = j. \end{cases} \quad (6)$$

In a spatial inverse-distance matrix $(\mathbf{W}^{ID})$, w between provinces i and j is calculated as the inverse of the distance between the two provinces.

$$w_{ij}^{ID} = \begin{cases} \dfrac{1}{d_{ij}^{\gamma}}, \ i \neq j, \\ 0, \ i = j, \end{cases} \quad (7)$$

where γ is the distance decay parameter (in this paper γ is set to 2), and the distance between two provinces is measured by the Haversine distance between the two provincial people's committee buildings (Elhorst and Vega, 2013). Specifically,

$$d_{ij} = 2R\arcsin\left(\sqrt{\sin^2\left(\frac{x_j - x_i}{2}\right) + \cos(x_i)\cos(x_j)\sin^2\left(\frac{y_j - y_i}{2}\right)}\right), \quad (8)$$

where R is the radius of the sphere, x_i, x_j are the latitudes of the two provincial buildings, and y_i, y_j are the longitudes of the two provincial buildings. A spatial contiguity matrix focuses on the impact from the neighboring provinces, whereas a spatial inverse-distance matrix highlights that this impact diminishes as distance increases. The third spatial weight matrix, called a spatial inverse-distance contiguity matrix $(\mathbf{W}^{IDC})$, combines the above two matrices where w of two non-contiguous provinces is set to 0 and that of two contiguous provinces is their inverse distance.

$$w_{ij}^{IDC} = \begin{cases} \dfrac{1}{d_{ij}^{\gamma}}, \text{ provinces } i \text{ and } j \text{ are neighbors,} \\ 0, \text{ provinces } i \text{ and } j \text{ are not neighbors or } i = j, \end{cases} \quad (9)$$

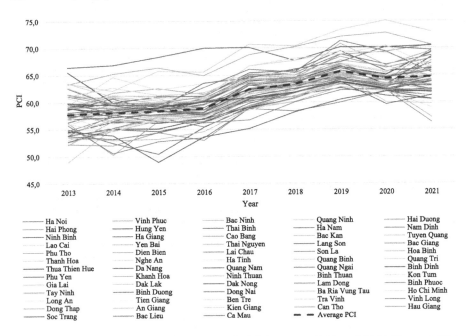

Fig. 1. Provincial Competitiveness Index (PCI) between 2013 and 2021 (Source: Vietnam Chamber of Commerce and Industry (2023).)

where the distance decay parameter is set to 2. Row normalization is applied to all three spatial weight matrices. These matrices are simultaneously used to test the robustness of the spatial regression models.

4   Data and Empirical Results

4.1   Data

Like many relevant studies, we use the Provincial Competitiveness Index (PCI) as a proxy for institutional quality. Business surveys using stratified random sampling are annually conducted by the Vietnam Chamber of Commerce and Industry (VCCI) to assess and rank economic governance of provincial authorities that affect private sector development. PCI comprises ten sub-indices: (1) entry costs, (2) land access and security of tenure, (3) transparency and access to information, (4) informal charges, (5) time costs and regulatory compliance, (6) policy bias, (7) proactivity of provincial leadership, (8) business support services, (9) labor and training policies, and (10) legal procedures for dispute resolution (VCCI, 2023).

Figure 1 shows PCI of all sixty three provinces in Vietnam from 2013 to 2021. The average PCI seemed to be stable for the first four years and improved steadily between 2017 and 2019 then slightly declined in 2020 and 2021 probably due to unprecedentedly devastating consequences of the COVID-19 pandemic.

In this period, Danang, Quang Ninh, and Dong Thap always showed up in the top five provinces. Long-standing members of the bottom five group were Ha Giang, Cao Bang, Bac Kan, Lai Chau, and Dak Nong. Provinces in Northen midlands and mountainous region normally had worse institutional quality.

Table 1. Variable definitions and data sources

Code	Definition	Data Source
PCI	Provincial Competitiveness Index	www.pcivietnam.vn
GRDPpc	Gross regional domestic product per capita (VND million)	General Statistics Office (GSO) of Vietnam
IIP	Industrial production index	GSO
PBT	Profit before tax of acting enterprises having business outcomes (VND trillion)	GSO
FDI	Accumulated registered capital of foreign direct investment projects (USD billion)	GSO
$L_{trained}$	Percentage of trained labor force	GSO
Spend	Provincial government spending (VND trillion)	GSO

Industrial development has played a central role in the economic growth of countries and of provinces. Industrialization often starts in manufacturing sectors and is believed to generate society transformation. The formation of a technostructure, including a highly trained labor force with modern technology, extends to all other sectors of a developing economy. However, the impacts of industrialization are subject to its performance, which is measured in this paper by profit before tax (PBT) of acting enterprises. We use three indicators, namely, gross regional domestic product per capita (GRDPpc), industrial production index (IIP), and PBT, to represent the development level of a province.

All variable definitions and data sources are presented in Table 1, and descriptive statistics in Table 2. Our panel data set consists of all sixty three provinces in Vietnam from 2013 to 2021. Gap in GRDP per capita between the richest and the poorest provinces was huge. A person in the wealthiest province on average made goods and services worth twenty six times as much as her counterpart in the poorest province in 2012. By the end of the last decade, GRDP per capita in the poorest province more than doubled while that in the rich one declined a little, thus reducing the gap by more than half to 10.8 times. Table 3 shows that GRDP per capita is strongly positively correlated with FDI. Advanced technology and better access to input and output markets of foreign-invested enterprises help produce not only more goods but also more profit before tax (PBT). FDI and PBT in turn are closely associated with local government spending, which tends to foster trained labor force ($L_{trained}$). This workforce has weaker positive correlations with PCI, GRDP per capita, FDI, and PBT. It should be noted that industrial production index (IIP) is not correlated or marginally significantly with all other variables. In terms of the coefficient of variation (the ratio of the standard deviation to the mean), all explanatory variables disperse a lot

Table 2. Summary statistics

	Mean	Maximum	Minimum	Std. Dev.	Observations
PCI	61.5	75.1	49.0	4.26	567
GRDPpc	51.3	361.7	12.7	41.51	567
IIP	110.4	322.8	43.3	15.83	567
PBT	9.5	205.3	−28.3	27.11	567
FDI	4.7	48.2	0.001	8.50	567
$L_{trained}$	17.9	48.5	5.1	7.29	567
Spend	20.1	288.2	5.3	21.37	567

Table 3. Unconditional correlation matrix among variables

	PCI	GRDPpc	IIP	PBT	FDI	$L_{trained}$	Spend
PCI	1						
GRDPpc	0.41***	1					
IIP	0.01	−0.09**	1				
PBT	0.23***	0.51***	−0.03	1			
FDI	0.28***	0.71***	−0.03	0.82***	1		
$L_{trained}$	0.45***	0.49***	−0.02	0.54***	0.55***	1	
Spend	0.32***	0.31***	−0.03	0.67***	0.64***	0.55***	1

and much more than PCI. It seems that smaller heterogeneity of PCI may be due to spatial spillover effect that will be analyzed in the next section.

4.2   Empirical Results

First, we compute the yearly global Moran's I to check the spatial autocorrelation of the dependent variable. Table 4 gives the estimates using three spatial weight matrices, i.e., a spatial contiguity matrix ($\mathbf{W}^C$), a spatial inverse-distance matrix ($\mathbf{W}^{ID}$), and a spatial inverse-distance contiguity matrix ($\mathbf{W}^{IDC}$). Most of the values are significant, implying that PCI of different provinces has a big spatial dependency. Then, we estimate the spatial autoregressive model (SAR), spatial Durbin model (SDM), and spatial error model (SEM) with these spatial weight matrices. There are a random-effects (RE) and a fixed-effects (FE) estimators for each model (and a matrix). The Hausman test is used to determine whether RE or FE is more appropriate. Estimation results are provided in Table 5.

Table 4. Estimated Moran's I for PCI

Spatial weight matrix $\mathbf{W}$	Year								
	2013	2014	2015	2016	2017	2018	2019	2020	2021
$\mathbf{W}^C$	0.25***	0.17**	0.09*	0.15**	0.17**	0.18**	0.17**	0.20***	0.26***
$\mathbf{W}^{ID}$	0.18***	0.03	0.02	0.04	0.08*	0.07*	0.09**	0.09**	0.10***
$\mathbf{W}^{IDC}$	0.24***	0.10	0.07	0.13*	0.18**	0.14*	0.19**	0.20**	0.28***

Notes: ***, **, and * denote significance at the 0.01, 0.05, and 0.1 level.
Source: Authors' calculation.

Table 5. Coefficient estimates of the SAR, SDM, and SEM models with a spatial contiguity, a spatial inverse-distance, and a spatial inverse-distance contiguity matrices

	Spatial contiguity			Spatial inverse-distance			Spatial inverse-distance contiguity		
	SAR	SDM	SEM	SAR	SDM	SEM	SAR	SDM	SEM
	(1)	(2)	(3)	(4)	(5)	(6)	(7)	(8)	(9)
GRDPpc	0.03***	0.02**	0.02**	0.03***	0.02***	0.02***	0.04***	0.03***	0.05
IIP	0.01**	0.01**	0.01	0.01**	0.01**	0.01*	0.01**	0.01**	0.05
PBT	−0.02	−0.01	−0.02**	−0.01	−0.01	−0.01	−0.02*	−0.01	0.05***
FDI	−0.003	0.04	0.04	−0.004	0.02	0.01	−0.02	0.02	0.28***
$L_{trained}$	0.23***	0.19***	0.18***	0.20***	0.18***	0.16***	0.26***	0.22***	0.01
Spend	0.01	0.01	0.005	0.01	0.01	0.01	0.01**	0.01	0.01*
WX									
GRDPpc		0.06***			0.02			0.03***	
IIP		0.01			0.02			0.01	
PBT		0.02			0.01			0.01	
FDI		−0.25***			−0.09			−0.10	
$L_{trained}$		−0.003			0.07			0.04	
Spend		0.01			0.01			0.01	
Constant	15.2***	19.3***	56.5***	10.4***	11.0***	56.3***	17.7***	20.0***	
Estimator	RE	RE	RE	RE	RE	RE	RE	RE	FE
$\hat{\rho}$	0.64***	0.54***		0.73***	0.66***		0.58***	0.51***	
$\hat{\lambda}$		0.77***			0.86***				0.66***
Obs	567	567	567	567	567	567	567	567	567

Notes: ***, **, and * denote significance at the 0.01, 0.05, and 0.1 level.
Source: Authors' calculation.

All estimates of ρ or λ are highly significant, indicating that spatial models should be used instead of the classical OLS model. Their positive numbers signify that when the institutional quality in surrounding provinces increases, so does the institutional quality in each province. In addition, Table 5 shows a quite uniform picture across the three spatial models employing the three spatial weight matrices with regard to their significance and sign. It proves that our models are reasonably robust. For ease of interpretation, the likelihood ratio test is conducted to select the best model which is the SAR model with a spatial inverse-distance matrix and a RE estimator. It results are in column 4 of Table 5, and its indirect and corresponding total effects are presented in Table 6.

GRDP per capita has not only a high unconditional correlation with PCI (in Table 3), but also a substantial ceteris paribus impact on institutional quality. Richer people are inclined to ask their local government to improve institutions. The significantly positive spillover effects of GRDP per capita could imply that these rich residents also boost institutional quality in neighboring provinces. Inter-provincial income comparison might strengthen motivation for institutional improvement in adjacent provinces as they would like to avoid the inferior feeling of being relatively poorer.

The same holds true for industrial production (IIP), which positively influences PCI. The chronically low productivity of the farming sector has cultivated a belief that industrialization is a (if not the only) key to economic growth and poverty reduction. Public officials are willing to execute adjustment require-

Table 6. Indirect and total effects for the SAR model with a random-effects estimator and a spatial inverse-distance matrix

	Coeff.	(Std. Err.)
Average indirect effects		
GRDPpc	0.067***	(0.015)
IIP	0.025**	(0.012)
PBT	−0.024	(0.022)
FDI	−0.014	(0.112)
$L_{trained}$	0.507***	(0.090)
Spend	0.027*	(0.016)
Average total effects		
GRDPpc	0.097***	(0.022)
IIP	0.036**	(0.017)
PBT	−0.035	(0.032)
FDI	−0.022	(0.160)
$L_{trained}$	0.731***	(0.115)
Spend	0.039*	(0.023)

Notes: ***, **, and * denote significance at the 0.01, 0.05, and 0.1 level.
Source: Authors' calculation.

ments made by enterprises. While manufacturing firms play an increasing role in a provincial economy, providing more jobs to local people, and paying more taxes to local governments, they tend to build up bigger pressure for better institutional quality. They also affect the institutional quality of adjacent provinces favorably. Their cumulative indirect effects equal to 0.025 more than double the magnitude of their cumulative direct effects of 0.01. However, profit before tax (PBT) does not have direct and spillover effects on PCI. It may be due to measurement errors in PBT caused by fraudulent financial reporting for tax evasion which is rampant in developing countries, including Vietnam.

Vietnam and its dynamic provinces have continually made various incentives and commitments to simplify administrative formalities towards international standards in order to attract FDI. It is interesting to note from Tables 5 and 6 that FDI in turn does not leverage the institutional quality of the host province and its neighbors. Its limited role could arise from its rather weak linkages with domestic firms. Most domestic enterprises have inadequate capacity to absorb technology and knowledge from foreign firms and meet their demands for high quality, competitive prices, and reliability. Therefore, Vietnam has not been successful in relying on foreign-invested firms for granting domestic firms entry into global value chains (World Bank, 2017). It seems that FDI has existed as an oasis to take advantage of concessions offered by central and local governments rather than as an important driver for finer institutional quality.

Table 5 endorses the positive impact of trained labor force on institutional quality. Skilled employees not only demand more transparent and dynamic institutions but also promote firm productivity, thus providing necessary means to establish good institutions. Each additional percentage of trained labor force would increase the provincial competitiveness index by 0.2% point. This positive impact spills over to contiguous provinces thanks to labor mobility among regions. Table 6 indicates that the average indirect effects are twice and a half as much as the average direct effects, with an average total of 0.73% point in PCI of relevant provinces given an extra percentage of trained work force. It seems unusual that provincial government spending does not exert a beneficial influence on the institutional quality of that province but enhances that of bordering provinces.

5   Conclusion

Good institutions, particularly those in the public sector, are instrumental to sustainable economic growth. This paper is to investigate the determinants of institutional quality at province level and its possible spillover effects among sixty three provinces in Vietnam. We use three spatial econometric models, i.e., spatial autoregressive model, spatial Durbin model, and spatial error model, with three spatial weight matrices, namely, a spatial contiguity matrix, a spatial inverse-distance matrix, and a spatial inverse-distance contiguity matrix. The data set covers the period from 2013 to 2021. GRDP per capita, manufacturing sectors, and trained labor force seem to have considerable positive impacts on the institutional quality of the host province and that of neighboring provinces. Local government spending affects the latter but not the former. Nevertheless, FDI plays no role in institutional quality probably due to its rather loose connections with domestic firms. For profit before tax, the problem might be measurement errors.

Acknowledgements. Chon Van Le acknowledges financial support from the Vietnam National University, Ho Chi Minh City under the research project B2022-28-05.

References

Acemoglu, D., Robinson, J.: Why Nations Fail: The Origins of Power, Prosperity, and Poverty. Crown Business, New York (2012)

Alesina, A., Rodrik, D.: Distributive politics and economic growth. Quart. J. Econ. **109**(2), 465–490 (1993)

Alonso, J.A., Garcimartín, C.: The determinants of institutional quality. more on the debate. J. Int. Dev. **25**(2), 206–226 (2013)

Borensztein, E., De Gregorio, J., Lee, J.-W.: How does foreign direct investment affect economic growth? J. Int. Econ. **45**(1), 115–135 (1998)

Campbell, J.L., Pedersen, O.K.: Institutional competitiveness in the global economy: Denmark, the United States, and the varieties of capitalism. Regul. Gov. **1**(3), 230–246 (2007)

Casella, A.: Free trade and evolving standards. In: Bhagwati, J.N., Hudec, R. E. (eds.) Fair Trade and Harmonization. Prerequisites for Free Trade?, vol. 1, pp. 119–156. MIT Press (1996)

Christensen, J.: Fair trade, formal equality, and preferential treatment. Soc. Theory Pract. **41**(3), 505–526 (2015)

Cutts, D., Webber, D.J.: Voting patterns, party spending and space in England and Wales. Reg. Stud. **44**(6), 735–760 (2010)

Elhorst, P., Vega, S.H.: On spatial econometric models, spillover effects, and W. ERSA conference papers ersa13p222. European Regional Science Association (2013)

Feng, T., Du, H., Lin, Z., Zuo, J.: Spatial spillover effects of environmental regulations on air pollution: evidence from urban agglomerations in China. J. Environ. Manage. **272**, 110998 (2020)

Fukumi, A., Nishijima, S.: Institutional quality and foreign direct investment in Latin America and the Caribbean. Appl. Econ. **42**(14), 1857–1864 (2010)

Goel, R.K., Nelson, M.A.: Corruption and government size: a disaggregated analysis. Public Choice **97**(1–2), 107–120 (1998)

Hasan, I., Wachtel, P., Zhou, M.: Institutional development, financial deepening and economic growth: evidence from China. J. Banking Finance **33**(1), 157–170 (2009)

Ho, T.B., et al.: Comprehensive reform to develop the country. New Age Magazine, 23 (2011)

Islam, R., Montenegro, C.E.: What Determines the Quality of Institutions? Policy research working paper 2764. World Bank (2002)

Javaid, M.N., Iftikhar, N.M., Ahmed, G.: What drives the quality of institutions in Asian economies? Directions for economic reforms. J. S. Asian Stud. **5**(3), 127–139 (2017)

Kimberley, A. (ed.): Corruption and the Global Economy. Peterson Institute for International Economics, Washington, D.C. (1997)

La Porta, R., Lopez-de-Silanes, F., Shleifer, A., Vishny, R.: The quality of government. J. Law Econ. Organ. **15**(1), 222–279 (1999)

Leebron, D.W.: Lying down with procrustes: an analysis of harmonization claims. In: Bhagwati, J., Hudec, R.E. (eds.) Fair Trade and Harmonization: Prerequisites for Free Trade?, vol. I: Economic Analysis, pp. 41–117. MIT Press (1996)

LeSage, J., Pace, R.K.: Introduction to Spatial Econometrics. Chapman and Hall/CRC, Boca Raton, FL (2009)

Lin, J.Y., Nugent, J.B.: Institutions and economic development. In: Chenery, H., Srinivasan, T.N. (eds) Handbook of Development Economics, vol. 3, Chapter 38, pp. 2301–2370. Elsevier (1995)

MacFarlan, M., Edison, H., Spatafora, N.: Chapter III: Growth and institutions. https://www.imf.org/~/media/Websites/IMF/imported-flagship-issues/external/pubs/ft/weo/2003/01/pdf/_chapter3pdf.ashx (2003)

Mukand, S.W., Rodrik, D.: In search of the holy grail: policy convergence, experimentation, and economic performance. Am. Econ. Rev. **95**(1), 374–383 (2005)

Nguyen, N.A., Nguyen, N.M., Tran-Nam, B.: Corruption and economic growth, with a focus on Vietnam. Crime Law Soc. Chang. **65**, 307–324 (2016)

Nguyen, V.B.: The role of institutional quality in the relationship between FDI and economic growth in Vietnam: empirical evidence from provincial data. Singap. Econ. Rev. **64**(3), 601–623 (2019)

North, D.C.: Institutions. J. Econ. Perspect. **5**(1), 97–112 (1991)

Qian, Y., Roland, G.: Federalism and the soft budget constraint. Am. Econ. Rev. **88**(5), 1143–1162 (1998)

Persson, T., Tabellini, G.: Democratic capital: the nexus of political and economic change. Am. Econ. J. Macroecon. **1**(2), 88–126 (2009)

Poel, D.H.: The diffusion of legislation among the Canadian provinces: a statistical analysis. Can. J. Polit. Sci./Rev. Can. de Sci. Politique **9**(4), 605–626 (1976)

Potrafke, N.: Economic freedom and government ideology across the German states. Reg. Stud. **47**(3), 433–449 (2013)

Rigobon, R., Rodrik, D.: Rule of law, democracy, openness, and income: estimating the interrelationships. Econ. Transit. **13**(3), 533–564 (2005)

Rodrik, D.: One Economics, Many Recipes: Globalization, Institutions, and Economic Growth. Princeton University Press, New Jersey (2007)

Rosett, A.: Unification, harmonization, restatement, codification, and reform in international commercial law. Am. J. Comp. Law **40**(3), 683–697 (1992)

Su, D.T., Nguyen, P.C., Schinckus, C.: Impact of foreign direct investment, trade openness and economic institutions on growth in emerging countries: the case of Vietnam. J. Int. Stud. **12**(3), 243–264 (2019)

Vietnam Chamber of Commerce and Industry: The Provincial Competitiveness Index. https://pcivietnam.vn/en/about-us.html (2023)

Wei, S.J.: How taxing is corruption on international investors? Rev. Econ. Stat. **82**(1), 1–11 (2000)

World Bank: Vietnam enhancing enterprise competitiveness and SME linkages: Lessons from international and national experience. https://hdl.handle.net/10986/30047 (2017)

Economic Condition, Multidimensional Poverty and NEETs in Thailand

Supanika Leurcharusmee[(✉)] and Piyaluk Buddhawongsa

Center of Human Resource and Public Health Economics, Faculty of Economics, Chiang Mai University, Chiang Mai, Thailand
supanika.l@cmu.ac.th

Abstract. This study examines how the economic situation and multidimensional poverty in Thailand impact the probability of young people in Thailand being in the NEET (Not in Education, Employment, or Training) status. The study employs the generalized structural equation model (GSEM) to explores the structural relationships between per capita gross provincial product (GPP) and poverty intensity in education, healthy living, living conditions, and financial security dimensions, in relation to the occurrence of NEET status. The results reveal significant associations between all dimensions of poverty to the NEET status, with education and healthy living poverty showing the strongest correlation. Regarding the impact of the economic condition on the NEET status, the findings indicate that youths in provinces with higher per capita GPP are less likely to experience poverty, leading to a reduced likelihood of being NEET. However, when controlled for the multidimensional poverty situation, it is observed that youths in provinces with a higher per capita GPP are more likely to becoming NEETs. These intricate findings suggest potential heterogeneity in the socio-economic backgrounds among NEETs. Overall, the results underscore the importance of providing support to youth from low education and poverty-stricken households who are susceptible to NEET status, particularly during economic downturns.

Keywords: Youths · NEET · economic condition · poverty · social exclusion · generalized structural equation

1   Introduction

The increasing number of NEETs (Youths not in Education, Employment, or Training) in Thailand is a major problem for society and the economy. It inhibits economic growth, productivity, and competitiveness. Thailand is also facing the challenge of becoming an aging society, while the number of young people aged 15–24 is declining from 9.6 million in 2016 to 9.2 million in 2021 [11]. This demographic shift leads to a higher old-age dependency ratio and a decline in labor force participation. These factors will have long-term implications for Thailand's socioeconomic development.

V.-N. Huynh et al. (Eds.): IUKM 2023, LNAI 14375, pp. 320–330, 2023.
https://doi.org/10.1007/978-3-031-46775-2_28

According to the International Labor Organization (ILO), approximately 1.4 million youth between the ages of 15 and 24 are NEET [11]. Youth who are NEET are more likely living in poor economic conditions and social exclusion because they often do not have the means to improve their skills and competencies. Improving employment, education and training opportunities and promoting the social inclusion of disadvantaged youth has therefore become a major policy concern for the government.

There are several studies examining risk factors for being in NEET. Tamesberger et al. [9] attempted to identify the influencing factors for young men and women separately. They found that risk was associated with early school leaving, health impairments, and first experience of unemployment in both groups. However, the additional factor for young women was associated with child care for children under age three. In South Korea, Noh and Lee [6] used data from the Youth Panel Study to examine how individual and family characteristics at age 16–17 affect the likelihood of being in NEET during 20–25. They found that individuals with low income, no career plan, and dissatisfaction with school during the high school years were more likely to be in NEET.

Macroeconomic factors also influence rates of NEET. Pesquera Alonso et al. [7] tested the hypothesis using data from countries in the European Union and confirmed that economic cycles are stronger factors than culture or other country-specific variables. Nevertheless, the NEET rate can be high even during economic upturns because young NEETs aged 19–24 had to manage the transition from school to work [2]. For a country, there are still regional impacts on the NEET rate. Using data from Austria, Bacher et al. [1] found that regional factors such as upper secondary completion, dual education, job vacancies, and spending on active labor market policies are also relevant in explaining regional differences in NEET rates.

Only a few research has been conducted on the factors affecting the situation of NEET particularly in Thailand. The recent research by the UNICEF and College of Population Studies, Social Research Institute, Chula Unisearch [11] found that the group NEET is diverse, e.g., school dropouts, unemployed youth, or those who drop out of the labor force for various reasons. Factors that lead to a young person becoming NEET include gender, education level, and socioeconomic conditions. This paper aims not only to contribute to the existing literature by quantitatively examining the direct impact of economic conditions on the NEET status of youth in Thailand but also to explore the indirect effect of economic conditions on NEET status through four dimensions of multidimensional poverty: education, healthy living, living conditions, and financial security. The findings of this study can provide policymakers with insights into the dual pathways through which the economic condition affect the likelihood of NEET status and assist in identifying NEET individuals based on their multidimensional poverty profiles. This, in turn, can facilitate the development of effective prevention policies.

2   Data

This study utilizes data from the 2019 Socio-economic Survey (SES) of Thailand, a comprehensive nationwide household survey conducted by the National Statistical Office (NSO). The dataset comprises a total of 124,874 individuals, including 11,801 youths aged 15 to 24. However, 14 observations were excluded due to the occupation of housemaids, which compromised the reliability of household characteristics as indicators of their socio-economic status. Additionally, 47 observations were excluded due to inconsistencies in their reported educational status. Therefore, the analysis in this study was conducted on a sample of 11,740 youths.

3   NEET Situation

Youths can be categorized into three primary groups: (1) in-school, (2) employed, and (3) NEETs. The in-school group comprises all youths registered in the formal educational system, irrespective of their employment status. The employed group includes youths who have secured employment but are not currently enrolled in school. The NEET group consists of individuals who are neither enrolled in school nor employed. Based on data from the 2019 SES sample, the majority of youths aged 15–24 years fell under the in-school category, accounting for 50.4% of the total. Around 37.4% of youths were employed, while the remaining 12.2% were classified as NEETs (Table 1).

Table 1. Classification of youths in Thailand

	Frequency	Percent
In-school	5,914	50.4%
Employed	4,392	37.4%
NEET	1,434	12.2%
Total	**11,740**	**100%**

Source: Calculated from the 2019 SES sample (without population weighting).

4   Multidimensional Poverty Situation

Numerous studies have provided evidence that social exclusion and poverty are significant contributing factors to the disengagement of young individuals, leading them to become a NEET [4,8]. Therefore, the present research aims to investigate the impact of social exclusion and poverty by utilizing indicators derived from Thailand's multidimensional poverty index, as reported by NESDC [5]. To determine the presence of poverty within each dimension, an individual residing in a household that falls under the criteria for a particular dimension of poverty is considered to be living in poverty. Thailand's multidimensional poverty index encompasses four primary dimensions, namely education, health and well-being, living conditions, and financial security. Each dimension consists of a set of three indicators, as illustrated in Table 2.

Based on the summary statistics derived from the 2019 SES sample[1], several factors contributed to the occurrence of multidimensional poverty among Thai youths. The factors that had the highest contributions were years of schooling (edu1), garbage management (live7), and pension (fin12), accounting for 40.4%, 33.6%, and 32.3% respectively. This indicates that some proportion of Thai youths reside in households characterized by low levels of education among some members, unproper garbage management practices, and the absence of pension support for older household members.

When comparing the likelihood of experiencing multidimensional poverty between NEETs and non-NEETs, the statistics suggest that NEETs experienced higher rates of poverty across all indicators, indicating varying degrees of social exclusion within the population. Notably, the dimensions that exhibit the most substantial disparities between NEETs and non-NEETs included consumption poverty (health6), years of schooling (edu1), and delayed schooling (edu2). Regarding consumption poverty, the proportion among NEETs was 20.6%, whereas among non-NEETs, it was only 9.8%, highlighting that NEETs were 10.8% more likely to belong to households categorized under consumption poverty compared to non-NEETs. In the education dimension, it was observed that NEETs were more inclined to originate from households characterized not only by individuals with low educational attainment but also by children experiencing delays in their education.

5   Generalized Structural Equation Model

To examine effects of the economic condition and social exclusion on the NEET status, this study modified the general framework from Ruesga-Benito et al. [8], which shows the effects of economic condition, social exclusion and poverty on the NEET rate. This study investigates the influence of the economic condition, as measured by the per capita gross provincial product (gpppc), and four dimensions of multidimensional poverty on the NEET status of each youth. Recognizing that the economic condition can impact NEET status both directly and indirectly through the four dimensions of multidimensional poverty, the study employs a Generalized Structural Equation Model (GSEM) to perform a path analysis with multidimensional poverty acting as mediators, as depicted in Fig. 1.

---

[1] It is important to note that the statistics presented in this study are derived from the 2019 SES sample described in the data section and have been calculated without using population weighting. Consequently, these statistics can only provide insights into the characteristics of the sample itself and do not necessarily reflect the population.

Table 2. Multidimensional poverty situation of youths in Thailand

Var	Description	Non-NEETs	NEETs	All youths
Education				
edu1	*Years of schooling*: Households with at least one member: (1) aged 15–29 years that did not complete lower secondary education or (2) aged 30–59 years that did not complete primary education.	0.393	0.487	0.404
edu2	*Delayed schooling*: Household that have at least one child aged 6–17 years who is either not attending school or has a delay in education of more than 2 years, except for those who have completed lower secondary education.	0.039	0.113	0.048
edu3	*Presence of parents*: Households that have at least one child aged 0–6 years who is not living with their father and/or mother (in the case where the father and/or mother are still alive).	0.034	0.037	0.034
Healthy Living				
health4	*Drinking water*: Households obtain water from (1) an indoor well within the house, or (2) an outdoor well located outside the house, (3) a river/stream/canal/ waterfall/mountain, or (4) rainwater, or (5) other sources.	0.248	0.264	0.250
health5	*Dependency*: Households with at least one member who is 15 years or older and unable to take care of themselves in their daily life without assistance and is unable to travel outside the living area without a caregiver present.	0.019	0.061	0.024
health6	*Consumption poverty*: Households living under poverty line.	0.098	0.206	0.111
Var	**Description**	**Non-NEETs**	**NEETs**	**All youths**
Living Conditions				
live7	Garbage management: Households manage garbage by (1) burning it, or (2) burying it, or (3) disposing of it in rivers or canals, or (4) littering in vacant public areas, or (5) other methods.	0.333	0.361	0.336
live8	Internet accessibility: Households with no member using the internet.	0.034	0.092	0.041
live9	Asset procession: Households that do not own at least 4 small items and 1 large item.	0.050	0.054	0.051
Financial Security				
fin10	Saving: Households with no financial assets for the purpose of saving.	0.069	0.086	0.071
fin11	Financial burden: Households with difficulties paying for home rent, water bills, electricity bills, or educational expenses during the past 12 months.	0.090	0.128	0.095
fin12	Pension: Households with at least one member who is 60 years or older without pensions or social assistance.	0.320	0.343	0.323

Source: The multidimensional poverty indicators were from NESDC [5] and the statistics were calculated from the 2019 SES sample (without population weighting).

Note: From NESDC [5], the indicator *health6* was *"Households where the expenses related to food are lower than the food poverty line, calculated based on the recommended calorie intake for each age group and gender, there is a requirement to meet the daily food consumption."*. However, with the information limitation on the food poverty line, this study uses general poverty line.

Using the GSEM method, the entire network of direct and indirect connections depicted in Fig. 1 can be simultaneously estimated via the system of five equations using the maximum likelihood estimation. Notably, the dependent variables across all regressions, namely the NEET status and the four dimensions of poverty intensity, exhibit a discrete nature, characterized by binary or ordinal variables. For this reason, the standard structural equation model (SEM) is not appropriate due to the assumption of normally distributed errors in all equations. Therefore, the GSEM is a more suitable estimation method as it accommodates different error distributions for each regression equation [3].

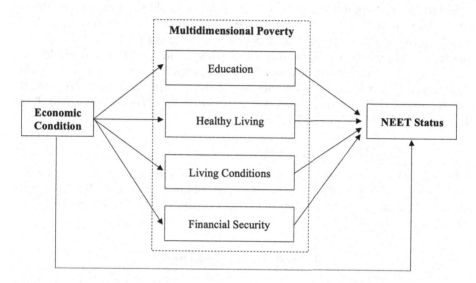

Fig. 1. The GSEM framework to examine the effects of economic condition and multidimensional poverty on NEET status

The system of five regressions equations for the GSEM structural framework for this study can be shown as follows:

Factors determining the NEET status. As the NEET status ($NEET$) is a binary variable that takes value 1 if the individual is NEET and 0 otherwise, the regression is assumed to follow the binary Logit model as shown in Eq. 1.

$$\Pr\{NEET = 1 \mid X\} = L(X\beta), \qquad \text{(Eq. 1)}$$

where $L(\cdot)$ is the logistic distribution, X is a vector of factors determining the NEET status including MP_{edu}, MP_{health}, MP_{live}, MP_{fin}, and $\ln(gpppc)$, and β

is a vector of corresponding path parameters. All variable descriptions and summary statistics are shown in Table 3.

Effects of the economic condition on the four dimensions of multidimensional poverty. Each poverty dimension j, namely education (edu), healthy living (health), living conditions (live), and financial security (fin), is represented by three indicators shown in Table 2. The multidimensional poverty intensity (MP_j) refers to the count of indicators that indicate a household's poverty status, determined by whether they fall below the respective thresholds. Therefore, the intensity of poverty ranges from 0 to 3, where 0 indicates that households do not experience poverty in a specific dimension, while 3 indicates that households are considered to be in poverty across all indicators within that dimension. As the MP_j variable representing each dimension of the intensity of poverty are ordinal, the regression for each dimension $j = edu, health, live, fin$ follows the ordinal Logit model as shown in Eq. 2 - Eq. 5 as follows.

$$\Pr\{MP_j = i \mid Z\} = \Pr\left(c_{i-1} < Z\gamma_j \le c_i\right), \qquad \text{(Eq. 2 - Eq. 5)}$$

where Z is $\ln(gpppc)$, and γ_j is the corresponding path parameter for poverty dimension j. In this study, there are four possible responses and, thus, there are three cut points c_1, c_2 and c_3. All variable descriptions and summary statistics are shown in Table 3.

Table 3. Variable description

Variables	Description	Mean	SD	Min	Max
$NEET$	NEET status (=1 if being NEET, =0 otherwise)	0.122	0.327	0	1
MP_{edu}	Intensity of poverty in education	0.486	0.615	0	3
MP_{health}	Intensity of poverty in healthy living	0.384	0.567	0	3
MP_{live}	Intensity of poverty in living condition	0.428	0.567	0	3
MP_{fin}	Intensity of poverty in financial security	0.489	0.614	0	3
$gpppc$	Per capita grossed provincial produce (GPP) (THB per year)	189,352	175,234	60,185	1,007,570

Source: Calculated from the 2019 SES sample (without population weighting).

6   Results

To examine the effects of the economic condition and the multidimensional poverty status on the NEET status of youths in Thailand, this study estimates a system of five structural equations with discrete dependent variables (shown in Eq. 1–Eq. 5) using the GSEM estimation[2]. Following the estimation, odd ratios

---

[2] It should be noted that this study estimates both Logit-based and Probit-based models. The results are only slightly different, but the Logit-based model yields lower AIC (85,254 for Logit and 85,283 for Probit) and BIC (85,416 for Logit and 85,445 for Probit). Therefore, this study reports the Logit-based model, which assume logistic distribution for all five equations.

were calculated by exponentiating the estimated coefficients and are presented in Table 4.

Table 4. Effects of the economic condition and multidimensional poverty on the NEET status from the GSEM estimation.

	(Eq. 1) $NEET$	(Eq. 2) MP_{edu}	(Eq. 3) MP_{health}	(Eq. 4) MP_{live}	(Eq. 5) MP_{fin}
MP_{edu}	1.409***				
	(0.063)				
MP_{health}	1.454***				
	(0.068)				
MP_{live}	1.168***				
	(0.058)				
MP_{fin}	1.148***				
	(0.052)				
$\ln(gpppc)$	1.098**	0.808***	1.080***	0.557***	0.809***
	(0.045)	(0.022)	(0.029)	(0.017)	(0.022)
Constant	0.028***				
	(0.014)				
Observations	11,740	11,740	11,740	11,740	11,740

Note: (1) Odd ratios are reported for all equations, (2) Standard errors are reported in parentheses, and (3) *** $p<0.01$, ** $p<0.05$, * $p<0.1$. Source: Calculated from the 2019 SES sample (without population weighting).

The findings presented in Table 4 can be presented in two distinct sections. Firstly, the study discusses the associations between the economic condition and the intensity of poverty in the four dimensions. Secondly, an analysis is conducted to examine the overall effects of the economic condition and the intensity of multidimensional poverty on the likelihood of youths being in the NEET status.

6.1   The Economic Condition and Multidimensional Poverty

The findings indicate substantial links between the economic condition and all multidimensional poverty dimensions. Odds ratios below 1 reveal a negative relationship, signifying that with higher per capita GPP, youth are less likely to experience elevated poverty levels in education, living conditions, and financial security dimensions. Specifically, for every 1% increase in the per capita GPP, youths are 0.808 times as likely to experience a high intensity of poverty in the education dimension, 0.557 times as likely in the living conditions dimension, and 0.809 times as likely in the financial security dimension.

The odds ratio is slightly above 1 for the healthy living dimension, suggesting that as per capita GPP increases, youths are more likely to experience higher poverty levels in this dimension. This contrasts with conventional economic theories. This may stem from specific indicators used to measure poverty intensity in healthy living, including drinking water (health4), dependency (health5), and consumption poverty (health6). Analysis in Table 2 reveals the drinking water indicator contributes most to poverty in this dimension, with 25% of youths falling below its threshold. Conversely, only 2.4% and 11.1% fall below thresholds for dependency and consumption poverty. Notably, the drinking water indicator, involving non-standard water sources, correlates positively with per capita GPP. This might arise from the indicator's omission of filtration methods, failing to capture true unhealthy living conditions among households.

6.2 The Economic Condition, Multidimensional Poverty, and NEET Status

For factors explaining the NEET status, both the per capita GPP and the poverty intensity across all dimensions exhibit a positive correlation with the NEET status, as evidenced by the odds ratios in Eq. 1 being greater than 1. Regarding the impact of poverty intensity, the study reveals that youths from households with higher poverty intensity in the education dimension are 1.409 times more likely to be NEET compared to those from households with lower poverty intensity in this dimension. The odds ratios for healthy living, living conditions, and financial security dimensions are 1.454, 1.168, and 1.148, respectively. These findings suggest that youths in impoverished households have an increased likelihood of being in the NEET situation, particularly emphasizing education and healthy living dimensions.

The influence of the economic condition on the NEET status yields intricate findings. The results show that, on one hand, the per capita GPP has a negative indirect effect on the NEET status through the mediating factor of poverty. This means that better economic condition lead to reduced poverty, which indirectly decreases the likelihood of NEET status. On the other hand, the per capita GPP simultaneously has a positive direct effect on the NEET status. This suggests that a higher per capita GPP is associated with an increased likelihood of being in the NEET situation directly. This observation suggests the presence of heterogeneity among NEETs, wherein certain groups are more likely to increase during favorable economic condition, while others are more likely to experience NEET status during economic downturns.

7 Conclusion and Implications

To understand the effects of the economic condition and the multidimensional poverty situation on the NEET status of youths in Thailand, this study employs the generalized structural equation model (GSEM) to estimate the structural relationships among relevant factors. Specifically, the study examines the connections between the per capita gross provincial product (GPP) and poverty intensity in the dimensions of education, healthy living, living conditions, and financial security, in relation to the occurrence of NEET status among youth.

The results show strong associations between all dimensions of poverty to the NEET status confirming the previous studies that the occurrence of NEET is more likely to happen among households with a high intensity level of poverty [4,8]. This study provides a deeper analysis to identify that the NEET status exhibits the strongest correlation with poverty in the education and healthy living dimensions of multidimensional poverty. These findings suggest that the social exclusion of youths leading to the NEET status is not solely determined by their financial circumstances. In the education dimension, the results reveal that the educational levels of other household members, educational delays, and the presence of parents in households with children significantly influence the likelihood of being NEET, which is associated with an increased probability of leaving the educational system prematurely. These findings demonstrate the intergenerational transmission of educational deprivation, underscoring the need for policy interventions aimed at breaking this cycle. Furthermore, the healthy living dimension, as measured by indicators such as access to drinking water, dependency, and consumption poverty, also exhibits strong significance. This dimension may be influenced by monetary poverty, which manifests in limited access to safe drinking water and inadequate consumption levels. Additionally, it may be influenced by the responsibility of assisting household members who are dependent and require daily care.

In addition to the household multidimensional poverty situation, this study also examines the relationship between the economic condition and the likelihood that youths become NEET. The results indicate that in provinces with a higher per capita GPP, the likelihood of youth experiencing poverty diminishes, consequently reducing their chances of being NEET. However, when accounting for the multidimensional poverty situation, it is observed that youths in provinces with a higher per capita GPP are more likely to becoming NEETs. The findings reveal that certain types of NEETs are closely associated with poverty, while other types tend to increase during periods of economic prosperity, indicating a potential link to higher socio-economic backgrounds. There are two key implications derived from this finding. Firstly, during economic downturns, there is a potential rise in the prevalence of NEETs, as a greater number of youths may experience a heightened risk of falling into poverty. Secondly, it highlights the necessity for conducting additional research aimed at understanding the various categories of NEETs and identifying their specific needs for support in successfully reintegrating into either the educational system or the labor market.

Although this study has utilized the extensive multidimensional poverty data obtained from Thailand's SES survey and investigated the influence of the economic condition and poverty on NEET status, the results highlight the necessity for further research that addresses practical inquiries to inform policy formulation and implementation. Particularly, inquiring about the types of support required by youths residing in low education and poverty-stricken households who are more likely to becoming NEETs, especially in the periods of economic downturn. Additionally, exploring the potential heterogeneity among NEETs, specifically examining the characteristics of the higher socio-economic group and determining the proportion they represent, would be beneficial for a comprehensive understanding of the issue.

References

1. Bacher, J., Koblbauer, C., Leitgöb, H., Tamesberger, D.: Small differences matter: How regional distinctions in educational and labour market policy account for heterogeneity in NEET rates. J. Labour Market Res. **51**, 1–20 (2017)
2. Caroleo, F.E., Rocca, A., Mazzocchi, P., Quintano, C.: Being NEET in Europe before and after the economic crisis: an analysis of the micro and macro determinants. Soc. Indic. Res. **149**(3), 991–1024 (2020)
3. Huber, C.: Generalized structural equation modeling using stata. In: Italian Stata Users Group Meeting, pp. 14–15 (2013)
4. Mussida, C., Sciulli, D.: Being poor and being NEET in Europe: are these two sides of the same coin?. J. Econ. Inequality **21**, 463–482 (2023). https://doi.org/10.1007/s10888-022-09561-7
5. NESDC: Poverty and Inequality Report. National Economic and Social Development Council, Bangkok (2019)
6. Noh, H., Lee, B.J.: Risk factors of NEET (Not in Employment, Education or Training) in South Korea: an empirical study using panel data. Asia Pacific J. Soc. Work Dev. **27**(1), 28–38 (2017)
7. Pesquera Alonso, C., Muñoz Sánchez, P., Iniesta Martínez, A.: Is there a uniform NEET identity in the European Union? Int. J. Adolesc. Youth **27**(1), 207–220 (2022)
8. Ruesga-Benito, S.M., González-Laxe, F., Picatoste, X.: Sustainable development, poverty, and risk of exclusion for young people in the European Union: the case of NEETs. Sustainability **10**(12), 4708 (2018)
9. Tamesberger, D., Leitgöb, H., Bacher, J.: How to combat NEET? Evidence from Austria. Intereconomics **49**, 221–227 (2014)
10. TDRI. Youth Employability Scoping Study. Bangkok: UNICEF (2020). https://www.unicef.org/thailand/media/4771/file/Youth%20Employability%20Scoping%20Study.pdf
11. UNICEF & College of Population Studies, Social Research Institute, Chula Unisearch. In-depth Research on Youth Not in Employment, Education or Training (NEET) in Thailand. Bangkok: UNICEF (2023). https://www.unicef.org/thailand/media/10746/file/In-depth%20research%20on%20youth%20NEET%20in%20Thailand.pdf

Author Index

Printed in the United States
by Baker & Taylor Publisher Services